# Climate Change Biology

# Contents

# Preface

We authors all feel that climate change is an important and pressing political issue and that action on our part is warranted and urgently needed, but this is not a book about policy, ethics, economics or environmental activism; it is a book about biology. We are biologists who study the impacts of climate change; this is a book about research into those impacts. Climate change biology is a big subject. Climate change influences all levels of biological organization from biochemistry to ecosystems. While we have tried to be comprehensive in concepts, we have certainly not been comprehensive in examples, in coverage of the variety of ecosystems, taxa, biogeographical areas, etc. We aimed to provide an introduction to the study of the biological impacts of climate change, leaving the more advanced topics for further study.

This book is organized around the basic outline of a 12-week undergraduate course that one of us (J.A.N.) designed in 2006. That course was difficult for undergraduates for two principal reasons. First, there was no suitable textbook available. There are many edited volumes written with a specialist audience in mind (see e.g. Newton *et al.*, 2007) and there are many texts about climate change aimed at undergraduates, but they have little or nothing to do with biology. Our book is intended to fill this gap. The second difficulty that undergraduates have with this topic is the general lack of synthesis. Climate change biology is a relatively new field of research and the questions asked within this field are large and difficult. We have tried to provide some synthesis of the field where that is possible, and to point out the open questions where it is not.

We hope that the book will provide a good survey introduction to climate change biology that can be covered comfortably in a single semester. We have aimed the text at third- and fourth-year undergraduates, but properly supplemented it should work well as a second-year undergraduate text or even a first-year graduate text. We generally assume students are familiar with first-year biology and general ecology, although we try to remind the reader of basic concepts where appropriate.

We appreciate all of the help and support we received from our colleagues throughout the process of writing and assembling this book. Colleagues provided us with images, photos and in some cases data so that we could re-graph them as needed. We also benefited from many discussions with our colleagues. The remaining faults are entirely our own. The literature on climate change biology is increasing at what seems to be an exponential rate, and it was impossible to keep up with everything. We could not provide a comprehensive literature review due to space constraints, and we apologize if we have not cited important papers.

# Glossary

**Acclimation.** When plants reduce their rate of photosynthesis in response to long-term exposure to elevated $CO_2$. Sometimes used more generically to mean a reduction in a plant's growth rate response to elevated $CO_2$. See also **Down Regulation**.

**Adaptation.** Used in two different contexts in this book. (1) *Evolutionary adaptation* is a population's change in gene frequency through time in response to a specific selection pressure. (2) *Climate change adaptation* refers to the process of altering management options or behaviour to reduce the impacts of climate change on a process or metric of interest.

**Bioclimatic Envelop Model.** A mathematical/statistical model where the observed distribution of a species is correlated with the observed climate. These models can be used to predict changes in the potential observed distribution in the future under climate change. Compare with **Ecological Niche Model**.

**$C_3$ Plants.** These plants have a form of photosynthesis where they use the Calvin Cycle directly, catalysing the primary uptake of $CO_2$ with the enzyme Rubisco. Compare with **$C_4$ Plants**.

**$C_4$ Plants.** These plants have a form of photosynthesis where they catalyse the primary uptake of $CO_2$ with the enzyme PEP carboxylase. $CO_2$ taken up in this way is then transported deeper into the leaf tissue where it enters the Calvin Cycle. Because of this carbon concentrating mechanism, $C_4$ plants generally have greater **Water Use Efficiency** than **$C_3$ Plants**.

**Carbon Sequestration.** Carbon can basically be in three places in a terrestrial ecosystem: the atmosphere, the soil or living organisms. In general sequestration occurs when the rate of carbon entering a carbon pool occurs faster than the rate of carbon leaving that pool. In particular, we only really use the term 'sequester' when referring to either the soil or some longer-lived organisms such as trees, in the stemwood. Some sub-pools of carbon in the soil turn over very slowly (e.g. humic substances) and so are able to keep that particular carbon out of the atmosphere for very long periods of time. Another term for a pool that sequesters carbon is a *carbon sink* (as compared to a *carbon source*).

**Climate Proxy.** A measure of something biological or physical that is correlated with contemporary climate. Measuring that variable is not a *direct* measure of climate, it is a *proxy* measure of climate. Classic examples of climate proxies are tree ring widths (see **Dendroclimatology**), deuterium concentrations, or fossilized organisms or parts such as diatoms and pollen. Climate proxies are used in climate reconstructions that allow us to infer what the climate would have been like in a particular place either where direct measurements were not taken, or before such measurements existed.

**$CO_2$ Fertilization Effect.** Sometimes referred to as simply the *fertilization effect*. Since photosynthetic rates of **$C_3$ Plants** are often not saturated at ambient $CO_2$, adding additional $CO_2$ tends to stimulate photosynthesis and hence plant growth. This effect is commonly exploited in the greenhouse industry where commercial greenhouses are flooded with additional $CO_2$ to produce better yields and faster plant growth. The effect is also commonly seen in climate change experiments where $CO_2$ is manipulated, at least for $C_3$ plants. **$C_4$ Plants** tend to be less limited by $CO_2$ concentrations at ambient $CO_2$ concentrations and are therefore less likely to show this effect.

**Community.** A group of **Populations** that coexist in the same geographic area and that interact with each other. Sometimes this term is used with a more restrictive adjective such as *grass community* or *bacterial community*. It is rare to see community used in its most inclusive forms in the context of experiments. See also **Succession**.

**Dendroclimatology/Dendroecology.** 'Dendro' refers to tree rings (from the Greek word *dendron*, meaning tree limb). *Dendroclimatology* is the process of using known relationships between tree-ring widths and contemporary climate to reconstruct past climates (i.e. the tree ring record is a form of climate proxy). *Dendroecology* is the process of using tree-ring widths through time to infer something about the local ecology at the time the rings were deposited. For example, tree rings can be used to reconstruct the pattern of fire disturbance in a forest.

**Diversity Index.** There are many *diversity indices*. They are ways of reducing diversity, which is essentially a multi-dimensional quality, into a single dimensional metric. In general, ecologists tend to think of diversity as a function of both the number of species present (*richness*) and the *evenness* with which individuals are distributed between different species. Diversity indices are mathematical means of combining these two dimensions (richness and evenness) into a single dimension. A common metric is *Shannon Diversity* (H).

**Down Regulation.** In the context of climate change this term is usually used synonymously with **Acclimation**. More broadly in biology, the term usually refers to when a gene is not as active as it usually is. There are occasions in the climate change biology literature where down regulation takes on this latter meaning.

**Ecological Niche Model.** A model that uses experimental information about the climate tolerances of a species, and then maps those tolerances to observed or future climate to identify geographical regions where a species might be found. These models are used in studies of species ranges. See also **Bioclimatic Envelop Model**.

**Ecosystem.** An ecosystem is an entire biotic **Community** along with its physical environment. It can be difficult to define the boundaries of an ecosystem since individuals interact across almost any boundary used as a definition. It is common to see ecosystems defined in terms of a watershed or body of water, both of which would seem to have relatively clear boundaries, but even in these cases energy and material certainly flow back and forth across the shoreline or the topological boundary of the watershed. In many cases the boundary of an ecosystem is not much more than a human construct used for descriptive and accounting purposes. Also notice that *ecosystem* has come to be used much more broadly over the past 10 years or so. It is sometimes used synonymously with *nature*, *habitat* and *community*.

**Ecosystem Functioning.** Ecosystem functioning usually refers to things that ecosystems do that benefit humans, although it need not be so restrictive. Typical functions include productivity (e.g. **NPP**), nutrient cycling, **Carbon Sequestration**, pest management, water purification, pollination services, and so on.

**Evapotranspiration.** Abbreviated ET, evapotranspiration is the sum of two different processes: evaporation and transpiration. ET is one of the important sources of moisture loss from an ecosystem or watershed. Along with surface runoff and ground water recharge, ET helps account for the level of soil moisture, which can be key in some ecosystems for plant productivity. See also **Transpiration**.

**External Validity.** A term that refers to the similarity between the conditions of an experiment and those of the real system to which the experiment is meant to be relevant. External validity must often be traded off against considerations of experimental control and experimental replication.

**Extirpation.** When a species becomes *locally* extinct. Also applies to subspecies and populations.

**FACE.** An acronym standing for *Free Air Carbon dioxide Enrichment*. FACE is a technique for experimentally enriching the $CO_2$ concentration over a plot of ground. It is widely considered to be the most **Externally Valid** method for conducting $CO_2$-enrichment experiments, but the method tends to suffer from low levels of experimental replication and so is vulnerable to a **Type II Error**.

**Functional Traits.** These are phenotypic characteristics possessed by individual organisms within an ecosystem. They usual distinguish those traits that are relevant for ecosystem functioning such as **NPP**, **Trophic**

**Level**, nitrogen fixation, and so on. Functional traits tend to ignore characteristics that we think have little to do with ecosystem functioning (e.g. colour). Functional traits are often discussed in the context of **Ecosystem Functioning**, where it is hypothesized that the greater the diversity of functional traits within an ecosystem, the higher the level of function. If functional trait diversity is to be a useful concept, it must do some work that 'species diversity' does not. Individuals from different species may share the same, or similar, functional traits, leading to the conclusion that, from an ecosystem perspective, these species are 'redundant'. Objectively defining functional traits is not easy. If traits are defined too narrowly, then there will be as many traits as species and the concept becomes useless. Defined too broadly, nearly all species will appear to be the same from the perspective of ecosystem functioning.

**GCM (AOGCM).** Stands for *Global Circulation Model*, although it is also commonly used in a more generic sense to mean Global *Climate* Model. AOGCM is the more technically correct term and stands for Atmosphere–Ocean coupled GCM. However, since these days nearly all GCMs are AOGCMs, it is common to see them all referred to simply as GCMs. GCMs are large mathematical models of the entire Earth that represent the physical processes that determine weather and hence climate. These models are told what the current and future concentrations of greenhouse gases will be, and they then project the climate into the future. See also **SRES Scenario** and **RCM**.

**GPP.** Stands for *Gross Primary Productivity*. This is the rate at which carbon ($CO_2$) is captured by an ecosystem. It is measured in kg/m$^2$/year. Much of this captured carbon is quickly respired by the autotrophs and this difference is known as **NPP**. Note: GPP also stands for Gross Primary *Production* where the units are simply kg/m$^2$ and the time period is understood to be 1 year.

**Holocene Epoch.** A geological time period that began about 12,000 years ago. It is part of the **Quaternary Period** that began about 2.5 million years ago.

**Interaction.** This term is commonly used when discussing the results of experimental studies. Two treatments or factors are said to interact if the overall effect of the two is greater or lesser than the sum of the two individual effects. Suppose that doubling the level of soil nitrogen doubled the growth of a particular plant species, and that doubling the level of $CO_2$ increased the growth of that same plant species by 50%. Then presented in combination we would expect that doubled nitrogen *and* doubled $CO_2$ would result in a 2.5-fold stimulation of growth. If we observed more than 2.5-fold stimulation then we could conclude that $CO_2$ and nitrogen interacted positively in this experiment. Less than a 2.5-fold increase would cause us to conclude that $CO_2$ and nitrogen were antagonistic (or perhaps growth becomes limited by some third factor) in their effects. Either conclusion would be an example of two factors or treatments interacting with each other.

**Invasive Species.** Sometimes also referred to as *non-native* species or *exotic* species, these are species that did not evolve in the particular habitat in which they are now present. Some ecologists make distinctions among these different terms, with *invasive* referring to species that are both non-native and in some sense harmful to either humans (usually in the sense of economically harmful) or native species. Other ecologists use the terms interchangeably.

**IPCC.** Stands for the *Intergovernmental Panel on Climate Change*. The IPCC is convened under the auspices of the United Nations Environment Program (UNEP) and the World Meterological Organization (WMO). The IPCC is not a policy-making organization. Its job is to provide information to policy makers about climate change and its consequences. The IPCC is the most authoritative body speaking about climate change. Since 1990 the IPCC has been producing assessment reports every 6 years (the most recent, fourth, assessment report, AR4, was published in 2007).

**Meta-analysis.** This is a statistical technique for combining the results of multiple experiments that all ask similar questions. It has become a widely used technique in ecology since the early 1990s and is now fairly common in the field of climate change biology. It can help overcome the problem of low **Statistical Power**, but can be biased by the general tendency for journals to not publish negative results.

**Multitrophic.** Many or most climate change biology studies concentrate on the response of organisms at a single **Trophic Level** to some climate change variable. *Multitrophic* refers to considerations regarding the response of multiple interacting trophic levels at the same time and to the same environmental variables. For example, how do $C_3$ grasses respond to elevated $CO_2$? Asking this question in isolation from those grasses' herbivores, parasites and viruses might give a very different view than if we examined the entire *multitrophic* response.

**NEP.** Stands for *Net Ecosystem Productivity*. NEP is **NPP** minus heterotrophic respiration (i.e. respiration by the organisms that eat the plants). It is measured in $kg/m^2/year$. Note: NEP also stands for Net Ecosystem *Production* where the units are simply $kg/m^2$ and the time period is understood to be 1 year.

**NPP.** Stands for *Net Primary Productivity*. NPP is **GPP** minus the rate of autotrophic respiration. It is measured in $kg/m^2/year$. Because NPP includes all the belowground growth, and because that growth can be very difficult to measure in practice, it is not uncommon to see estimates of NPP restricted to aboveground NPP, either reported as such, or reported as total NPP after multiplying by a constant fraction used as a stand-in for actual estimates of belowground NPP. Note: NPP also stands for Net Primary *Production* where the units are simply $kg/m^2$ and the time period is understood to be 1 year.

**NUE.** Stands for *Nutrient Use Efficiency*. Sometimes refers to the more specific *nitrogen use efficiency* with the same abbreviation. NUE can be a more general measure, but sometimes refers to the more specific **PNUE**. There are many definitions of NUE and so the precise meaning in any given circumstance can sometimes be difficult to determine. Generally NUE means mass gained per unit mass of the nutrient. The confusion arises from both halves of this concept. Where is the mass being gained? Is it leaf mass, whole plant mass, grain mass, etc.? Where is the nutrient? Is it in the plant, in the soil, or in the form of fertilizer, etc.? For any specific application of the concept, one or more of these definitions might make more sense.

**Phenological/Phenology.** Phenology refers to the timing of critical life history events such as seed set, germination, moulting, egg laying, etc.

**Phenotypic Plasticity.** A *phenotype* is an observable characteristic of an organism that results from both its *genotype* and its environment. Phenotypic plasticity refers to the range of values the phenotype can display as the environment changes. Phenotypic plasticity is what allows individual organisms to tolerate (or perhaps thrive) across a range of some environmental variable.

**PNUE.** Stands for *Photosynthetic Nutrient Use Efficiency*. This is the photosynthetic rate per unit of leaf nitrogen. See also **NUE**.

**Population.** A group of organisms all belonging to the same species that coexist in the same geographic area and interact with each other.

**ppm.** Stands for 'parts per million'. ppm is a dimensionless measure of concentration commonly used to talk about the concentration of $CO_2$ in the atmosphere. Commonly used too is the mole fraction expression (number of moles of carbon per mole of air, expressed in $\mu mol/mol$, where micro ($\mu$) is $10^{-6}$, hence parts per million). Both expressions are commonly seen in the literature. We have standardized on ppm throughout this text.

**Quaternary Period.** This is the current geological period that began approximately 2.5 million years ago. It encompasses two geological epochs, the Pleistocene (2.5 million years ago to about 12,000 years ago) and the **Holocene** (12,000 years ago to present) Epochs.

**RCM.** Stands for *Regional Climate Model*. RCMs cover small parts of the globe and simulate climate at a much finer level of detail than **GCMs**.

**Rubisco.** Stands for *Ribulose-1,5-bisphosphate carboxylase oxygenase*. Also commonly abbreviated RuBisCO. Rubisco is the enzyme responsible for catalysing the uptake of $CO_2$ at the start of the Calvin Cycle. Rubisco and its activity is the subject of much research in the field of climate change biology because of its role in the primary uptake of $CO_2$ for $C_3$ **Plants**, and because it often represents a significant amount of the total nitrogen in a plant and is therefore a major player in our understanding of **NUE**.

**Scenario.** Also called *climate change scenario* and **SRES Scenario**. This is a coherent, self-consistent 'story' about what levels of future greenhouse gas emissions might look like. They are stories about how population growth will change, the pace of economic development, the pace of technological change, the mixture of fuels used in the future, and so on. These assumptions are then 'interpreted' through mathematical models that produce greenhouse gas emissions. These emissions are then fed into **GCMs** that are then used to project climate into the future. Scenarios are not *predictions*, since they depend on processes that can only be predicted with great uncertainty; so, for example, there are no *most likely* or *least likely* scenarios. Equally, scenarios are meant to be *value neutral*; for example, there is no *best case* or *worst case* scenario.

**Species Diversity.** See **Diversity Index**.

**Species-specific Responses.** The conclusion from many climate change experiments. Species-specific responses refer to the situation where similar species did not respond in the same way to one or more climate change variables. Such a conclusion is often the same as saying that there was not a general response observed across all species. Sometimes the term *idiosyncratic* will be used, which can mean the same thing, but more commonly means that the responses are not only species-specific, but also specific to the particular set of species combinations. That is, not only do species X and species Y not respond to higher temperatures in the same way, their responses also depend of whether species A, B, C, etc. are present and, in which combinations.

**SRES Scenario.** The subset of climate scenarios developed under the **IPCC**'s Special Report on Emissions Scenarios. This report developed a standard set of scenarios used by all of the climate modelling groups for the IPCC's third and fourth assessment reports. New emissions scenarios are being developed for the IPCC's fifth assessment report, due out in 2013. These new emissions scenarios and related climate change projections will probably start to appear in the scientific literature by early 2012.

**Successional/Succession.** The change in **Community** membership through time. Most areas have one or a few possible *climax* communities that would arise if not for disturbance, either anthropogenic or natural. We can talk about a successional stage (e.g. early, mid, late) or type (e.g. primary, secondary, bog, etc.).

**Trophic Level.** Refers to the position occupied by a species within the foodweb of a specific ecosystem. For example, producers (autotrophs, e.g. plants), primary consumers (herbivores) and secondary and tertiary consumers, etc. (predators, parasitoids, detritovores) are descriptions of different trophic levels.

**Trophic Structure.** One way of describing how a biological community is structured. Trophic structure can refer to the number of organisms at different **Trophic Levels**, or the ratio between the numbers of organisms at two different trophic levels (e.g. predators to herbivores).

**Type II Error.** A Type II Error occurs when we conclude that an experimental treatment has no effect when it actually does. The smaller our sample size or the smaller the effect size we are trying to detect, the more vulnerable will our experiments be to the occurrence of Type II Errors.

**WUE.** Stands for *Water Use Efficiency*. In general, WUE is a measure of the number of kilograms of biomass produced by a plant per kilogram of water. Confusion can arise as to whether we are talking about water available in the soil or water in the leaf/plant.

# PART I
# Preliminaries

Open-topped chambers for studying the impacts of elevated $CO_2$. These chambers are located in a scrub oak ecosystem at the Kennedy Space Center in the Merritt Island National Wildlife Refuge. They were established by Bert Drake, of the Smithsonian Environmental Research Center, in 1996. This is an example of one of several methods for manipulating the concentration of $CO_2$ in the atmosphere to study the potential impact of climate change. This and other methods are discussed in Chapter 3. (Photo courtesy of the Smithsonian Environmental Research Center.)

# Overview

In this section we explore the following questions: How do we know about past climates? How do we know about future climates? How do we study the biological impacts of climatic change? In Chapter 1 we introduce the methods of studying past climates and climatic changes. We also provide some insight into the climatic history of the Earth and the causes of non-anthropogenic climatic variation.

In Chapter 2 we explain that future climatic change is projected in three main steps. First we construct stories about how the Earth may change in the coming years in terms of things like population growth, availability and use of various sources of energy, the pace of economic development, and the pace of technological change. Next, these stories are interpreted via 'scenario models'. These models yield quantitative projections of the magnitude and dynamics of greenhouse gas emissions. Finally, these quantitative greenhouse gas emissions are used in large computer simulations of atmospheric and ocean physics which yield not only the climate projections that we readily encounter in the popular press, but much more detailed projections by region and year for many more variables than simply temperature.

In Chapter 3 we turn our attention to the methods that we use to study and predict the biological impacts of climatic change. These methods range from the laboratory, to the greenhouse, to the field. They involve theory, observation and experimentation. We have come a long way in our study of the biological impacts of climatic change, but there are limits to what we can show and what we have shown. In this chapter we identify and discuss some of these limitations. To become intelligent consumers of such impact studies, we need to appreciate their limitations. This chapter is particularly important as the ideas within apply to nearly all of the remainder of the book.

# 1 Putting it in Perspective: The Palaeorecord and Climate Reconstructions

The climate of the Earth has not been static at any scale of variability. Variations in the brightness of the Sun, fluctuations in the Earth's orbital properties, the changing geometry of the continents, the impacts of meteorites and the eruptions of volcanoes, and the activities of organisms all interact to produce a constantly changing climate. By examining the history of the Earth's climate we can answer questions such as: How unusual are the recent and anticipated changes in the Earth's climate? What have the effects of climatic changes been on species' distributions? How sensitive is the climate system to changes in atmospheric chemistry? How have patterns of variability and climatic extremes varied with changes in mean temperature? Because instrumental records of climate are available for barely 100 years of the Earth's approximately 4.5 billion year history, to see anything more than the briefest snapshot of its climate history it is necessary to make use of natural archives of climate information. In this chapter we examine the types of proxy records available for reconstructing the pre-instrumental climate of the Earth; we explore the history of the Earth's climate over the past 300 million years, with a particular focus on the climate of the last 15,000 years.

## 1.1 Methods of Palaeoclimatic Reconstruction

Many biological and geophysical processes are sensitive to variations in climate and thus record in their structures information that may be used to reconstruct the climate at their time of formation. These sources of information are generally termed 'proxy records of climate', since they are not direct measurements of the climate system itself, but rather indirect measurements of associated processes. There is a wide range of potential sources of proxy climate data (Table 1.1), but they all share a few common characteristics. First, they all contain non-climatic signals (or noise) as well as a climate signal. Second, they are sensitive only to particular features of the climate system, such as growing season moisture or mean annual water temperature. Lastly, they are sensitive to climate only at specific frequencies – above which it is not possible to resolve a signal and below which the climate signal may be confounded by other, non-climatic signals.

To extract the climate signal from these proxy records four main steps are necessary.

1. The chronology of the archive must be determined. That is, dates must be assigned to the components of the system that will be measured. There is a wide range of tools and techniques for doing this, some of which are addressed in more detail below, but two of the more commonly used techniques include radiocarbon dating – which allows the dating of organic material, within some range of uncertainty, that is up to 60,000 years old – and tree-ring cross-dating, which uses unique patterns of wide and narrow growth rings to assign specific calendar years to individual tree-ring measurements.

2. The parameter of interest must be measured. For some proxy records this step is relatively straightforward – as with the width of a tree ring or the particle size distribution for a sediment sample – but very often the measurement of proxy records is technically challenging, expensive and time-consuming.

©CAB International 2011. *Climate Change Biology* (J.A. Newman *et al.*)

**Table 1.1.** Common sources of palaeoclimatic data, their typical temporal properties and their commonly extracted climate signals.

| Proxy record | Minimum resolution (years) | Typical time span (years) | Optimal frequency (years) | Main climate signals |
|---|---|---|---|---|
| Historical records | 0.001–1 | 100s–1,000s | Variable | T, P, D, E, V, S |
| Tree rings | 0.1–1 | 100–1,000 | 1–100 | A, D, E, L, T, P |
| Sedimentary pollen | 1–10 | 1,000–10,000 | 10–100 | D, T, P, V |
| Sedimentary charcoal | 1–10 | 1,000–10,000 | 10–100 | D, E, V |
| Lake sediments | 1–10 | 1,000–10,000 | 10–100 | E, L |
| Diatoms | 1–10 | 1,000–10,000 | 10–1,000 | T, P, D, W |
| Chironomids | 1–10 | 1,000–10,000 | 10–1,000 | T, P, D, W |
| Ice cores | 1–100 | 1,000–10,000+ | 10–100 | A, D, E, P, T, S |
| Bivalve shells | 0.1–1 | 10–100 | 1–10 | T, W |
| Speleothems | 1–10 | 100–1,000+ | 10–1,000+ | T, P, W |
| Coral chemistry | 0.1–1 | 10–100 | 1–100 | T, L, P |
| Packrat middens | 10–100 | 100–10,000 | ~100 | D, V |
| Borehole temperature profiles | ~10 | 500–1,000 | 10–100, decreasing with age/depth | T |
| Marine sediments | 10–100 | 10,000–1,000,000 | 100–1,000 | T, W, E, L |

A, atmospheric composition; D, drought; E, events (e.g. volcanic eruptions, wildfires, storms); T, temperature; P, precipitation; V, vegetation change; S, solar variability; L, sea level; W, water chemistry.

For example, the measurement of gas concentrations in air bubbles trapped in ice cores requires that the core remains solidly frozen, isolated from potential contaminants (including the air in the laboratory itself), and the use of highly specialized equipment.

3. Once the parameter of interest has been dated and measured it must be calibrated to the climate signal being reconstructed. For some types of proxy record there is a direct physical relationship between the climate system and the measured properties of the proxy record. This is the case with many isotopic measurements made on geophysical proxy records (although they present other challenges to climate reconstruction). More commonly, a statistical relationship is calculated between the measured variable and the climate signal of interest, using the modern period for which instrumental data are available to calibrate the reconstruction model.

4. The resulting reconstruction must be verified against independent sources of climate information in order to establish its quality. For highly resolved proxy records such as tree rings, coral skeletal chemistry and varied sediments (see Section 1.3), the instrumental record may be used. A portion of the available data is usually withheld from the calibration and is used to quantify the quality of the reconstruction. For more poorly resolved proxies, and for those that do not overlap with the instrumental record, other means must be found to verify them, such as comparisons with other palaeoclimatic reconstructions or with global circulation models (GCMs; Chapter 2).

While there is an almost bewildering array of biological and geophysical records that can be used to reconstruct elements of the climate system, these records can generally be placed into one of four categories (Bradley, 1999): glaciological, geological, biological or historical. Within each of these categories there are many potential records available. For example, glaciological records include such measures as maximum extent (as evidenced by the formation of moraines), rates of ice accumulation (measured from core samples), atmospheric composition (measured from air bubbles trapped in these cores), the isotopic composition of the ice itself, the amount and source area of dust entrained within glaciers, and structural properties of the ice itself. Because many palaeoclimatic records are assigned dates using similar techniques, we begin our examination of palaeoclimatic records with a discussion of the main tools available for dating these samples.

## 1.2 Methods of Chronology Determination

One of the greatest challenges facing palaeoclimatologists lies in assigning dates to samples. Accurate dating is essential for comparing different records with each other, for determining the sequence of changes around the Earth, for evaluating rates of change and for disentangling the many chains of cause and effect in the Earth's climate system. A wide range of tools and techniques is available for dating samples of interest. They vary in their precision and accuracy, the materials to which they may be applied and the time scales over which they are useful. There are three main groups of techniques that are most commonly used for dating palaeoclimatic data: radioisotopic, palaeomagnetic and biological.

Radioisotopic techniques rely on the fact that under certain conditions unstable isotopes of a given element may be formed. If these initial conditions occur as a discrete event in time, and the decay rate of the element is known and appropriate to the age of the material being dated, these relationships can be used to determine the age of the material. The most common radioisotopic technique is radiocarbon dating, but there are other tools available to palaeoclimatologists including potassium–argon dating, uranium-series dating and cosmogenic dating. Palaeomagnetic techniques make use of episodic reversals in the Earth's magnetic field and movement of the position of the magnetic poles and the Earth's plates that result in predictable variations in the orientation of magnetic particles in rocks and sediments. Biological dating techniques make use of the growth rates and growth patterns of organisms to estimate their age and, by extension, the age of the surfaces on which they are growing. Very often these organisms also contain natural archives that can also be used to reconstruct properties of the climate system directly (see Section 1.3).

### Radiocarbon dating

Radiocarbon dating is undoubtedly the workhorse of palaeoclimatic dating techniques. It can be applied to any organic compound, is useful for material up to approximately 50,000 years in age and is relatively inexpensive to perform. Radiocarbon dating is possible because carbon atoms occur in three distinct forms or isotopes.

Carbon-12 atoms (denoted $^{12}C$) contain six protons and six neutrons, and are the most common carbon isotope, accounting for about 98.9% of the Earth's atmosphere. The other important isotope is carbon-14 ($^{14}C$), which represents only 0.0000000001% of the Earth's atmosphere. The remaining isotope, carbon-13 ($^{13}C$), accounts for the remaining approximately 1.1% of the Earth's atmosphere, but is not used for dating purposes. $^{14}C$ is formed in the Earth's upper atmosphere when a nitrogen atom is struck by a neutron energized by cosmic radiation. This neutron displaces one proton, leaving the atom with only six protons but eight neutrons – converting it from $^{14}N$ to $^{14}C$. A hydrogen atom is also produced during this process. $^{14}C$ is an unstable isotope of carbon, though, and episodically a neutron in the atom will decay into a proton and an electron, converting the carbon atom back into nitrogen. The rate of this decay is relatively constant, such that, in a given sample, half of the $^{14}C$ will decay in approximately 5730 years. This interval, termed the 'half-life' of the isotope, means that, for example, a 100 g sample of wood might initially contain $5.0 \times 10^{-9}$ g $^{14}C$, and would contain only $9.8 \times 10^{-12}$ g $^{14}C$ after ten half-lives.

Newly produced carbon atoms oxidize rapidly to form carbon dioxide ($CO_2$) and are readily mixed throughout the atmosphere. Because plants assimilate $CO_2$ during photosynthesis, plant tissues contain a proportion of $^{14}C$ similar to the atmosphere. This radioactive carbon cascades through the food web as herbivores consume plants, and as carnivores consume herbivores. Thus, all living organisms contain a small proportion of $^{14}C$ in their tissues. The fact that $^{14}C$ is assimilated only by living organisms, coupled with the fact that it decays over time, allows scientists to measure the relative abundance of $^{14}C$ to $^{12}C$ to determine the length of time that has passed since the organism died. The accuracy of dates determined by radiocarbon dating is complicated by several factors. First, the decay rate of $^{14}C$ is slightly variable, meaning that the half-life is more correctly stated as $5730 \pm 40$ years. Second, as samples age, the amount of $^{14}C$ present decreases to the point where it is difficult to measure the decay rate accurately. Several other issues are worth discussing in more detail.

One of the main assumptions behind radiocarbon dating is that the total amount of $^{14}C$ in the atmosphere has been constant over time. It is now widely recognized that this assumption is not valid and so corrections must be made to account for these differences. The main causes of long-term variability in $^{14}C$ abundance include variations in cosmic radiation, enrichment of $^{12}C$ in the atmosphere due to volcanic eruptions, and effects of air temperature and ocean circulation on rates of $CO_2$ exchange between oceans and the atmosphere. More recently, atmospheric nuclear testing caused a temporary spike in atmospheric $^{14}C$ during the 1950s and 1960s. Most importantly, the burning of fossil fuels since the early 19th century has reduced the relative abundance of $^{14}C$ over time. Because coal and oil are composed of extremely old organic matter they contain effectively no $^{14}C$. Consequently, as fossil fuel-derived $CO_2$ in the Earth's atmosphere has increased, the ratio of $^{14}C$ to $^{12}C$ has decreased.

The net result of these fluctuations is that the dates determined by radiocarbon dating need to be corrected, or converted to calibrated radiocarbon dates, to account for differences between the radiocarbon age of the sample and the calendar age of the sample. Calibration curves have been developed based on the $^{14}C$ present in independently dated samples such as tree rings and speleothems (see Section 1.3). For samples that post-date the industrial revolution, calibration is often not possible because there are multiple possible matches. In these cases, other information about the sample may allow the correct age to be assigned, or other dating techniques may be necessary.

Another issue affecting the accuracy of radiocarbon dating depends on the source of the carbon that is assimilated by the organism being dated. For terrestrial plants it is generally safe to assume that the carbon comes directly from the atmosphere – although natural $CO_2$ vents do exist in some parts of the world (see Chapter 3). However, in aquatic and especially marine environments, the carbon assimilated by plants may be very old and consequently relatively depleted of $^{14}C$. One example of this is known as the 'hard water effect', and occurs when aquatic plants derive $CO_2$ from water that contains dissolved carbonate rocks. This problem is compounded in marine environments. First, when $CO_2$ is absorbed into the ocean, $^{14}C$ – being slightly heavier than $^{12}C$ – is preferentially retained in surface waters. This leads to an approximately 1.5% enrichment in $^{14}C$ in surface water. $CO_2$ can remain dissolved in ocean water for very long periods of time, though, and circulation processes move water

unequally around the world's oceans. Consequently, the apparent age of seawater varies geographically, depending on the residence time of dissolved $CO_2$ and the degree of mixing with surface water. In general, seawater from the tropics has a younger apparent age than water from high latitudes (although the North Atlantic Ocean is an exception to this rule due to the Gulf Stream and very strong downwelling). While calibration curves have been developed to correct marine radiocarbon dates, they do introduce additional causes of uncertainty. The question of how stable the observed patterns of apparent age have been over time remains an important line of scientific inquiry.

### Other radioisotopic methods

Although radiocarbon dating is the most widely used radioisotopic dating technique, other isotopes are available and can be used to date samples using the same basic principles. Here we describe three of the more commonly used radioisotopic methods. For recently deposited sediments, lead-210 dating provides better dating control than radiocarbon dating. Lead-210 ($^{210}Pb$) decays to $^{206}Pb$ with a half-life of 22.3 years. Thus, the ratio of $^{210}Pb$ to $^{206}Pb$ can be used to date materials up to about 150 years in age. This technique is used most commonly for dating of the upper portions of sediment cores (Fig. 1.1) and ice cores. Uranium-series dating techniques represent a suite of approaches, all based on the various daughter isotopes of uranium ($^{238}U$ and $^{235}U$). Uranium ultimately decays to a stable form of lead (either $^{206}Pb$ or $^{207}Pb$), but during this process it passes through a number of unstable intermediate daughter isotopes, several or which are useful for dating materials (Table 1.2). Of these, the most commonly used are the $^{230}Th/^{238}U$ ratio, which is useful for dating materials up to about 350,000 years old, and the $^{231}Pa/^{235}U$ ratio, which is useful for materials from 5000 to 150,000 years old. For much older deposits, potassium–argon dating can be used. Argon gas ($^{40}Ar$) has a half-life of $1.31 \times 10^9$ years. This gas is forced out of molten rock, so that newly deposited volcanic sediments have no $^{40}Ar$ contained within them. Over time, radioactive decay causes small amounts of $^{40}Ar$ to accumulate, which can be measured by heating the rock in the laboratory and measuring the $^{40}Ar$ released. This technique is not useful for dating samples less than about 100,000 years old, but has been a very useful tool for dating seafloor basalts, allowing the use of palaeomagnetic dating techniques (see 'Palaeomagnetism' section).

**Fig. 1.1.** Colin Mustaphi of the Carleton University Paleoecology Laboratory extracts a surface core from a montane lake near Nelson, British Columbia. Lead-210 dating is used to determine the chronology and accumulation rate of recently deposited sediments. (Photo: Z. Gedalof, Department of Geography, University of Guelph.)

## Palaeomagnetism

Molten lava, baked clay, and lake, ocean and aeolian (i.e. windblown) sediments all contain iron compounds that orient parallel to the Earth's magnetic field as they cool or settle. Because the position of the magnetic north pole changes over time, and because the polarity of the Earth's magnetosphere reverses episodically, the orientation of these compounds provides information about the age of the deposits. Sediments that are deposited (or lava that cools) during periods of normal magnetism are oriented northward, and during periods of reversed magnetism they are oriented southward. The period of time that the polarity of the Earth remains fixed is termed a 'chron', and has been highly variable over the Earth's history (Merrill *et al.*, 1990). Within a given chron there may be several subchrons – brief (typically less than $10^5$ years) intervals of polarity reversals. The precise cause of magnetic reversals is still debated, but the most likely mechanism relates to a 'tangling' of the Earth's magnetic field relating to different behaviour of the solid core and the liquid mantle.

Because the Earth's magnetosphere is global in nature, the chronology of magnetic reversals, termed the 'geomagnetic polarity time scale', can be used to date iron-bearing deposits anywhere in the world. As long as an approximate age can be determined, the sequence of normal and reverse polarity sediments can be correlated to this time scale to assign a date to the deposits. This time scale was developed mainly through potassium–argon dating of volcanic flows and basalts, but it also allows dating of marine sediments. The resolution and accuracy of palaeomagnetic dating are variable, but it can be used to date deposits up to about 76 million years in age.

**Table 1.2.** Important daughter isotopes of $^{238}$U and $^{235}$U. There are a number of shorter-lived intermediate daughters not included here because they are too short-lived to be useful for uranium-series dating.

| Isotope | Half-life (years) | Isotope | Half-life (years) |
|---|---|---|---|
| $^{238}$U (uranium) | $4.51\times10^9$ | $^{235}$U (uranium) | $7.13\times10^8$ |
| $^{234}$U (uranium) | $2.50\times10^5$ | $^{231}$Pa (protactinium) | $3.24\times10^4$ |
| $^{230}$Th (thorium) | $7.52\times10^4$ | $^{207}$Pb (lead) | Stable |
| $^{226}$Ra (radium) | $1.62\times10^3$ | | |
| $^{210}$Pb (lead) | 22 | | |
| $^{206}$Pb (lead) | Stable | | |

## Dendrochronology

As well as being a source of palaeoclimatic information itself (see Section 1.3), dendrochronology can also be a tool for dating other sources of palaeoclimatic data. Dendrochronological methods make use of the fact that in temperate regions, or in regions with reliably seasonal variability in precipitation, trees produce annual growth rings. Tree-ring dating is used in three main ways to assign dates to data of unknown age (Luckman, 2000).

First, the age of trees provides a minimum age for the surface on which they are growing. This technique has been widely used to develop chronologies of moraine formation associated with Little Ice Age (about 16th–19th centuries CE) glacier fluctuations (Luckman, 1995; Lewis and Smith, 2004). A limitation of this approach is that the time lag between moraine formation and colonization by trees is typically unknown, and can be decades to centuries. This technique is also only useful for dating events that occurred in the last few centuries to perhaps a millennium, as it is limited by the longevity of trees.

A second application of dendrochronology to date palaeoclimatic events relies on the direct impact of the event of interest on the structure of the trees. When trees are tilted, for example by snow avalanches, hurricanes or glacier advances, they respond by producing eccentric growth rings in order to restore an upright posture. The onset of this eccentric growth allows an exact calendar year to be assigned to the cause of tilting (Heusser, 1960; Luckman, 1995; Lewis and Smith, 2004).

Probably the most useful application of dendrochronology is based on the principle of cross-dating (Stokes and Smiley, 1968; Fritts, 1976). Variability in growth rates associated with climatic fluctuations leads to a unique sequence of wide and narrow rings that can be matched between samples to assign exact calendar years to wood of unknown age (Fig. 1.2). Samples from living trees have an outer-ring date that corresponds to the year that the tree was sampled. By measuring growth rates back through time a precise chronology of wide and narrow rings can be determined. Tree-ring samples of unknown age (e.g. from fallen logs or pieces of buried wood) can be compared with this known chronology and their rings assigned to specific years. Often, the inner rings in these subfossil samples will predate the earliest date of the living chronology, and the total cross-dated chronology can be extended further back in time. Using this technique of incorporating progressively older wood into tree-ring chronologies produces continuous records spanning up to 12,460 years.

Cross-dating has been widely used to determine dates of glacial advances over the past several thousand years. Glaciers that advanced into forests killed trees, which were preserved *in situ*. Recent glacial retreats have exposed these buried forests and the calendar year in which they were killed can be determined (e.g. Reyes *et al.*, 2006). The same approach has been used to date trees killed by lake level rise (Woodhouse and Overpeck, 1998) and sand dune advances (Bégin *et al.*, 1995; Wiles *et al.*, 2003). Cross-dating can also be combined with the two techniques above to allow their application to trees that are dead at the time of sampling.

## 1.3 Sources of Palaeoclimatic Information

In this section we examine the scope and variety of sources of historical climatic information. These range from organisms and parts of organisms to ice, boreholes and other physical sources of information.

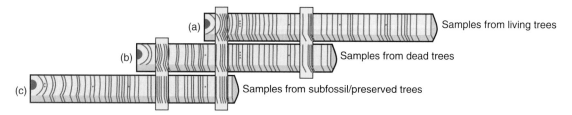

**Fig. 1.2.** Cross-dating matches the sequence of wide and narrow rings between samples of known age, such as those derived from living trees, with those of unknown age, such as those sampled from dead trees and subfossil material of unknown age, making it possible to precisely determine the age of these samples.

## Marine sediments

Oceans occupy more than two-thirds of the Earth's surface, but until recently very little was known about their climate – either in the overlying atmosphere or through the water column. Even with satellite observation systems, subsurface conditions are still poorly understood. Ocean circulation is a key component of the Earth's climate system, helping to redistribute energy from the tropics to the higher latitudes and reducing seasonal fluctuations in maritime regions. For these reasons it is especially critical that the climate of the world's oceans be better understood. Ocean floor environments are also often more stable and less affected by geomorphic processes (such as glaciation or erosion) than terrestrial environments, meaning that in most cases their preserved record is much longer.

Marine sediments accumulate at the ocean floor through the deposition of sediments that originated in continental regions, along with marine organisms. As long as the seafloor is not disturbed sediments accumulate over time, so that the youngest sediments will always be at the surface and progressively older sediments will be buried with increasing depth. Digging down into the sediments therefore amounts to digging further back in time. Two main types of information are stored in marine sediments. Inorganic sediments delivered by rivers, wind, ice or ocean currents can be used to infer properties about the source region as well as atmospheric and oceanic circulation. Organic sediments include a wide range of marine organisms that settle to the seafloor when they die, or that live directly on the seafloor. Often the organic sediments that are preserved do not directly reflect the local community due to transportation by water currents and the selective dissolution of some types of organisms.

The main way that inorganic sediments are used to infer past climate is through determination of their source region and measurement of their accumulation rate and size distribution. The main sources of terrestrial sediments are the Earth's deserts (Prospero, 1979). During periods of drought, fine particles are mobilized and can be transported thousands of kilometres from their source region. Research has shown that while the size of these particles is related to wind velocity, the total amount accumulated is proportional to aridity in the source region (Rea *et al.*, 1985). Analyses of marine sediment cores downwind from arid regions in Central Asia and Central America, for example, show reduced accumulation rates during glacial periods over the past 700,000 years that likely reflect greater humidity in these regions during periods of glaciation (Janecek and Rea, 1985).

Organic sediments account for the bulk of most marine sediments. They are composed of a wide variety of species, but only a few are useful for palaeoclimatic reconstructions. The foraminifera (amoeboid protists) are by far the most widely used marine organisms, although coccolithophores (single-celled algae, protists and phytoplankton) are also commonly used. Both groups of species share the same basic properties: they have a hard skeleton that preserves well in the seafloor environment; they can usually be identified to species; individual species have fairly precise environmental tolerances; and they have globally cosmopolitan distributions. Two main approaches are used to reconstruct climate from deposits of marine organisms: (i) when species can be accurately identified from the tests and when their behavioural attributes and environmental tolerances are precisely known, the presence or absence of individual species can be used to infer ocean conditions (most commonly temperature and salinity); and (ii) oxygen isotopes can be extracted from the carbonates in the test and their relative abundances used to infer past ocean conditions. This technique is described in more detail in the section 'Polar ice caps' (p. 13).

## Lake sediments

Like marine sediments, lake sediments contain a wealth of information about the climate system. The main differences between marine and freshwater lake sediments lie in the specific organisms that are preserved, the abundance of biotic and abiotic materials derived from terrestrial environments, the generally much higher sedimentation rates, and the typically shorter timespan that is preserved in them. As a source of information about the late Quaternary (approximately the last 40,000 years in unglaciated regions, the last 12,000 years in glaciated regions), lake sediments are unparalleled in their resolution, the quality of information offered and the spatial distribution of records.

As sediments accumulate in lakes, a wide variety of biotic and abiotic components are entrained that may be used to reconstruct past climates. There are two main sources of these components: autochthonous materials are produced within the lakes itself and allochthonous materials are transported to the lake from external sources. Many of these components provide information not only about the climatic history of the lake, but also about terrestrial and aquatic ecological responses to climatic variability and change. We examine the more commonly used records individually for now, although in practice they are typically undertaken in varying combinations with each other. There are many other less commonly used sources of information that are not discussed here; readers wishing to learn more on these topics should consult the comprehensive work of Last, Smol and colleagues (Last *et al.*, 2001, 2002; Smol *et al.*, 2002a,b).

### Pollen grains

One of the richest sources of palaeoenvironmental information over the late Quaternary is derived from reconstructions of vegetation based on the analysis of pollen grains preserved in lakes, as well as bogs and estuaries. Wind-pollinated plants require chance dispersal events to deliver pollen from the male flower to the female flower to enable reproduction. In order to maximize the likelihood of a successful pollination occurring, these plants produce huge quantities of pollen – most of which does not reach its intended destination. Some of this pollen finds its way into waterways, and ultimately is incorporated into lake sediments. The pollen grains themselves are coated in a highly resistant outer layer that preserves them in aquatic environments for exceptionally long periods of time. The morphology of pollen is highly distinctive; most grains can be identified to the genus they belong to and many can be identified to species (Fig. 1.3).

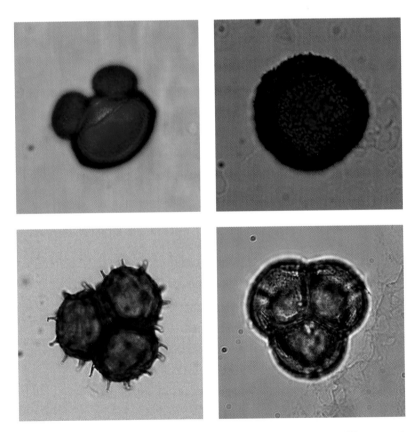

**Fig. 1.3.** Shown here (clockwise from top left) are pollen grains from mountain hemlock (*Tsuga mertensiana*), western hemlock (*Tsuga heterophylla*) and Pacific rhododendron (*Rhododendron macrophyllum*) and a spore from spike clubmoss (*Selaginella selaginoides*). (Photos: M. Pellatt, Parks Canada, Western and Northern Service Centre.)

The field of analysing pollen is called 'palynology'. Palynologists extract pollen grains from sediment cores in order to reconstruct the history of vegetation in the region of the coring site. Because the pollen contained in the core is wind dispersed, the record represents a regional perspective on the vegetation history. While pollen can be dispersed incredibly long distances, analyses have shown that the majority of pollen found in a given core is derived from vegetation growing within about 5 km, depending on the size and density of the pollen grain, as well as the size of the lake and its catchment (Bradshaw and Webb, 1985). Small lakes (<9 ha) in particular capture their dominant signal from vegetation within 1–4.5 km.

The reconstruction of climate from pollen is similar to the use of foraminifera in marine sediments. Many species (or genera) of plants occur naturally under a fairly narrow range of climatic conditions (see Chapters 4 and 5). The presence of their pollen in lake sediments therefore indicates that the climate at the time of deposition must have been within the climatic tolerances of those plants. Pollen abundance has often been used qualitatively to infer climate. For example, the presence of *Artimesia* (sagebrush) in the pollen record is indicative of cold, dry conditions; *Quercus* and *Pinus* (oak and pine) generally indicate warm, dry conditions; and *Tsuga* and *Acer* (hemlock and maple) cool, moist conditions. Because plants have these individualistic relationships with climate it is often possible to develop a transfer function (a type of mathematical model) to reconstruct the climate of the past quantitatively based on the precise assemblage and abundance of species present. Several quantitative approaches to climate reconstruction are possible, including simple ratios of key species (Brown and Hebda, 2002), multivariate statistics (Guiot, 1987) and response surfaces (Huntley *et al.*, 1993).

All of these approaches are predicated on a number of key assumptions, the most important of which is that climate is the fundamental factor driving vegetation dynamics at these spatial scales. This assumption is probably valid to a first approximation. While other factors undoubtedly affect the presence or absence of individual species (or groups of species), climate ultimately determines the range of conditions, or bioclimatic envelope, that most species can tolerate. Within this bioclimatic envelope there is a lot of variability in species presence or absence due to biotic interactions, such as insect

outbreaks, wildfire, anthropogenic activity, and abiotic factors such as processes of soil development, geological constraints and stochastic dispersal events (see Chapters 4 and 5).

Several factors interact to make the direct reconstruction of climate from pollen problematic. First, there are often significant lags between periods of climatic change and the resulting responses in the pollen spectrum. This lag is due to the time it takes for vegetation to migrate away from its previous range and to become established in its new range, and the time that established vegetation may persist on the landscape and continue to produce pollen. This situation is especially problematic in the case of conifers, which can survive as mature trees in conditions that are well outside their ideal bioclimatic envelope and continue to produce pollen even though conditions are unsuitable for their regeneration.

A second set of challenges can be generally termed the 'no analogue problem'. In its more common meaning it refers to the situation where vegetation assemblages of the past do not have modern analogues, meaning that there is no direct modern equivalent for the fossil assemblage that can be used to infer climatic conditions. This situation turns out to be more often the rule than the exception, as pollen reconstructions have consistently shown that vegetation assemblages are transient and seldom persist for more than 2000 to 5000 years (Brubaker, 1988). A related problem occurs when the climate of the past has no modern analogue. In the early Holocene this may have occurred due to fundamentally different atmospheric circulation patterns caused by presence of the retreating ice sheet (Lamb and Woodroffe, 1970). Under both of these scenarios climate reconstructions are necessarily extrapolations rather than inferences, and must be interpreted with some caution.

### Diatoms

Diatoms are a group (Division *Bacillariophyta*) of unicellular aquatic algae that are characterized by siliceous cell walls that preserve well in anaerobic environments (Battarbee *et al.*, 2001). These cell walls have complex features that allow identification of most samples to the species level once the surrounding organic material has been removed (Fig. 1.4). Optimal environmental conditions for individual species of diatoms are typically narrow, and the precise assemblage and relative abundance

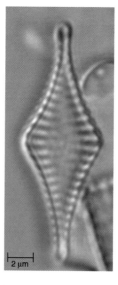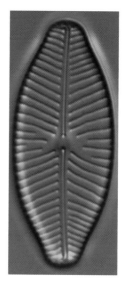

**Fig. 1.4.** Diatoms' distinctive ornamentation and narrow environmental optima make them ideal environmental indicators that allow palaeolimnologists to infer properties of the aquatic ecosystem at the time of their deposition, including temperature, pH, salinity and light regime. Shown here, from left to right, are *Aulacoseira islandica*, *Fragilaria parasitica* and *Navicula gastrum*. (Photos: K. Rühland, Palaeoecological Environmental Assessment and Research Laboratory (PEARL), Queen's University.)

present in a lake at any given time is a function of water temperature, light regime, lake chemistry (e.g. nutrient loadings, pH and salinity) and patterns of seasonal variability in these properties. As with pollen, the identity and relative abundance of diatoms present at a given point in time can be used to infer climatic conditions using a number of qualitative and quantitative techniques.

There are two important differences between diatoms and pollen. First, diatoms provide a more localized signal in both time and space than pollen. This higher resolution is a result of the fact that the diatoms entrained in the sediment core are derived almost exclusively from the lake from which the core was extracted. Diatoms also have much shorter life cycles than trees and even shrubs, and so the abundance record responds more quickly to environmental changes. Second, variability in the taxa present reflects changes to the aquatic environment rather than changes to the terrestrial environment. While mean lake temperature is often a good proxy for air temperature, a number of other factors may influence the lake environment independent of the terrestrial environment. Most importantly, changes in nutrient loadings and lake turbidity associated with geomorphic processes or

disturbance in the catchment can cause important changes in the diatom assemblage that may be unrelated to climate.

### Zoological indicators

There are a number of zoological indicators preserved in lake sediments, of which chironomids are the most widely used. The Chironomidae are collectively a group of insects known as the 'non-biting' midges and their remains are the most abundant of any insect in aquatic sediments. Chironomids complete the early stages of their life cycle in aquatic environments and, as they undergo metamorphosis, they shed body parts that may be preserved. Of these, the most important parts are the head capsules shed by the third and fourth instar larvae (Walker, 2001). These head capsules preserve well, and are sufficiently distinctive to allow identification of the taxa to species level. Cladocera are a diverse order of crustaceans, generally called 'water fleas', that can be used similarly to diatoms to reconstruct water chemistry and climate in freshwater lakes (Fig. 1.5; Bredesen *et al.*, 2002). As with other biological records, chironomids and cladocerans are sensitive to properties of the

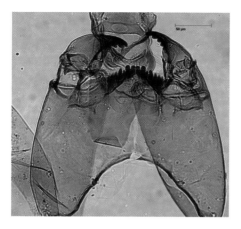
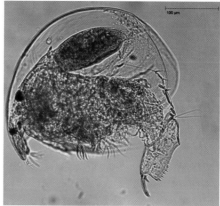

**Fig. 1.5.** Chironomids and cladocerans are important zoological proxy records of past aquatic environmental conditions. Their distinctive morphologies and precise environmental tolerances allow palaeolimnologists to use their presence or absence to infer properties of the aquatic ecosystem at the time of their deposition, including temperature, pH, salinity and light regime. Shown here are the head capsule of the chironomid *Sergentia* sp. (left) and the cladoceran *Chydorus* sp. (right). (Photos: courtesy of J. Sweetman, Parks Canada, Western and Northern Service Centre.)

aquatic environment, including temperature and water chemistry, and their analysis and interpretation are similar to diatoms.

### Polar ice caps

Probably no source of palaeoclimatic data has been as informative as the ice cores taken from the polar ice caps in Greenland and Antarctica. Because the temperature of these polar regions is permanently below freezing, snow that falls on glacier surfaces does not melt and is ultimately incorporated into the glacial ice as subsequent snow accumulates at the surface. Deeper ice therefore represents older snow, and provides a snapshot of environmental conditions at the time of deposition (Fig. 1.6). Palaeoclimatic information is available from a range of sources within ice cores: the relative abundance of stable isotopes in the ice, air bubbles trapped in the ice, particulate matter entrained in the ice and the structure of the ice itself.

In addition to the radioactive isotopes used to date material described above, there are also commonly occurring isotopes that do not decay over time but can change in relative abundance due to climatic changes. The most important of these isotopes for climatic reconstruction are $^2H$ and $^{18}O$. $^2H$ is an isotope of hydrogen that contains an extra neutron. It is also called deuterium, and is often identified using the symbol D. Oxygen occurs in

**Fig. 1.6.** Ice cores, both from polar ice caps and from high-elevation glaciers, contain a wealth of palaeoenvironmental information. Here, scientists from the GISP2 drilling project in Greenland extrude a core from its barrel. In preparing the core for analysis it is critical that the sample is not exposed to modern contaminants such as dust or water, and that the core remains solidly frozen. (Photo: M. Twickler, Climate Change Research Center, University of New Hampshire.)

three stable forms: $^{16}O$ is the most common isotope, but $^{17}O$ and $^{18}O$ also occur naturally. Hydrogen and oxygen are of interest to palaeoclimatologists because they are the constituents of water. Three forms of water are of particular interest: (i) $^1H_2{}^{16}O$ is the most common type of water molecule; (ii) $^1H^2H^{16}O$ (also identified as

HDO) contains one normal hydrogen atom, one deuterium atom and one normal oxygen atom; and (iii) $^1H_2^{18}O$ contains two normal hydrogen atoms and one oxygen-18 atom. Other combinations are possible, but are so rare that they are not used for palaeoclimatic reconstruction.

Palaeoclimatic reconstructions using the relative abundances of these water molecules are possible because their different masses cause them to evaporate and condense at different rates depending on air temperature. $^1H_2^{16}O$ is the lightest form of water and is evaporated most readily. This has the effect of producing water vapour that is relatively depleted in the two heavier forms of water and concentrating heavier forms in the liquid water body. This process of separating heavy and light isotopes from each other is termed 'fractionation'. The degree of fractionation is, with all other factors being held constant for now, an inverse function of temperature. That is, higher temperatures result in relatively more of the heavy isotopes being evaporated. The opposite effect occurs when water vapour condenses to form precipitation. Heavier water molecules condense more readily and thus are removed by precipitation more quickly than lighter molecules. Consequently, precipitation that occurs at

higher temperatures is relatively enriched in heavier isotopes and precipitation that occurs at lower temperatures is relatively depleted (Fig. 1.7).

There are a number of other factors that affect the relative abundance of heavier isotopes in precipitation. Most critically, variability in the source water region, the distance from the source region, elevation and the amount of water stored in glaciers contribute to variability in the relative abundances of the various forms of water molecules. At a global scale, the tropics are less depleted of heavier isotopes than are polar regions. Continental interior regions also receive precipitation that is more depleted than coastal regions. As air masses are advected towards the continental interior heavier isotopes condense and are lost to precipitation, and water vapour originating in terrestrial regions (e.g. transpiration from plants, evaporation from lake surfaces) forms a larger component of subsequent precipitation. Similarly, higher elevations also receive precipitation depleted in heavier isotopes due to orographic lifting, which preferentially removes the heavier isotopes during precipitation at lower elevations.

At very long time scales, the above factors combine to cause the oceans to be relatively enriched in

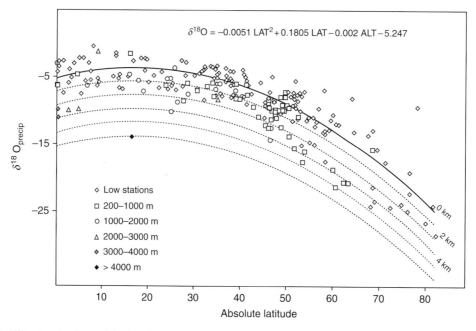

**Fig. 1.7.** $\delta^{18}O$ of meteoric precipitation shown as a function of latitude and elevation. (Redrawn from Bowen and Wilkinson, 2002.)

Chapter 1

heavier isotopes during times of glaciation. Because glaciers form at high latitudes and high elevations, they are composed of water that is highly depleted of heavy isotopes. As more water is removed from the oceans and stored in glacial ice, lighter isotopes accumulate in that ice and heavier isotopes are concentrated in the oceans.

In addition to the isotopic chemistry of the water from which the glacial ice is composed, ice cores also contain trapped air bubbles that provide a snapshot of the atmosphere at the time that they were entrained in the ice. Of particular importance, reconstructions of atmospheric composition using trapped air bubbles have extended the instrumental record back in time several hundred thousand years, providing a context for recent observed changes. These reconstructions have shown that radiatively important gases such as $CO_2$, methane ($CH_4$) and nitrous oxide ($N_2O$) have fluctuated in their abundance considerably over glacial and interglacial time periods. Most critically, these 'greenhouse' gases have co-varied with temperature over at least the last 650,000 years (Siegenthaler et al., 2005), confirming a strong link between greenhouse gas concentrations and climate (see Box 1.1).

A third source of palaeoclimatic data that can be derived from ice cores derives from the analysis of dust and volcanic ash. As with analysis of marine sediment cores, the dust can be analysed to determine its source region. Increased dust concentrations in ice cores correspond to periods of drought and consequent wind erosion. Similarly, aerosol particles may reflect changes in global cloudiness that have important implications for the Earth's energy balance and climate system. Volcanic eruptions produce sulfur gases and particles that are transported very long distances from the site of the eruption and are recorded in the chemical and particulate composition of glacial ice.

## Boreholes

Another recently developed source of palaeoclimatic data is geothermal temperature profiles, the study of which is generally called 'borehole climatology' (Bodri and Čermák, 2007). Scientists have been studying the temperature profiles of the Earth's outer crust since the 1920s, but only recently have these profiles been used to reconstruct the temperature history of the past several centuries. Borehole climatology is based on the fact that rock is a very poor conductor of heat. The temperature at the Earth's core is very high and decreases towards the surface. Heat flow from the Earth's core is assumed to be constant, so that under a constant climate the subsurface temperature profile would be a fixed function of depth. Variations in temperature at the Earth's surface propagate downward very slowly. Diurnal temperature fluctuations typically affect only the top 1 m of rock, and annual variations are barely detectable below 10 m. Prolonged periods of colder or hotter temperature than normal will propagate slowly downward from the Earth's surface and appear as deviations from the fixed function of depth. Once the geothermal temperature profile has been measured, the thermal conductivity of the rock can be determined and the temperature history can be reconstructed using inversion techniques.

There are two distinct advantages of boreholes as a source of palaeoclimatic data over other data sources. First, because subsurface temperatures are being measured directly, they provide a direct measurement of past temperature regimes, rather than an indirect measure. Second, the method can be applied anywhere in the world, allowing truly global reconstructions to be developed. There are also limitations to the method, though. Most critically, the further back in time we are interested in examining, the more diffuse the signal is. For this reason, reconstructions typically extend back only about 500 years (Pollack et al., 1998; Beltrami et al., 2003; Mann et al., 2003).

## Tree rings

In addition to being a tool for dating, tree rings also contain a lot of information directly related to the climate system. Dendroclimatology is the science of using the growth rings of trees to understand variability in the climate system. Tree-ring data for climate reconstructions can be derived from measurements of features of the growth rings and from the chemical and isotopic composition of the wood within the growth rings themselves (Cook and Kairiukstis, 1990). Trees have many desirable qualities as natural archives of climatic information: they are widely distributed in regions with strong seasonal variability in climate; they have annually resolved growth rings; they are sessile; they typically live for centuries and some

## Box 1.1. CO$_2$ and climate.

The reason we care about CO$_2$ is that it is a radiatively active gas. That is, it is transparent to visible light from the Sun, but is opaque to energy in ranges of the thermal infrared portion of the spectrum. The fact that CO$_2$ is a greenhouse gas has been known since the Victorian scientist John Tyndall conducted experiments on it in 1859. Increasing amounts of CO$_2$ in the atmosphere have the potential to raise the Earth's mean temperature. It is therefore of interest to know how temperature and CO$_2$ have varied over the Earth's history. Reconstructions of temperature and CO$_2$ concentrations from the Vostok ice core make this comparison possible (Petit *et al.*, 1999; Fig. 1.8).

An examination of the history of CO$_2$ and temperature reveals at least three important observations: (i) temperature and CO$_2$ have co-varied very closely over the past 420,000 years; (ii) modern CO$_2$ levels are higher than they have been at any point in that period; and (iii) of some scientific interest, rises in temperature *precede* rises in CO$_2$. This last point has been emphasized by climate change deniers who argue that it is temperature that determines CO$_2$, not the other way around. There are several flaws with this argument and it is worth spending some time addressing them.

The fact that CO$_2$ increases lag behind temperature increases was predicted by scientists long before the evidence was available to show that the relationship does, in fact, work this way (Lorius *et al.*, 1990). The explanation for this apparent paradox can be summarized fairly simply. Previous glacial periods have been ended because of changes in orbital forcing (see Section 1.4) that have warmed the Earth's surface by a small amount. This warming releases CO$_2$, CH$_4$ and other greenhouse gases that were being stored in oceans, soils and biomass. These greenhouse gases cause warming to continue for many thousands of years. We now examine the evidence for this relationship in a little more detail.

Before we can address *why* temperature changes have preceded CO$_2$ changes, it is worth looking at the time length by which the two records differ. This question is not as easy to answer as it might appear from examining Fig. 1.8. The reason for this is that while [18]O (which is used to reconstruct temperature records) is derived directly from the ice that forms the glacier, and thus corresponds directly to the temperature at the time of snow deposition, the air bubbles that are used to reconstruct the composition of the atmosphere are trapped at some

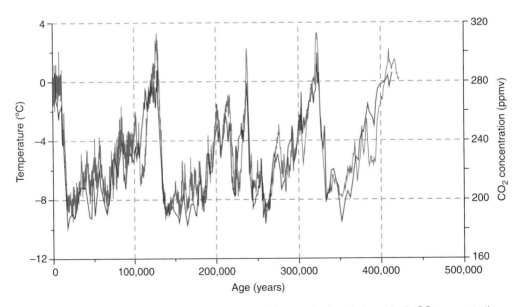

**Fig. 1.8.** Temperature anomalies relative to the 1960–1990 mean (red) and atmospheric CO$_2$ concentrations in parts per million by volume (ppmv; blue) reconstructed from the Vostok Dome C ice core (Petit *et al.*, 1999). For reference, the 2009 mean atmospheric CO$_2$ concentration at Mauna Loa was 387.35 ppm.

later date. This disparity occurs because snow is porous and until there is sufficient mass accumulation that the snow is compacted into ice, there is an exchange of air between the upper portions of the glacier and the atmosphere. There is, therefore, a lag between the two records that has nothing to do with their relationship to each other, but simply reflects differences in how they are recorded. While interpretations vary, it is likely that in the Vostok core the air bubbles are 4000 to 6000 years older than the ice in which they are embedded (Barnola *et al.*, 1991). Even accounting for this difference, though, a lag of approximately 800 years remains between initial rises in temperature and rises in $CO_2$ (Monnin *et al.*, 2001). This lag can be seen most readily as the termination of glacial periods, such as occurred at the start of the Holocene.

This disparity occurs because it was not the rising $CO_2$ levels specifically that initiated warming at the end of the last glacial period. Instead, two changes in the orbital properties of the Earth coincided to result in increased insolation in the northern hemisphere. Specifically, the Earth's tilt on its axis was maximized, such that the northern hemisphere received more direct sunlight in the summer. Enhancing this effect were changes in the Earth's precession, which meant that the Earth was closest to the Sun during the boreal summer. These factors combined to initiate melting of continental ice sheets and reduce areas of snow cover, which in turn decreased the surface albedo and would have contributed to additional warming. Modelling efforts show, however, that these effects are insufficient to warm the planet as much as has been observed and additional mechanisms are needed to cause the magnitude of warming that has been seen. This additional warming can be explained only by including forcing from $CO_2$ and other greenhouse gases.

During the last ice age atmospheric $CO_2$ levels were about 190 ppm – substantially lower than modern levels. This carbon was stored in a wide range of reservoirs, including the southern oceans, terrestrial soils and of course fossil fuels. As the Earth's temperature rose, some of this $CO_2$ would have been released to the atmosphere, changing the radiative balance of the Earth and amplifying the original warming signal. In short, while the first 800 years of warming cannot be explained by $CO_2$ increases, the more than 4000 years of warming that followed may well be related to $CO_2$. Indeed, most climate models suggest that between half and two-thirds of the observed warming is probably a response to increasing greenhouse gases.

In any case, what matters now is not whether increases in $CO_2$ *initiated* warming in the past but whether they can *cause* warming. The evidence from the ice core record suggests that they can. To suggest that because temperature increases can cause $CO_2$ increases the opposite is impossible is akin, as one anonymous online user noted, to concluding that chickens cannot lay eggs because they have been observed to hatch from them. It does not matter what caused either temperature increases or $CO_2$ increases in the past; we know what the source of current elevated $CO_2$ concentrations is, and we know that it is a greenhouse gas and is capable of warming the planet.

species can live for millennia. Wood also preserves well in many locations, allowing very long chronologies to be constructed by cross-dating subfossil material. In terms of understanding the climate of the last millennium, dendroclimatology has probably been the most important source of proxy data (Jansen *et al.*, 2007), although its use is not without controversy (see Box 1.2).

In most temperate regions trees produce alternating bands of light and dark wood at different times of the growing season (Fig. 1.10). There are many properties of growth rings that can be measured, many of which can be related to climatic variability. The most commonly used features of tree rings that are used for climatic reconstruction include total ring width, earlywood and latewood width, and maximum wood density (measured using X-ray machines). These features all vary in response to a wide range of biological and environmental factors leading up to the timing of cell division and growth, including the species of tree, its age and size, the availability of carbohydrate reserves, its total leaf area, available nutrients, the effects of competition with neighbours, seasonal variability in sunlight, temperature, precipitation, wind and damaging agents.

The goal in dendroclimatology is to sample trees in which the climatic signal is likely to be the

## Box 1.2. Dendroclimatology: the hockey stick debate.

*The 20th century was the warmest century in the last 600 years – and the warmest years in all of that period were 1990, 1995 and 1997. 'Our conclusion was that the warming of the past few decades appears to be closely tied to emission of greenhouse gases by humans and not any of the natural factors', said Dr Michael Mann.*

This was how the press reported what has come to be known as the 'hockey stick' result (Fig. 1.9). This reconstruction was originally published in the journal *Nature*, but eventually it was featured prominently in the Third Assessment Report of the Intergovernmental Panel on Climate Change (IPCC), appearing in both the Summary for Policymakers (Watson and Team, 2001) and the Scientific Basis report (Houghton *et al.*, 2001). Over the several years following its publication, a debate developed within the peer-reviewed literature, spilled over to the Internet and into the popular press, was investigated by the US House of Representatives' Committees on Science, and Energy and Commerce, and ultimately resulted in a US National Academies of Science hearing and report (NAS, 2006). The controversy began when Ross McKitrick and Stephen McIntyre attempted to replicate the result of Mann *et al.* and were unable to do so. In 2003 they published the first of several

papers in what has turned into a protracted debate about the veracity of the hockey stick result, the 'truth' about past warming, the necessity for greenhouse gas abatement, professional ethics and, it would seem, personal animosity (Pearce, 2006).

The controversy surrounding the reconstruction derives from several related issues. First, in attempting to replicate the Mann *et al.* (1998) analysis, McIntyre and McKitrick discovered that the reconstruction in general and the hockey stick itself depend critically on the inclusion of a particular set of bristlecone pine (*Pinus longaevea* and *Pinus aristata*) tree-ring chronologies. These chronologies arguably show 'excessive growth' during the 20th century. This growth could not be explained by instrumental temperature measurements and was thought to reflect the fertilization effect of the additional $CO_2$ that has been accumulating in the atmosphere since about 1850 (Lamarche *et al.*, 1984). Mann *et al.* (1999) linearly removed the growth component that they believed was attributable to elevated $CO_2$, and demonstrated that the hockey stick result was robust at least to this assumption. However McIntyre and McKitrick (2005b) pointed out that: (i) this correction was not applied to the Mann *et al.* (1998) reconstruction; (ii) the correction

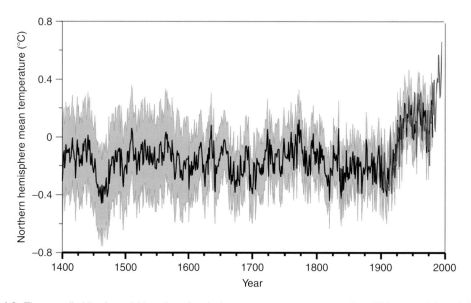

**Fig. 1.9.** The so-called 'hockey stick' northern hemisphere temperature reconstruction of Mann *et al.* (1998). The grey shading indicates the 95% confidence limits on the temperature estimate, and suggests that the 1990s were the warmest decade of the last 600 years, and that 1998 was the warmest year of the reconstruction.

suggests a stronger $CO_2$ fertilization effect in the 19th than in the 20th century (which Mann et al. (1999) attribute to $CO_2$ saturation); and (iii) it is possible that the excess growth is not related to $CO_2$, but to either increased precipitation or increased grazing pressure – both of which could enhance growth by reducing moisture stress. If this were true, they argue, the 'correction' could have 'further distorted' the proxy record (McIntyre and McKitrick, 2005b). A recent analysis by Salzer et al. (2009) suggests that neither of these explanations is correct, and that the excessive growth may in fact be related to a unique warming at particularly high elevations. This finding suggests that the bristlecone pine chronologies are consistent in their relationship to climate and that no correction is necessary.

A second concern raised by McIntyre and McKitrick relates to the application of principal components analysis (PCA) used by Mann et al. PCA is a statistical technique used to take a large number of variables that co-vary and reduce these variables to a smaller set of principal components (PCs), all of which are independent of each other. Mann et al. used an 'unconventional' method of data pre-processing that biased the solution towards hockey stick-shaped results (McIntyre and McKitrick, 2005a). When alternative pre-processing methods favoured by McIntyre and McKitrick are used, the hockey stick does not emerge as the dominant PC – but instead is extracted as the fourth most important pattern (PC4). If PC4 is left out of the resulting reconstruction then no hockey stick results; if it is included then the hockey stick returns. Wahl and Ammann (2007) argue that in order to represent the full structure of the tree-ring data, more PCs are needed and PC4 should be included. Additionally, if the data are standardized as part of the pre-processing (an uncontroversial treatment and one that was part of the Mann et al. original method), the hockey stick comes out as PC2 and would be used in either approach.

## Is this controversy proof that climate change is all bunk?

In order to make rational decisions about climate change, we would ideally like to know the answers to three big questions: (i) Can humans change the climate? (ii) How much, by when, and with what confidence do we think that climate will change? (iii) What are the likely impacts of this degree of climate change? Mann et al. (1998) had apparently answered (i) in the affirmative. Even if we conclude that their findings are unreliable, it does not necessarily mean that the alternative is true. Despite the sensitivity of their findings to the presence or absence of the bristlecone pine data, there are other hockey stick-shaped, proxy-based reconstructions that use neither the bristlecone pine nor the contentious pre-processing technique. There are also other lines of evidence that suggest the observed warming is historically unusual, such as instrumental records (Jones, 1999), changes in plant and animal phenology (Visser and Both, 2005), borehole records (Pollack et al., 1998; Huang et al., 2000), temperature reconstructions based on diatoms (Smol et al., 2005), and a rather strong argument from first principles about the physics of greenhouse gases and the determinants of climate (Houghton et al., 2001).

And even if we are utterly unable to show that humans have not yet caused any appreciable change in climate, this does not mean that humans are incapable of causing such change in the future. The change predicted for the next 100 years is much larger and more rapid than any period of warming in the last 1000 years. The veracity of this claim does not rest on whether or not we accept Mann et al.'s result.

strongest and the other causes of variability in ring features are minimized (or at least well enough understood to be mathematically filtered out of the ring-width series). The normal approach to achieving this goal is to sample trees growing near the limits of their ecological amplitude – specifically those growing near the high-elevation treeline, near the edges of their geographical distribution or at the forest–non-forest ecotone. Because these trees are growing near to the climatic limits of what they can tolerate, they are most likely to produce ring properties that co-vary with the lim-iting climatic factor. In dendroclimatology, this concept is known as the 'principle of limiting factors' (Speer, 2010).

## Other records

Here we have presented only the most commonly used sources of palaeoclimatic information that are relevant to understanding how the climate system has behaved in the past and how ecosystems have responded to these changes. There are many tools and techniques we have not discussed,

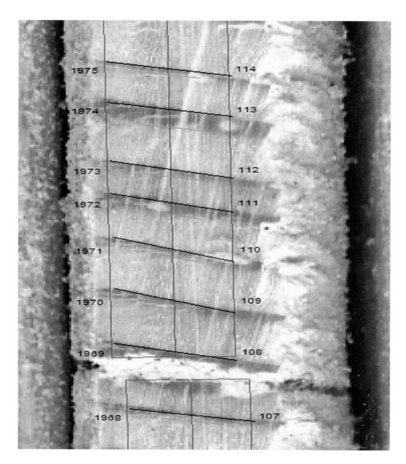

**Fig. 1.10.** A core sample from a Douglas-fir (*Pseudotsuga menziesii*), with ring boundaries shown using the WinDendro image analysis system. Early during the growing season trees growing in temperate regions typically produce cells that are large, thin-walled and light in colour, termed 'earlywood'. Towards the end of the growing season they produce smaller, denser cells with thicker walls, termed 'latewood'. The boundary between earlywood and latewood is shown here by the thin blue lines, and between annual rings the boundary is shown by the wider black lines. A small crack in the core is visible between 1968 and 1969. This variability allows the precise identification of the timing and magnitude of growth over the tree's life. (Photo: Z. Gedalof, Department of Geography, University of Guelph.)

for various reasons. For example, palaeoclimatic modelling studies use first principles of general circulation to simulate what the climate of the past should have been like given the Earth–Sun geometry, the geography of the continents, solar brightness and other fundamental controls on the climate system (Cane *et al.*, 2006). These models provide critically important insights into the physics of the climate system that guide the development of models that project future climate (see Chapter 2). However, they do not provide information that is directly relevant to understanding the effects of climatic change on

biological processes, so we have not addressed them here.

Similarly, there is a bewildering array of emerging tools and techniques that will undoubtedly prove useful in years to come in terms of understanding the climate history of the Earth and the response of ecosystems to climatic change. One obvious example is in the field of sclerochronology. Several recent breakthroughs have shown that the hinges of bivalves have annual features that are highly sensitive to climatic variability. For example, geoduck clams (*Panopea generosa*) can live to be 100 years old and have annual shell-growth increments that are very strongly controlled by seawater temperatures (Strom *et al.*, 2004; Fig. 1.11). Freshwater mussels live for over 80 years, and appear to be similarly sensitive to water temperature. Rockfish otoliths (ear bones) have annual increments, are climatically sensitive, and rockfish live to over 100 years (Black *et al.*, 2008).

Alongside new sources of palaeoclimatic information, new methods of analysis are being developed. Studies are adopting multi-proxy, multi-scale approaches that allow contributing proxy data to be used to their best advantage. Palaeoclimatic models are being used to assess the strengths and limitations of proxy records

and proxy networks. The science of palaeoclimatology is relatively young. While much has been learned already, it is an exciting time to be active in the field!

## 1.4 Causes of Climatic Variability

Now that we have examined how it is that we have learned about the climate of the past, it is worth spending some time examining the Earth's history of climatic variability. We examine this history first by looking at the causes of climatic variability in some detail. We then turn our attention to the history of the Earth's climate, examining climate at various scales, but with a focus on the climate of

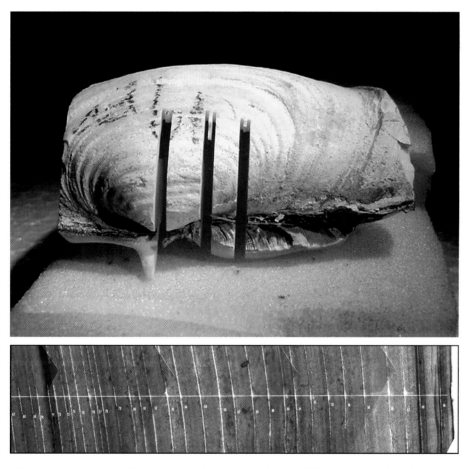

**Fig. 1.11.** The giant geoduck clam (*Panopea abrupta*) can live to be over 100 years old (top), and produces annual bands in its shell (bottom), analogous to the growth rings of a tree, that are sensitive to variability in water temperature. (Photos: Arë Strom, State of Washington Department of Fish and Wildlife.)

the Holocene (i.e. approximately the last 12,000 years). We end the chapter with a short summary of lessons for the future that might be derived from the palaeorecord.

## Tectonic-scale climatic change

At the longest time scales, the climate of the Earth has been affected by processes that are driven by the Earth's internal heat. Most critically, plate tectonics has altered the geometry of the continents and oceans and has affected how much carbon is stored in carbonate rocks versus other components of the carbon cycle. One remarkable feature of the Earth's climatic history is that over its 4.55 billion year history the climate has never been so cold or so hot that liquid water was not present at the surface. This remarkable quality is one of the features that distinguishes Earth from our nearest neighbours, Mars and Venus, and makes life possible.

The relative stasis in the Earth's climate at these very long time scales is most likely the result of a feedback system that exists between the rock cycle and the hydrologic cycle. During warm intervals, weathering of carbonate rocks occurs more rapidly due to the presence of organic acids associated with plant growth and the hydrologic cycle is accelerated due to the higher temperatures. So how does weathering of terrestrial rocks remove carbon from the atmosphere? To answer this question precisely requires a good deal of chemistry, but the arm-waving explanation is that small amounts of atmospheric $CO_2$ get dissolved in rainwater, which forms a weak carbonic acid ($H_2CO_3$). This acid is particularly effective at weathering carbonate rocks through hydrolysis, producing carbon compounds that are eventually incorporated in the shells of marine organisms and then ultimately are incorporated into sedimentary rocks. During cold intervals rates of weathering are reduced, causing $CO_2$ to accumulate in the atmosphere and leading to warming over very long time scales. This feedback cycle seems to have kept the Earth's climate temperate over most of its 4.55 billion year history, in spite of considerable variability in the composition of the atmosphere and the brightness of the Sun.

## Orbital-scale climatic change

From a human perspective the Earth–Sun geometry seems to be very stable: the Earth rotates on its axis once per day – giving us diurnal variability in climate – and it is tilted on its axis by 23.5° relative to the plane of its revolution around the Sun, giving rise to seasonal variability. However, at longer time scales, variations in the Earth–Sun geometry have important effects on climate. Three features of the Earth–Sun system are particularly important: (i) the tilt of the Earth on its axis determines the magnitude of seasonal variability in the mid-latitudes; (ii) the eccentricity of the Earth's orbit around the Sun affects the distance between the Earth and the Sun; and (iii) precession of the Earth's axis determines how close it is to the Sun at different times of the year. The role of these factors in regulating the Earth's climate was originally hypothesized by the Serbian scientist Milutin Milanković. Milanković developed his hypothesis while he was a prisoner of war in Belgrade during World War I, and the impact of these mechanisms on climate is now generally termed 'Milankovitch forcing'.

The angle of the Earth's tilt on its axis is measured relative to its plane of revolution around the Sun. Presently, that tilt is 23.5°. Consequently, at the summer solstice, the northern hemisphere is tilted towards the Sun by 23.5° and northern polar regions (i.e. those within 23.5° of the north pole, or northward of 66.5°N) receive continuous sunlight. The southern hemisphere is tilted away from the Sun, and the southern polar regions receive no sunlight at all. Thus, to a large degree, the tilt of the Earth on its axis determines the magnitude of seasonal variability. This tilt is not constant, though, and as the Earth wobbles on its axis it oscillates between about 22.2° and 24.5°. When the tilt is less than it is now, differences between summer and winter climate are reduced, and when it is greater seasonal differences are enhanced. The time it takes for one complete 'wobble' (i.e. for the tilt to go from 22.2° to 24.5° and back to 22.2°) is about 41,000 years.

The eccentricity of the Earth's orbit refers to how circular or elliptical the path followed by the Earth is as it revolves the Sun. At present the Earth's orbit is relatively circular, but in the past it has been even more circular and also much more elliptical. When the Earth's orbit is more elliptical, the Earth–Sun distance varies as the Earth revolves around the Sun. Not surprisingly, the climate is warmer when the Earth is close to the Sun and cooler when it is further away. Perhaps less intuitively, though, the eccentricity of the orbit interacts with the tilt to determine at what time of the year the climate is warmer or cooler. For example, when the Earth's orbit is highly elliptical and the Earth is furthest from the Sun during the boreal summer and closest during the boreal winter, then the northern hemisphere will experience relative cold summers and relatively warm winters. The southern hemisphere will experience the opposite conditions, with particularly hot summers and cold winters. The eccentricity of the Earth's orbit varies at a range of frequencies, but the strongest occur at scales of approximately 100,000 years, 413,000 years and 2.1 million years.

The final attribute of the Earth's orbit that affects climate is its precession on its axis. Much like the spindle on a spinning top, the direction that the axis points relative to the stars 'wobbles' over time. The effect that this has on climate results from interactions with the two other motions described above, such that the seasonality associated with the time of year that the Earth is closest and furthest from the Sun varies. This type of precession is called 'axial precession', and one complete cycle takes 25,700 years. A second type of precession occurs in the ellipse of the Earth's orbit itself. The effect of this precession is similar to the axial precession, due to its effect on the seasonal differences in the Earth–Sun distance. The two precessions together are known as the 'precession of the equinoxes' and have a strong periodicity of approximately 23,000 years, with other weaker periods also affecting the Earth's orbit.

Milankovitch forcing has a clearly evident impact on the climate of the Earth (Berger, 1988). Most of the Pleistocene glaciations coincide with intervals when the northern hemisphere received less total insolation during summer months. There is a common conception that the Earth is on the brink of entering another ice age and that global warming has saved humanity from this fate. While the first part of this statement is arguably factually correct, the brink on which the climate system is perched is geological in nature – calculations show that northern hemisphere insolation will, in fact, increase gradually over the next 25,000 years. Declines sufficient to cause an ice age are not expected for at least 50,000 years, and possibly twice that long.

## Solar brightness

The final 'external' factor affecting the Earth's climate that we consider relates to variability in solar brightness. While a full discussion of the history of solar variability and its contribution to climate is beyond the scope of this book, two features of the Sun's output are worth examining in a little more detail. First, it is now generally understood that early in the history of the solar system's development the Sun was likely much dimmer than it is today – perhaps emitting 25% less energy when life first developed 3.8 billion years ago (Sagan and Mullen, 1972). How is it that the Earth did not enter a deep freeze like our neighbouring planet Venus appears to have done? The available evidence suggests that the early Earth atmosphere was likely enriched in $CO_2$, ammonia ($NH_3$) and other greenhouse gases that kept the early Earth warm enough to support liquid water (Sheldon, 2006). As the Sun brightened, interactions with the tectonic cycle and the effects of plants on the Earth's atmosphere reduced the total amount of greenhouse gases at almost exactly the rate needed to keep a moderately stable climate on the Earth's surface. This finding is so remarkable that creationists have pointed to it as evidence that the Earth must have been created. Others, such as James Lovelock (1987), have used it as evidence that the Earth is a self-regulating system. Does this mean that the Earth will respond to current increases in $CO_2$ by somehow extracting excess greenhouse gases from the atmosphere? Perhaps, but the evidence from these previous eras suggests that these changes will take tens of thousands to millions of years to occur. We will be affected by the changing composition of the atmosphere more immediately than that!

While the Sun has got brighter over the scale of billions of years, superimposed on this are shorter-scale fluctuations. For example, the 11-year sunspot cycle is associated with alternating intervals of slightly reduced and slightly enhanced solar activity. Of most interest to climate change research, the Sun was likely dimmer during the coldest intervals of the Little Ice Age and from the 1960s to 1980s, a time of global cooling. However, other periods of decreased and increased solar activity have not corresponded to climatic changes, and there is still much to learn about the link between solar activity and climate. Nevertheless, reconstructions of the Sun's brightness suggest that overall it has got brighter over the 20th century (Foukal *et al.*, 2006) and it is important to determine what proportion of the observed warming is attributable to solar variability in order to understand the sensitivity of the climate system to changing levels of greenhouse gases.

Determining the link between solar activity and climate has proved to be surprisingly difficult. In part this is because measuring solar brightness turns out to be exceptionally difficult. Surface measurements are problematic because of interference by clouds, aerosols and atmospheric pollutants. Satellite measurements are more reliable, but extend back in time only a few decades. For this reason proxy measurements must be used, such as observations of sunspot activity and the relative abundance of radioactive isotopes (formed by cosmic rays) that are found in ice cores and tree rings (see Section 1.2). Additional complications derive from the relatively short duration of the instrumental record of climate. Two main tools have been used, then, to understand the link between solar brightness and climate. First, a wide range of proxy measurements of temperature has been used to reconstruct the link between solar brightness and climate. These analyses vary in their findings, but none supports a strong link between solar variability and climate, and none suggests that the increased brightness seen over the 20th century is sufficient to cause the observed warming. The second approach is to use GCMs (see Chapter 2) and to force them with solar brightness alone, greenhouse gases alone and the two in combination. Again, results differ in the particulars, but none suggests that solar brightness can explain the total observed increase in temperature alone; indeed it is probably responsible for only about 0.1°C of the observed warming (e.g. Lean, 2010). Similarly, the IPCC Fourth Assessment Report concluded that the total radiative forcing that could be attributed to increasing solar brightness was only about 0.12 W/m². For comparative purposes, the radiative forcing attributable to greenhouse gases is about 2.3 W/m² (Fig. 1.12). Thus, while changing solar brightness is an important contributor to climatic variability at both short and very long time scales, it cannot be responsible for the current warming trend.

## 1.5  The History of the Earth's Climate

The history of the Earth's climate could easily fill a number of textbooks by itself. Interested readers might find William F. Ruddiman's book *Earth's Climate: Past and Future* (Ruddiman, 2001) an engaging and accessible starting point. Here we focus on the climate of the Quaternary Period, looking at some important intervals and scales of climatic change as a lens through which we might contextualize anticipated future changes. The Quaternary Period is divided into two epochs: the Pleistocene, which began 2.59 million years BP, and the Holocene, which began approximately 10,000 years BP. In geological terms, the Quaternary Period represents only the briefest snapshot of the Earth's history. If the 4.55 billion year history of the Earth were scaled to a single day, the Quaternary would not even begin until less than one minute before midnight. Yet in terms of climate change biology it is of particular interest for a number of reasons: the Earth–Sun geometry, atmospheric chemistry and the distribution of the continents have consistently been more or less as they are now; 2.59 million years is long enough to have had an evolutionary impact on most species and so examining how species have responded to climatic change over this period is of relevance to future climatic change; and there are relatively well preserved archives of climate and ecosystems for many parts of the world, so we can examine the effects of climatic change on biota.

### Climate of the Pleistocene – glacial and interglacial periods

The Pleistocene epoch initiated with the uplifting of the Tibetan Plateau, the closing of the Panama Isthmus and the formation of glaciers in the northern hemisphere. The climate system in general became very unstable, with rapidly fluctuating transitions between glaciated and unglaciated conditions, rising and falling sea levels, and rapidly changing ecosystems (Webb and Bartlein, 1992). Mean temperatures were also generally cooler than in the preceding Pliocene epoch. These climatic changes were likely a consequence of orbital forcing, solar brightness, atmospheric composition and internal feedbacks. A high-resolution reconstruction of climate over the last 650,000 years shows six well-defined glacial intervals punctuated by relatively short-lived interglacial periods (Siegenthaler *et al.*, 2005). A distinctive feature of this particular reconstruction is that it provides an opportunity to look at the relationship between atmospheric chemistry and climate. As it turns out, temperature and $CO_2$ have closely tracked each other, as have other greenhouse gases, for at least the last 650,000 years (see Box 1.1).

There are at least two important lessons that biologists should take from observing the climate

Radiative forcing of climate between 1750 and 2005

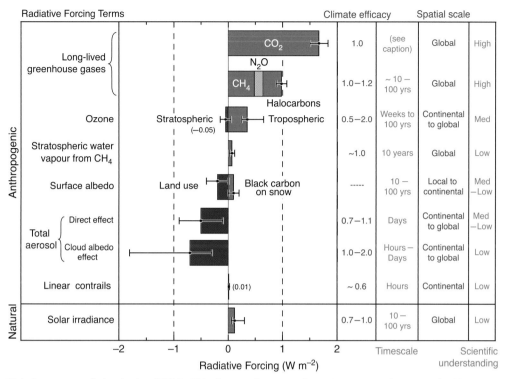

Global average radiative forcing (RF) in 2005 (best estimates and 5 to 95% uncertainty ranges) with respect to 1750 for $CO_2$, $CH_4$, $N_2O$ and other important agents and mechanisms, together with the typical geographical extent (spatial scale) of the forcing and the assessed level of scientific understanding (LOSU). Aerosols from explosive volcanic eruptions contribute an additional episodic cooling term for a few years following an eruption. The range for linear contrails does not include other possible effects of aviation on cloudiness. (From *Climate Change 2007: Synthesis Report. Contribution of Working Groups I, II and III to the Fourth Assessment Report of the Intergovernmental Panel on Climate Change*, Figure SPM.5. IPCC, Geneva, Switzerland.)

**Fig. 1.12.** The magnitude of global average radiative forcing (RF) estimates and uncertainties (as of 2005) for anthropogenic carbon dioxide ($CO_2$), methane ($CH_4$), nitrous oxide ($N_2O$) and other important agents and mechanisms, together with the typical geographical extent (spatial scale) of the forcing and the assessed level of scientific understanding. What is clear from this figure is that $CO_2$ is clearly the most important of the so-called 'long-lived greenhouse gases'. In particular, its effects on radiative forcing are probably more than ten times greater than those attributable to increases in solar irradiance.

of the Pleistocene. First, over most of the past 3 million years, the climate of the Earth was very different from what we experience today. Evolutionary selection has largely taken place under a colder and drier climate, with substantially different circulation patterns. Second, although climate has changed substantially and abruptly in the past, it is very possible that climate in the next several decades to centuries will change both more rapidly and by a greater magnitude than anything that has occurred in the last 3 million years (see Chapter 2).

**Climate of the Holocene**

The Pleistocene epoch ended approximately 10,000 years BP with the retreat of the last continental glacier. It was followed by the Holocene epoch, which continues to this day. In the context of the Quaternary, the Holocene has been a period of relative climatic stability – which has probably contributed to the rise of human civilization. In broad terms, glacial retreat began approximately 18,000 years BP. As glaciers retreated the Earth's albedo decreased, and $CO_2$ and

CH$_4$ were released from warming soils and ocean waters. These gases accelerated the rate of warming, and between 18,000 and 14,000 years BP the mean surface temperature in Antarctica warmed by about 6°C (Petit *et al.*, 1999). This long gradual warming trend was interrupted by a relatively brief cold interval starting approximately 13,000 years ago, lasting approximately 1500 years (Fig. 1.13). This interval is called the Younger Dryas, after the Arctic and alpine flower *Dryas octopetala*, whose pollen is considered an indicator of the transition back to cold conditions. The causes of the Younger Dryas are a matter of some debate, but one leading candidate is that a flood of fresh water into the North Atlantic Ocean caused by the catastrophic draining of glacial Lake Agassiz led to a shutdown of thermohaline circulation (Broecker, 2006). During the Younger Dryas glaciers re-advanced, although none approached its former extent. The Younger Dryas ended abruptly with a period of rapid warming approximately 11,600 years ago. Records from both Greenland and Antarctica suggest that temperatures increased within as little as 20 years (Alley *et al.*, 1993). Following the Younger Dryas glaciers continued retreating, and North America and Europe were effectively ice free by 10,000 years BP.

Temperatures continued to rise in the early Holocene and a long period of relative warmth, known as the Holocene Climatic Optimum, or the Hypsothermal, persisted from approximately 9000 to 5000 years BP. This period is of particular interest, as temperatures were comparable to modern conditions or possibly warmer, and therefore may present a useful analogue for observing the effects of climatic warming on ecosystem processes. Additionally, because there have been no periods of widespread glaciation, most lakes contain sediment records that span this period, so vegetation and aquatic ecosystem reconstructions are widely available (see Section 1.3).

Following the Holocene Climatic Optimum a long gradual period of cooling was initiated. This interval is often called the Neoglacial, and is associated with a series of alternating glacial advances and retreats in Arctic and alpine environments. Evidence for these advances is elusive, since a more extensive period of glacial advances lasting from about 1400 to 1900 CE obliterated the landforms associated with previous advances. Nevertheless, evidence from lake sediments and forests overridden by ice indicates frequent glacial advances and retreats during the Neoglacial.

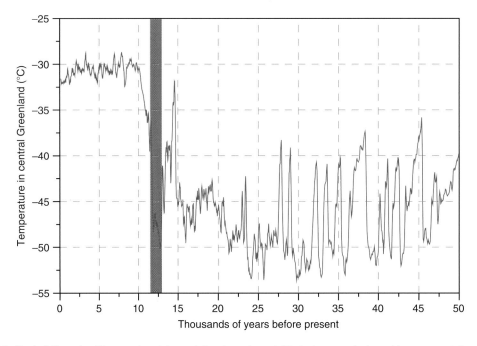

**Fig. 1.13.** Central Greenland temperature interpretation based on stable isotope analysis and ice accumulation data from the GISP2 ice core (Alley, 2000). The Younger Dryas interval is highlighted with the blue shaded bar.

The Neoglacial interval was interrupted by a short interval of relatively warm conditions that lasted from approximately 950 to 1250 CE, known as the Medieval Warm Period. It is not clear if it was a global phenomenon or was restricted to the North Atlantic sector, and this issue is a matter of considerable debate among climate change scientists and (especially) their deniers. The Medieval Warm Period corresponds to the time that the Vikings settled Greenland and wine grapes were grown in northern England. Like the Holocene optimum it provides a possible analogue for examining the response of ecosystems to warm conditions. Perhaps more importantly, though, it provides us with a natural experiment – if the climate of the Medieval Warm Period was warmer than today, it would suggest that the magnitude of warming we are currently experiencing is not unprecedented and can be explained without invoking greenhouse gas explanations. We are convinced that modern warming is unprecedented in the late Holocene by a wide range of evidence, including borehole data, high- and low-latitude ice core data, diatom and chironomid reconstructions, and palaeoclimate models. Nevertheless, the relative warmth of the Medieval Warm Period remains a matter of scientific (and non-scientific) debate (see Box 1.2).

The most recent period of glacial advances occurred during the Little Ice Age, which lasted from about 1400 to 1900 CE. During the Little Ice Age glaciers worldwide advanced, many to their maximum extent during the Holocene (Luckman, 2000). Since 1850, global mean temperatures have increased by approximately 1.5°C, although there is considerable regional variation in trends and some parts of the Earth have even cooled. Because it ended so recently, the termination of the Little Ice Age offers the best opportunity to examine how ecosystems respond to climatic warming. For example, Daniels and Veblen (2004) looked at the effects of climatic variability and change on the establishment and growth rates of trees at the high-elevation treeline in the South American Andes. They found that warmer temperatures are conducive to growth increases in adult trees, but that seedling establishment is associated with periods of increased wetness. Upslope migration of the treeline has occurred in response to alternating intervals of moisture (conducive to establishment) and warmth (conducive to growth of established trees). Linear increases in temperature will therefore not necessarily lead to upslope migration of trees in these environments.

Many climatologists argue that we have now left the Holocene and entered a new epoch: the Anthropocene. Since about 1850 humans have had a profound impact on the Earth at a global scale. In this book we focus principally on climatic changes, but there are a number of other confounding factors: species introductions, land-use change, habitat fragmentation, resource extraction and nutrient deposition, that will all interact with climate change to affect the capacity of organisms and ecosystems to respond to climate change. One particularly important feature of the Anthropocene that is both directly and indirectly related to climate change is the substantial increase in $CO_2$ that has occurred over the last 150 years. This increase can be attributed to the combustion of fossil fuels due to its isotopic signature (see Section 1.2). This increase will affect plants directly as well as indirectly through its effect on climate (see Chapter 4).

## 1.6 Conclusions

The last 50 years have seen huge advances in our understanding of the climate system. Critically important advances have come from newly developed sources of palaeoenvironmental information, better techniques for dating palaeoenvironmental data (allowing for comparisons between various reconstructions), and more sophisticated analytical tools and techniques. The main lessons that should be taken from this examination of the palaeorecord are the following.

1. Abrupt changes in climate are possible, and surprises are likely. In particular, the Younger Dryas began quite abruptly and ended even more abruptly. It was likely caused by a climatic threshold being crossed, and it had clear impacts on both climate and ecosystems in the late Pleistocene.
2. The climate system is complex and feedbacks exist. Climate is affected by a wide range of processes acting across a range of scales. At the longest time scales, tectonic activity and orbital variability have regulated the climate of the Earth. At shorter time scales, variability in solar brightness and atmospheric chemistry exert important controls on the Earth's climate. At the shortest time scales, internal dynamics are most important. The current release of greenhouse gases (see Chapter 2) is happening at a much shorter time scale than has happened at any point in the past, and its likely effect on climate has no observable precedent or analogue.

**3.** Similarly, current levels of $CO_2$ and $CH_4$ are unprecedented in at least the last 650,000 years – probably longer. If we look to deeper history, such as the age of dinosaurs, we find that periods with higher $CO_2$ have also had much warmer climates. However, the palaeorecord shows that in the past $CO_2$ increases were not the initial cause of climatic warming, but enhanced that initial warming. The concern now is that while external processes have not acted to initiate warming and release $CO_2$, $CO_2$ is increasing nevertheless and based on the palaeorecord will act to warm the climate system.

**4.** With the possible exception of the Younger Dryas, the climate change expected over the next century will be greater in magnitude or more abrupt than anything that has been seen in the palaeorecord (see Chapter 2). We are truly entering uncharted territory!

# 2 Projecting Future Climates

In order to consider the biological impacts of climatic change we need to know what these climatic changes will look like. In Chapter 1 we learned how we know about past climate, but how do we know what our future climate will be? Future climates are projected using mathematical models of our climate system, based on atmospheric physics. Given the mixture of greenhouse gases in the atmosphere at any given point in time, these models can deduce what the climate will be like. It is important to understand that these models do not predict the levels of greenhouse gases. Greenhouse gas concentrations are climate model inputs, not model outputs. So where do the greenhouse gas concentrations come from? They come from another type of mathematical model called a 'scenario model'. This two-step process, using first scenario models and then climate models, is the focus of this chapter. Having some understanding of this process is important if we as biologists are to make intelligent choices of climate projections for use in our studies of climatic impacts.

Throughout the remainder of this book we will, from time to time but particularly when we discuss future climatic change, refer to the 'IPCC'. This is the Intergovernmental Panel on Climate Change (www.ipcc.ch). The IPCC was established by the United Nations Environment Programme (UNEP) and the World Meteorological Organization (WMO). According to their website:

> The IPCC is a scientific body. It reviews and assesses the most recent scientific, technical and socio-economic information produced worldwide relevant to the understanding of climate change. It does not conduct any research nor does it monitor climate related data or parameters. Thousands of scientists from all over the world contribute to the work of the

IPCC on a voluntary basis. Review is an essential part of the IPCC process, to ensure an objective and complete assessment of current information. Differing viewpoints existing within the scientific community are reflected in the IPCC reports.

## 2.1 What are Scenarios?

Emissions scenarios are stories about the future. They are *not* predictions; they are descriptions of plausible alternative futures. To quote the IPCC's Special Report on Emissions Scenarios (SRES):

> Future levels of global GHG [greenhouse gas] emissions are the products of a very complex, ill-understood dynamic system, driven by forces such as population growth, socio-economic development, and technological progress; thus to predict emissions accurately is virtually impossible.

Predicting future greenhouse gas emissions with any degree of accuracy would require us to know things that we cannot possibly know, like the pace of technological change, the rate of economic development, and the political and cultural will to abate greenhouse gas emissions.

Equally, the scenarios used by the IPCC are also *not* statements of value. There are no 'preferred' or 'favourite' scenarios. This position may seem paradoxical. If anthropogenic greenhouse gas emissions are the cause of global warming, and global warming is undesirable, then why do we not prefer emissions scenarios that result in lower future greenhouse gas concentrations? The reason is that growth in greenhouse gas emissions is related to things like the pace of economic development, and economic development has moral justification, because it meets the needs of a growing population. That is, there is human and moral value in at least some of the activities that lead to higher greenhouse gas emissions. To some extent it is a matter of taste as to whether we prefer improved human development even if it comes with a cost to the environment. In any case, the IPCC is not in the business of telling governments what they should do. The IPCC's mandate involves the science around climatic change, not policy making.

So what are 'scenarios'? Figure 2.1 shows how scenario construction involves making logical connections between known drivers of greenhouse gas emissions and the emissions themselves. Why do we need this? Surely it would be possible to simply run the climate models under a whole range of greenhouse gas emissions and then draw conclusions about the consequences of rising greenhouse gas emissions, without ever considering where those emissions come from. Of course this is possible, but what scenario construction does is to provide tools for policy makers to think about how

changes in policy might alter future emissions. By attempting to elaborate on the connection between, say, the mixture of fuels used in the future and greenhouse gas emissions, scenarios provide the policy makers with a tool to think about how their energy policy will impact future climatic change.

Scenarios involve two parts, qualitative storylines and quantitative interpretations of those storylines. They use historical and contemporary information to generate some assumptions about the relationships between driving forces and greenhouse gas emissions. These assumptions are reproducible and internally consistent. Sometimes these assumptions are qualitative, sometimes they are quantitative and sometimes there are explicit mathematical models behind each of them. The qualitative and quantitative assumptions are fed into scenario models that then generate the time course of emissions over the next 100+ years. Figure 2.2 shows the annual emissions for one particular scenario as an example.

Scenario construction starts with something called a 'storyline'. Storylines are narratives that

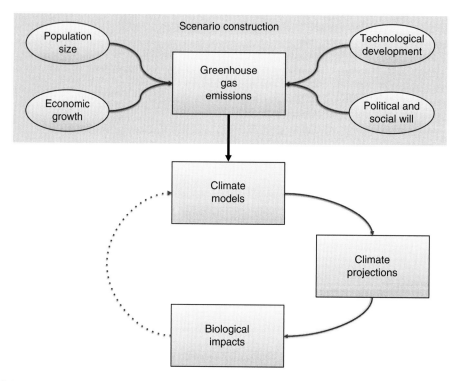

**Fig. 2.1.** Schematic diagram indicating how climate change scenarios are constructed and their role in projecting future climate conditions. An example of the greenhouse gas emissions projections produced from the scenario construction is seen in Fig. 2.2.

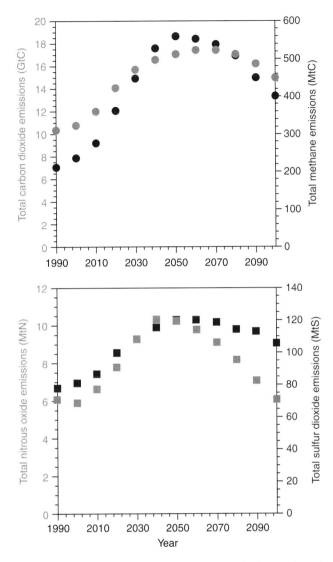

**Fig. 2.2.** The time series projections of greenhouse gas emissions for the A1C scenario as interpreted by the IMAGE scenario modelling framework (for more details, see http://tinyurl.com/ch2-image-model). The A1C scenario is described in Table 2.1. These greenhouse gas emissions outputs are used as inputs to the climate models described later in this chapter.

describe broad pictures of how the future might unfold. The storylines differ in the way that each conceives of how the major geopolitical regions of the world interrelate, how new technologies spread, how different economic regions develop, how environmental protection is conceived and implemented, and how population growth changes globally and regionally. These storylines contain no explicit policies to limit greenhouse gas emissions or to adapt to climatic change.

The IPCC settled on four storylines. These storylines are to be thought of as equally *plausible* views of the future. There are no 'business as usual', 'best case' or 'worst case' scenarios. In order to avoid *any* value implications, the IPCC decided to give the storylines neutral names: A1, A2, B1 and B2. The differences between the four storylines are summarized in Table 2.1.

Within each storyline there is a 'family' of scenarios. Storylines are *qualitative* views of the

**Table 2.1.** The basic qualitative differences among the four families of scenario storylines. (From Nakicenovic and Swart, 2000, section 4.2.1.)

| Family | A1 | | | | A2 | B1 | B2 |
|---|---|---|---|---|---|---|---|
| Scenario group | A1C | A1G | A1B | A1T | A2 | B1 | B2 |
| Population growth | Low | Low | Low | Low | High | Low | Medium |
| GDP growth | Very high | Very high | Very high | Very high | Medium | High | Medium |
| Energy use | Very high | Very high | Very high | High | High | Low | Medium |
| Land-use changes | Low/medium | Low/medium | Low | Low | Medium/high | High | Medium |
| Resource availability | High | High | Medium | Medium | Low | Low | Medium |
| Pace of technological change | Rapid | Rapid | Rapid | Rapid | Slow | Medium | Medium |
| Change favouring | Coal | Oil and gas | Balanced | Non-fossils | Regional | Efficiency and dematerialization | Dynamics as usual |

GDP, gross domestic product.

future, while the scenarios can be thought of as *quantitative* interpretations of the each storyline. The scenarios themselves are developed by interpreting the qualitative storylines through a scenario modelling framework. There are six frameworks used by the IPCC:

- Asian Pacific Integrated Model (AIM) from the National Institute of Environmental Studies in Japan;
- Atmospheric Stabilization Framework Model (ASF) from ICF Consulting in the USA;
- Integrated Model to Assess the Greenhouse Effect (IMAGE) from the National Institute for Public Health and Environmental Hygiene (RIVM), used in connection with the Dutch Bureau for Economic Policy Analysis (CPB) WorldScan model, the Netherlands;
- Multiregional Approach for Resource and Industry Allocation (MARIA) from the Science University of Tokyo in Japan;
- Model for Energy Supply Strategy Alternatives and their General Environmental Impact (MESSAGE) from the International Institute of Applied Systems Analysis (IIASA) in Austria; and
- Mini Climate Assessment Model (MiniCAM) from the Pacific Northwest National Laboratory (PNNL) in the USA.

The same storyline is run through more than one model framework. This exercise then results in different quantitative descriptions of each of the qualitative stories, and hence in different quantitative projections of greenhouse gas emissions. Figure 2.3 shows the differences between the model interpretations of each of the storylines for total anthropogenic emissions of $CO_2$. The error bars show the range of ±1 standard deviation in the distribution of results. So for example, the five model interpretations of the A2 storyline result in total $CO_2$ emissions in 2100 ranging between 28.19 GtC (MESSAGE) and 34.47 GtC (AIM). A1C and A1G (not shown in Fig. 2.3) are usually combined to form a scenario called A1FI (FI denotes *fuel intensive*) and A1FI is often referred to as the 'high emissions scenario', for obvious reasons. A2 is often referred to as the 'medium high emissions scenario', B2 as the 'medium low emissions scenario' and B1 as the 'low emissions scenario'.

## 2.2 From Emissions to Climate Projections: General Circulation Models

Formally, future climates are projected using a type of mathematical model called 'Atmosphere–Ocean General Circulation Models' or AOGCMs. It is still common to see them referred to simply as 'General Circulation Models' or GCMs. GCM is also used to mean 'Global Circulation Model' and 'Global Climate Model', but for our purposes these terms are all synonymous. Technically, the term GCM harkens back to a period when the ocean and atmosphere portions of climate models were either fully or partially 'uncoupled'. Today, GCM is used synonymously with AOGCM.

GCMs are computer models that represent the world by a large number of grid cells, typically representing a horizontal scale of between 1° and 5° latitude by longitude. These grid cells cover the

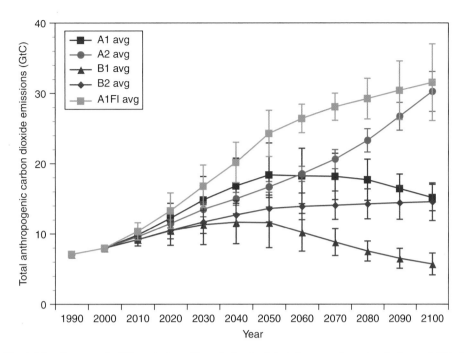

**Fig. 2.3.** The total anthropogenic $CO_2$ emissions. The ±1 standard deviation error bars denote the variability in how the various scenario modelling frameworks interpret the various storylines.

entire surface of land, water and ice. The grid cells are stacked on top of each other so that processes are modelled in layers extending from the top of the atmosphere to the bottom of the oceans. Most models used in the IPCC assessments include from 20 to 30 layers. What happens in one grid cell affects the conditions in surrounding grid cells. The 'atmosphere' comprises information about the density, in terms of temperature and pressure, the motion of the air mass up and down and horizontally, and the composition of the air in terms of its concentrations of greenhouse gases (e.g. $CO_2$ and water vapour). There are exchanges of heat and water between the atmosphere and the land and oceans (Houghton, 2004).

GCMs are based on well-established principles of physics, and are able to reproduce recently observed features of our current climate and can reproduce past climate. And as such, 'There is considerable confidence that Atmosphere-Ocean General Circulation Models (AOGCMs) provide credible quantitative estimates of future climate change, particularly at continental and larger scales' (Randall *et al.*, 2007: Executive Summary). Such spatial scales are, of course, problematic for biologists who rarely work at such large scales. There is some

relief from this limitation in the form of statistical downscaling and regional climate models, considered later in this chapter.

There are currently 18 modelling centres around the world, and 25 GCMs were used in the IPCC's Fourth Assessment Report (AR4). These modelling centres and model names are shown in Table 2.2.

We might reasonably ask: 'Why use so many models?' The fact is that climatologists know more about the mechanisms controlling climate than can be represented in a mathematical model and still have the model run, in a reasonable length of time, on some of the world's fastest computers. As a consequence, climate modellers must make trade-offs between speed and process representation. Some processes are represented mechanistically, while some are represented phenomenologically. To translate this into terms that biologists can understand, suppose we were modelling plant growth. In general, it will be true that if we add more mineral nitrogen to the soil, the plant will grow more. We could represent this mechanistically by modelling N-uptake by the roots, N-transport, N-use, N-recycling and so on. This representation would take many mathematical equations and demand a considerable amount of

**Table 2.2.** Modelling centres and model names for GCMs used in the IPCC's Fourth Assessment Report.

| Originating group(s) | Country | Names of models | Further information |
|---|---|---|---|
| Beijing Climate Center | China | BCC-CM1 | http://tinyurl.com/ch3-CSMD |
| Bjerknes Centre for Climate Research | Norway | BCCR-BCM2.0 | http://tinyurl.com/ch3-BCCR |
| National Center for Atmospheric Research | USA | CCSM3 | http://tinyurl.com/ch3-CCSM |
| Canadian Centre for Climate Modelling and Analysis | Canada | CGCM3.1(T47), CGCM3.1(T63) | http://tinyurl.com/ch3-CGCM |
| Météo-France/Centre National de Recherches Météorologiques | France | CNRM-CM3 | http://tinyurl.com/ch3-CNRM |
| CSIRO Atmospheric Research | Australia | CSIRO-Mk3.0, CSIRO-Mk3.5 | http://tinyurl.com/ch3-CSIRO |
| Max Planck Institute for Meteorology | Germany | ECHAM5/MPI-OM | http://tinyurl.com/ch3-echam5 |
| Meteorological Institute of the University of Bonn, Meteorological Research Institute of KMA, and Model and Data group | Germany/ Korea | ECHO-G | http://tinyurl.com/ch3-echo-G |
| LASG/Institute of Atmospheric Physics | China | FGOALS-g1.0 | http://tinyurl.com/ch3-fgoals |
| US Department of Commerce/National Oceanic and Atmospheric Administration (NOAA)/Geophysical Fluid Dynamics Laboratory | USA | GFDL-CM2.0, GFDL-CM2.1 | http://tinyurl.com/ch3-gfdl |
| NASA/Goddard Institute for Space Studies | USA | GISS-AOM, GISS-EH, GISS-ER | http://tinyurl.com/ch3-giss (Note: the FORTRAN code for these models can be downloaded from this site) |
| Istituto Nazionale di Geofisica e Vulcanologia | Italy | INGV-SXG | http://tinyurl.com/ch3-ingv-sxg |
| Institute for Numerical Mathematics | Russia | INM-CM3.0 | http://tinyurl.com/ch3-inm-cm3 |
| Institut Pierre Simon Laplace | France | IPSL-CM4 | http://tinyurl.com/ch3-ipsl-cm4 |
| Center for Climate System Research (The University of Tokyo), National Institute for Environmental Studies, and Frontier Research Center for Global Change (JAMSTEC) | Japan | MIROC3.2 (hires), MIROC3.2 (medres) | http://tinyurl.com/ch3-miroc |
| Meteorological Research Institute | Japan | MRI-CGCM2.3.2 | http://tinyurl.com/ch3-MRI-CGCM2 |
| National Center for Atmospheric Research | USA | PCM | http://tinyurl.com/ch3-NCAR-pcm |
| Hadley Centre for Climate Prediction and Research/Met Office | UK | UKMO-HadCM3, UKMO-HadGEM1 | http://tinyurl.com/ch3-HadCM3 |

computing power. On the other hand, we could forgo a mechanistic representation and simply capture the 'phenomenon' of higher soil nitrogen concentrations and greater plant growth with a single, in this case statistical, function. Such a phenomenological representation would demand far less computing time but it would sacrifice a mechanistic understanding.

We would have to consider our goals carefully before choosing one or the other approach. Climate modellers make many different decisions about which processes to represent mechanistically and which to represent phenomenologically. In other words, modellers decide which abstractions to make and which not to make.

Another trade-off that climate modellers must consider is computing time and spatial and temporal resolution. Early climate models from the 1970s had very little spatial resolution. They represented the globe with a lattice of grid cells that were 600 km×600 km, and these grid cells would be stacked six cells high to represent the atmosphere (Pope, 2007). To give a sense of scale, a grid cell this size is about the same size as Germany (357,114 km²). By comparison, some of the most modern climate models have a spatial resolution of 135 km×135 km, stacked 38 cells high to represent the atmosphere (see Fig. 2.4 for more details). These models take a minimum of 256 times the computing power required for the early 1970s models (Pope, 2007). Why does the spatial and temporal resolution matter? Bear in mind that space and time are actually *continuous*, but they must be represented as *discrete* units in order to be modelled in a computer. The smaller the discrete units, the better the approximation is to the continuous features of the model (space and time).

With that as background, we can now return to the question of 'why so many models?'. Since climate modellers have to make these various abstractions and compromises, there would always be the fear that the resulting projections are highly dependent on the particulars of the model. So the philosophy is to have many groups, all working essentially independently, making different abstractions and different compromises and using different assumptions. We then take the projections from all of these groups and models and compare them. In this comparison we look for 'robust' projections. A projection is robust if it does not depend strongly on the particular climate model. For example, one robust projection is that the global mean annual temperature will increase. All of these models, despite their differences in assumptions, abstractions and compromises, project an increase in mean annual global temperature. None of the models projects cooling. On the other hand, the particular degree of warming is less robust. Figure 2.5 shows that the particular degree of warming is still very much dependent on the assumptions, abstractions and compromises of each model.

We take up the issue of robust projections in greater detail in Box 2.1. Besides such robust solutions, there are also reasons to believe that averaging over the various differences in the GCMs yields the best model projections. The recent US Climate Change Science Program Synthesis and Assessment Product 3.1 (Bader *et al.*, 2008) concluded:

The CMIP3 [Coupled Model Intercomparison Project 3] 'ensemble-mean' model performs better than any individual model by this metric and by many others. This kind of result has convinced the

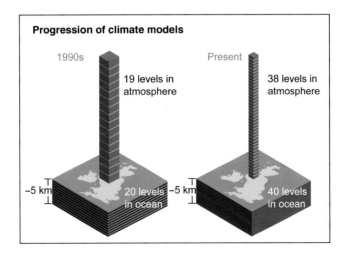

**Fig. 2.4.** Schematic showing the spatial scale of climate models in the 1990s (the IPCC's Second Assessment Report) and present (the IPCC's Fourth Assessment Report). (Figure from the Hadley Centre. © British Crown Copyright 2010, the Met Office, redrawn by the Southwest Climate Change Network, www.southwestclimatechange.org.)

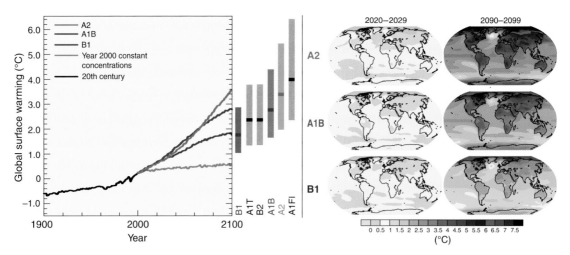

Left panel: Solid lines are multi-model global averages of surface warming (relative to 1980–1999) for the SRES scenarios A2, A1B and B1, shown as continuations of the 20th century simulations. The orange line is for the experiment where concentrations where held constant at year 2000 values. The bars in the middle of the figure indicate the best estimate (solid line within each bar) and the likely range assessed for the six SRES marker scenarios at 2090–2099 relative to 1980–1999. The assessment of the best estimate and likely ranges in the bars includes the Atmosphere Global–Ocean General Circulation Models (AOGCMs) in the left part of the figure, as well as results from a hierarchy of independent models and observational constraints. Right panels: Projected surface temperature changes for the early and late 21st century relative to the period 1980–1999. The panels show the multi-AOGCM average projections for the A2 (top), A1B (middle) and B1 (bottom) SRES scenarios averaged over decades 2020–2029 (left) and 2090–2099 (right). (From: *Climate Change 2007: Synthesis Report. Contribution of Working Groups I, II and III to the Fourth Assessment Report of the Intergovernmental Panel on Climate Change*, Figure 3.2, IPCC, Geneva, Switzerland.)

**Fig. 2.5.** Mean annual change in global surface temperatures predicted from climate models. The bars on the right-hand side show the range of projections for each scenario that are produced by all the different climate models. So we can see that a robust projection is that the mean annual global surface temperature will rise, but that the rise could be anywhere from 1 to 6.5°C depending on the emissions scenario and the particular climate model.

community of the value of a multimodel approach to climate change projection. Our understanding of climate is still insufficient to justify proclaiming any one model 'best' or even showing metrics of model performance that imply skill in predicting the future.

The report goes on to point out that different models perform differently on a host of different metrics (i.e. methods of measuring how 'good' they are). No single model does better on all metrics, and without knowing which metric makes a model better or worse at projecting future climate, when forcing variables will be very different from what they are now, it is impossible to say which model is 'best'.

## 2.3 Regional Models and the Problem of Downscaling

As we mentioned previously, even today GCMs simulate climate on fairly large spatial scales, so large that their ecological relevance may be questionable for many applications. Even the GCMs with the finest spatial resolution, such as the HadGEM1, use grid cells 135 km on a side. While considerably better than the early models that used grid cells up to 600 km on a side, even these finer-scale models are still quite coarse for our purposes. There are, however, two possible solutions to this problem: (i) the use of regional climate models (RCMs); and (ii) the use of statistical downscaling.

## Box 2.1. The issue of consensus.

The media has a lot to say about the issue of 'consensus' among climate change 'experts', on both sides of the issue. Some argue that there is substantial consensus among experts that the Earth is warming and that we humans are to blame. Others argue that such 'consensus' is a fiction, that there really is considerable disagreement among experts, and, besides, science doesn't work by consensus. So who is right? First, it is worth noting that the most authoritative body speaking about climate change, the IPCC, rarely uses the word 'consensus', and when it does it is not about such grand pronouncements but rather about small details. Instead, the IPCC talks about 'levels of uncertainty' in our understanding of different aspects of climate change.

There are four big questions when it comes to climate change and GCMs: (i) Has there been detectable warming in the recent past? (ii) Is this warming unusual against the background of historical climate change? (iii) Are humans to blame for this warming? (iv) How much warming will there be in the future if emissions continue to increase? There are many other important questions we might ask, but these are the really big ones. These four questions all require different sorts of evidence but, more importantly, we need to understand that science never provides definitive answers to any questions, let alone these. Science is not about 'absolute truths', it is about 'very likely maybes'. All of science, and all applications of science, are the same in this regard – we don't deal in absolutes. Non-trivial questions in science are 'weight of evidence' arguments. For some questions we can give answers for which we have very little uncertainty, and for others there may be a great deal of uncertainty. But notice that there are no answers that we can give with absolute certainty. To feel comfortable with scientific explanations of natural phenomena, one has to feel comfortable with uncertainty. It is a misunderstanding of how science works, by politicians and lay-people, which causes them to seek definitive answers from scientists. There are no definitive answers, there are only 'very likely maybes'. We expand on these points in Chapter 14.

We mentioned that the IPCC deals with levels of uncertainty, not with consensus and not with absolute definitive answers. This quote from the IPCC AR4 Synthesis Report (IPCC, 2007: 27) explains how it considers uncertainty:

> Three different approaches are used to describe uncertainties each with a distinct form of language. Choices among and within these three approaches depend on both the nature of the information available and the authors' expert judgment of the correctness and completeness of current scientific understanding.
>
> Where uncertainty is assessed qualitatively, it is characterised by providing a relative sense of the amount and quality of evidence (that is, information from theory, observations or models indicating whether a belief or proposition is true or valid) and the degree of agreement (that is, the level of concurrence in the literature on a particular finding). This approach is used by WG III [Working Group III: Mitigation of Climate Change] through a series of self-explanatory terms

such as: high agreement, much evidence; high agreement, medium evidence; medium agreement, medium evidence; etc.

> Where uncertainty is assessed more quantitatively using expert judgment of the correctness of underlying data, models or analyses, then the following scale of confidence levels is used to express the assessed chance of a finding being correct: very high confidence at least 9 out of 10; high confidence about 8 out of 10; medium confidence about 5 out of 10; low confidence about 2 out of 10; and very low confidence less than 1 out of 10.
>
> Where uncertainty in specific outcomes is assessed using expert judgment and statistical analysis of a body of evidence (e.g. observations or model results), then the following likelihood ranges are used to express the assessed probability of occurrence: virtually certain >99%; extremely likely >95%; very likely >90%; likely >66%; more likely than not >50%; about as likely as not 33% to 66%; unlikely <33%; very unlikely <10%; extremely unlikely <5%; exceptionally unlikely <1%.
>
> WG II [Working Group II: Impacts, Adaptation and Vulnerability] has used a combination of confidence and likelihood assessments and WG I [Working Group I: The Physical Science Basis] has predominantly used likelihood assessments.

The problem with the term 'consensus' is that it implies an 'all or nothing' judgement – we either have consensus or we do not. But clearly there can be degrees of consensus. Better to just stop using the word at all, and stick to these operational definitions of the degrees of uncertainty.

So what does the IPCC actually say about any 'consensus' projection of climate change? Here is its statement on questions (i) and (ii), about whether there has been detectable warming in the recent past and whether this warming is unusual against a background of historical climate change:

> Average Northern Hemisphere temperatures during the second half of the 20th century were *very likely* higher than during any other 50-year period in the last 500 years and *likely* the highest in at least the past 1300 years. (WGI 6.6, SPM) (IPCC, 2007: 30)

Its answer to question (iii), on whether humans are to blame, is:

> There is *very high* confidence that the global average net effect of human activities since 1750 has been one of warming, with a radiative forcing of +1.6 [+0.6 to +2.4] W/m$^2$. (WGI 2.3, 6.5, 2.9, SPM) (IPCC, 2007: 37)

And one way that it answers question (iv), regarding how much climate change will there be in the future, is:

> climate sensitivity is *likely* to be in the range of 2 to 4.5°C with a best estimate of about 3°C, and is *very unlikely* to be less than 1.5°C. Values substantially higher than 4.5°C cannot be excluded, but agreement of models with observations is not as good for those values. (WGI 8.6, 9.6, Box 10.2, SPM) (IPCC, 2007: 38)

Climate sensitivity is the equilibrium change in surface temperature that results from a doubling of atmospheric $CO_2$ concentrations. More details on this measure are provided elsewhere in this chapter.

RCMs are dynamic models in the same way that GCMs are, but they cover only a specific region (often of land surface). They use either observed climate or that generated by a relevant GCM to provide the RCM with so-called 'boundary conditions'. RCMs have considerably finer spatial scales, on the order of 50–100 km on a side, with some of the best having grid cells of just 20 km on a side. This is still large by ecological standards, but considerably better resolution than GCMs can provide. Like GCMs, RCMs are based on physical principles of the climate system and they are capable of reproducing contemporary climate for many regions of the globe. This ability gives us some confidence in their ability to downscale future climates, too.

The chief disadvantage of RCMs is that despite the fact that they cover a much smaller portion of the globe than GCMs, RCMs can be as computationally demanding or even more so than GCMs. Because of the finer spatial scale, a given region will have many more grid cells than the same region in a GCM. More grid cells mean more calculations; but also because of the finer spatial scale, a shorter time step is required for the integration in order to achieve numerical stability (Bader *et al.*, 2008). Partly because of the computational costs, these models are rarely ever run for as long as GCMs. While GCMs are used to project climate for the next 100–200 years, RCMs are rarely used for periods longer than a few tens of years.

Statistical downscaling is much simpler. With statistical downscaling, we take contemporary relationships between observed large-scale climate features and observed smaller-scale climate features and we use these relationships to translate large-scale projections into smaller-scale projections. The principal disadvantage with statistical downscaling is that it assumes the relationships between large- and small-scale climate features remain constant through time. The principal advantage to statistical downscaling is that it is computationally cheap. Where comparisons have been made between statistical downscaling and RCMs, neither has been found to be consistently superior. Both have strengths and weaknesses and there is currently a place for both approaches (Bader *et al.*, 2008).

A word of caution is probably useful at this point. There is a temptation to think of downscaled or regional climate projections, particularly those generated by RCMs, as being in some sense 'more accurate' than projections from GCMs. This temptation probably stems from the higher spatial resolution that tends to produce a 'smoother' picture of projected climate. The finer spatial scale allows more accurate representation of topography for example, and topography can influence local climate. Sometimes this confidence is justified. For example, there is reason to believe that RCMs do a better job of capturing the spatial variability in precipitation, something that often varies on a spatial scale much finer than GCMs operate at. Notwithstanding such examples, improved fit does not always happen and in some cases the fit can actually be worse (see Bader *et al.*, 2008: 32 for further discussion). RCM projections are also totally dependent on the source of their boundary conditions. These conditions usually come from GCMs and, as we see below, GCMs vary a good deal in their abilities, strengths and weaknesses.

## 2.4  Hindcasts and Model Validation

Model validation is the process of determining if a model is, in some sense of the word, 'good'. Model validation differs a bit between different kinds of models and models built for different purposes. In general, model validation would involve using the model to make a series of predictions and then comparing those predictions with what is actually observed. However, because what we are actually trying to predict with GCMs is the future climate, model validation must necessarily utilize surrogate data.

There are three general methods for validating GCMs: (i) the model is run backward through time to generate recent past climate and then these simulations are compared with historical climate data collected from weather stations all over the world; (ii) the models are run for a period in the Earth's history when climate forcings and other key variables were substantially different, and we see if the model can adequately reproduce the climate we think was present at these more distant times (see Chapter 1); and (iii) the models are used to predict the effects of large perturbations on the formation of major circulation events, such as El Niño. We concentrate only on the first method in this section. The others are interesting, but perhaps more esoteric than necessary for our purposes.

In any GCM, the equations are integrated to see how conditions change in time. To predict future climate we integrate forward in time but, as mentioned above, we have nothing to compare these results with, so they will not do for model validation. Mathematically, there is no reason that the integration needs to be carried out forward in time; these equations can be integrated backward from the present time, and this produces 'hindcasts' (the opposite of forecasts) that can be compared with past climate (recall our discussion in Chapter 1 of how we know about past climates).

Let us start by looking at the 'anomalies' in global mean temperature for the past century. When talking about climate change (past or future) it is common to express the measure in terms of an anomaly, i.e. a difference between the measure's value at one point in time and its value during a particular reference period of time. Figure 2.6 shows the results for 58 simulations of 14 GCMs (shown in yellow). The mean of all 58 runs is shown in red. The observed temperature anomalies are shown in black. Vertical lines denote major volcanic eruptions (Randall *et al.*, 2007).

The validation shown in Fig. 2.6 is nice, but non-spatial. Do the models capture the spatial variation with any degree of accuracy? Figure 2.7 shows the multi-model mean simulated contemporary temperatures minus the observed temperatures. The IPCC's Working Group I report (Randall *et al.*,

2007: 608) summarizes the correspondence between simulated and observed climate as follows:

With few exceptions, the absolute error (outside polar regions and other data-poor regions) is less than 2°C. Individual models typically have larger errors, but in most cases still less than 3°C, except at high latitudes. Some of the larger errors occur in regions of sharp elevation changes and may result simply from mismatches between the model topography (typically smoothed) and the actual topography. There is also a tendency for a slight, but general, cold bias. Outside the polar regions, relatively large errors are evident in the eastern parts of the tropical ocean basins, a likely symptom of problems in the simulation of low clouds. The extent to which these systematic model errors affect a model's response to external perturbations is unknown, but may be significant. In spite of the discrepancies discussed here, the fact is that models account for a very large fraction of the global temperature pattern: the correlation coefficient between the simulated and observed spatial patterns of annual mean temperature is typically about 0.98 for individual models. This supports the view that major processes

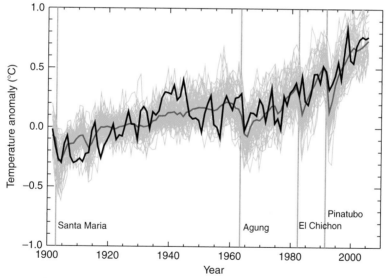

Global mean near-surface temperatures over the 20th century from observations (black) and as obtained from 58 simulations produced by 14 different climate models driven by both natural and human-caused factors that influence climate (yellow). The mean of all these runs is also shown (thick red line). Temperature anomalies are shown relative to the 1901 to 1950 mean. Vertical grey lines indicate the timing of major volcanic eruptions. Figure adapted from IPCC WGI Chapter 9, Figure 9.5. Refer to corresponding caption for further details: http://tinyurl.com/fig2-6-supp. (From: *Climate Change 2007: The Physical Science Basis. Working Group I Contribution to the Fourth Assessment Report of the Intergovernmental Panel on Climate Change*, Figure 8.1, Cambridge University Press.)

**Fig. 2.6.** The match between GCM-simulated global temperature anomalies (yellow lines, red line shows the mean) and observed anomalies (black line).

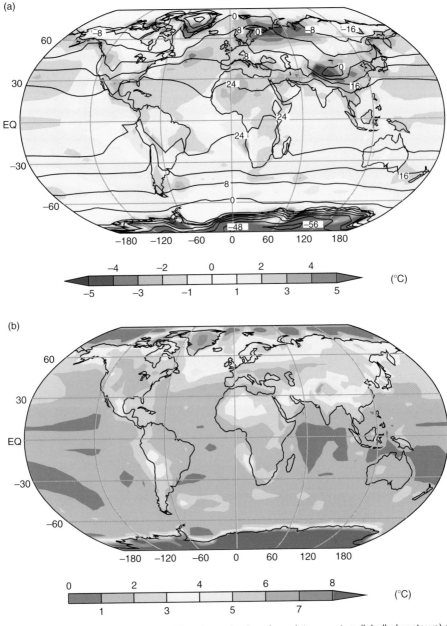

(a) Observed climatological annual mean SST and, over land, surface air temperature (labelled contours) and the multi-model mean error in these temperatures, simulated minus observed (colour-shaded contours). (b) Size of the typical model error, as gauged by the root-mean-square error in this temperature, computed over all AOGCM simulations available in the MMD at PCMDI. The Hadley Centre Sea Ice and Sea Surface Temperature (HadISST; Rayner *et al.*, 2003) climatology of SST for 1980 to 1999 and the Climatic Research Unit (CRU; Jones *et al.*, 1999) climatology of surface air temperature over land for 1961 to 1990 are shown here. The model results are for the same period in the 20th-century simulations. In the presence of sea ice, the SST is assumed to be at the approximate freezing point of seawater (−1.8°C). Results for individual models can be seen in the Supplementary Material, Figure S8.1. (From: *Climate Change 2007: The Physical Science Basis. Working Group I Contribution to the Fourth Assessment Report of the Intergovernmental Panel on Climate Change*, Figure 8.2, Cambridge University Press.)

**Fig. 2.7.** (a) The sea-surface temperature or the surface air temperature (contour lines) and (b) the difference between the multi-model mean predictions and the observed values (shaded contours).

governing surface temperature climatology are represented with a reasonable degree of fidelity by the models.

These models then seem to capture global mean temperature, at least for the recent past, with reasonably good accuracy. What about other aspects of climate? Let us consider precipitation. Figure 2.8 shows the observed pattern of annual precipitation (Fig. 2.8a) and the multi-model mean hindcast (Fig. 2.8b). The IPCC (Randall *et al.*, 2007: 611–612) summarizes the correspondence between simulated and observed precipitation as follows:

> Models also simulate some of the major regional characteristics of the precipitation field, including the major convergence zones and the maxima over tropical rain forests, although there is a tendency to underestimate rainfall over the Amazon. When considered in more detail, however, there are deficiencies in the multi-model mean precipitation field. There is a distinct tendency for models to orient the South Pacific convergence zone parallel to latitudes and to extend it too far eastward. In the tropical Atlantic, the precipitation maximum is too weak in most models with too much rain south of the equator. There are also systematic east–west positional errors in the precipitation distribution over the Indo-Pacific Warm Pool in most models, with an excess of precipitation over the western Indian Ocean and over the Maritime Continent. These lead to systematic biases in the location of the major rising branches of the Walker Circulation and can compromise major teleconnection pathways, in particular those associated with El Niño (e.g., Turner *et al.*, 2005). Systematic dry biases over the Bay of Bengal are related to errors in the monsoon simulations. Despite the apparent skill suggested by the multi-model mean, many models individually display substantial precipitation biases, especially in the tropics, which often approach the magnitude of the mean observed climatology.

So these models would seem to produce pretty accurate reconstructions of recent past climate, when we know with reasonable accuracy the relevant climate forcings, at least for the global mean temperature. In and of itself, this is not too surprising because the models actually go through a parameter 'tuning' phase to produce better matches in the energy balance parts of the models. It is also possible for a model to perform well on average, even if it performs poorly in detail. A better test of accuracy is how the models capture the spatial and temporal variability around the mean (Figs 2.7 and 2.8). But even here the models perform quite well, with overall correlations between hindcast and observed temperature being ≥95%. There still exist some local errors and

these errors can be quite large, but generally the models do a good job at reconstructing temperatures. On the other hand, hindcasts of monthly mean precipitation are not as well correlated with observed pattern; overall correlations are typically 50–60%, and the fit tends to be worse in the tropics than at higher latitudes. Nevertheless, GCMs do a pretty good job at matching the observed large-scale patterns and inter-annual variability in precipitation (Bader *et al.*, 2008).

## 2.5 Model Results and Projections

Figure 2.9 shows the so-called 'time-evolving' change in global mean surface temperature and percentage change in global mean precipitation for the various GCMs under a few different scenarios. The first thing that is evident from Fig. 2.9 is that there is considerable inter-annual variability in the projection for any given model. This is something little understood by the media, who made such a hoopla over the observed downturn in global average temperatures in 2008. The same is true for the mean model projections, although much of the inter-annual variability is cancelled out when such a mean is calculated. This becomes even more evident when we look at Fig. 2.10, in which the global mean temperature anomaly is summarized. A robust conclusion from these models is that global mean temperatures will increase by between 1.5 and 3.5°C by the end of this century (barring the increasingly unlikely commitment scenario being realized). Also note that there is little difference between the scenarios until around the middle of the present century.

Global mean values obscure seasonal and spatial variability. Compare Fig. 2.9 with Fig. 2.11, where we put back in the spatial (but not seasonal) variability. We see that the global mean increase in temperature is not distributed uniformly in space. In Fig. 2.12 we see the seasonal variation for the middle right-hand column result from Fig. 2.11. But even Fig. 2.12 hides variability on even shorter time scales, particularly for precipitation. One topic that is not well developed in AR4 but is potentially really important for ecologists predicting the likely impacts of climatic change is the distribution of extreme weather events. In AR4 (Meehl *et al.*, 2007: 778) the authors make reference to two studies and conclude that there will be: 'a decrease in temperature variability during the cold season in the extratropical NH

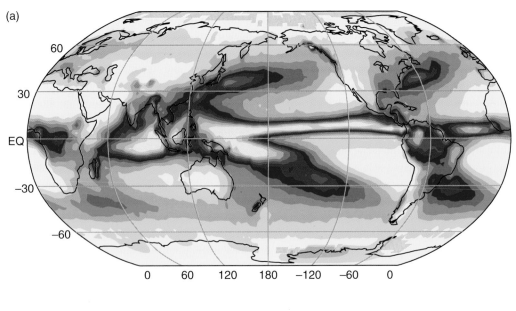

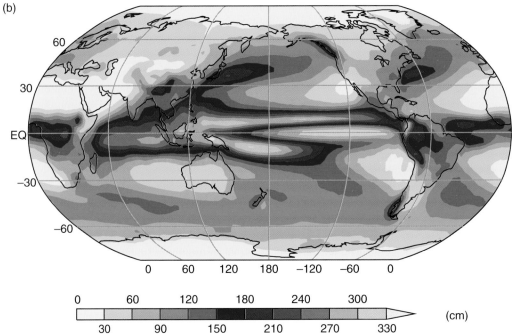

Annual mean precipitation (cm), observed (a) and simulated (b), based on the multi-model mean. The Climate Prediction Center Merged Analysis of Precipitation (CMAP; Xie and Arkin, 1997) observation-based climatology for 1980 to 1999 is shown, and the model results are for the same period in the 20th-century simulations in the MMD at PCMDI. In (a), observations were not available for the grey regions. Results for individual models can be seen in Supplementary Material, Figure S8.9. (From: *Climate Change 2007: The Physical Science Basis. Working Group I Contribution to the Fourth Assessment Report of the Intergovernmental Panel on Climate Change*, Figure 8.5, Cambridge University Press.)

**Fig. 2.8.** The observed pattern of annual precipitation (a) and the hindcast pattern of annual precipitation (b).

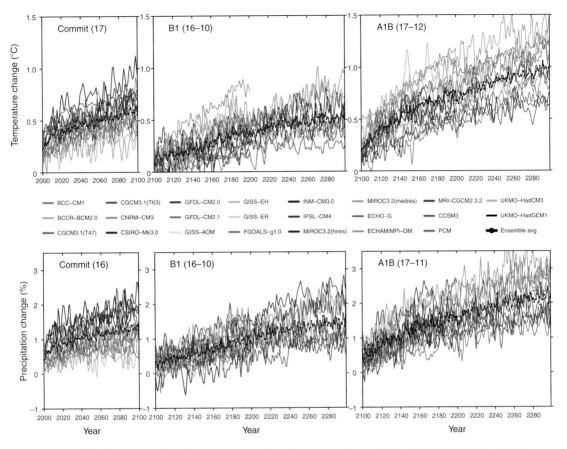

Top left: Globally averaged surface air temperature change relative to 1980–1999 for the 20th century commitment experiment; Top centre: Same as left except for the B1 commitment experiment computed with respect to the 2080–2099 average; Top right: Same as centre except for the A1B commitment experiment; Bottom row: Same as top row but for percentage change in globally averaged precipitation. The numbers in the panels denote the number of models used for each scenario and each century. (From: *Climate Change 2007: The Physical Science Basis. Working Group I Contribution to the Fourth Assessment Report of the Intergovernmental Panel on Climate Change*, Figure 10.3 supplemental material, Cambridge University Press.)

**Fig. 2.9.** The individual model projections for global mean temperature anomaly (relative to 1980–1999) and the percentage change in global mean precipitation. The numbers in parentheses denote the number of models used for each century (e.g. B1(16–10) denotes that 16 models were used for 2100–2200 and 10 models were used for 2200–2300).

[Northern Hemisphere] and a slight increase in temperature variability in low latitudes and in warm season northern mid-latitudes'. While for precipitation they conclude that there will be: 'an increase in monthly mean precipitation variability in most areas, both in absolute value (standard deviation) and in relative value (coefficient of variation)'. Figure 2.13 shows the changes in extreme precipitation events based on the multi-model mean anomalies. 'Precipitation intensity' is defined as the annual total precipitation divided by the number of wet days. We can see from Fig. 2.13a that variability in precipitation intensity increases quite substantially over the course of the century, regardless of the scenario. We also see from Fig. 2.13b that this variability is not uniformly distributed across the globe. 'Dry days' is

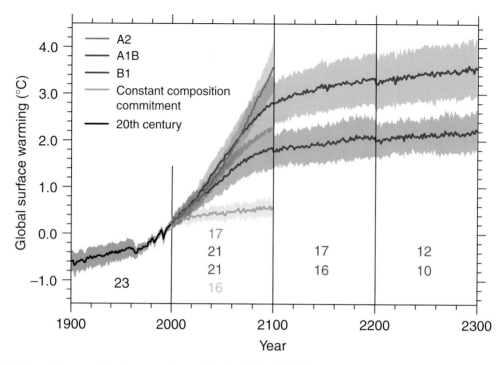

Multi-model means of surface warming (relative to 1980–1999) for the scenarios A2, A1B and B1, shown as continuations of the 20th-century simulation. Values beyond 2100 are for the stabilisation scenarios (see Section 10.7). Linear trends from the corresponding control runs have been removed from these time series. Lines show the multi-model means, shading denotes the ±1 standard deviation range of individual model annual means. Discontinuities between different periods have no physical meaning and are caused by the fact that the number of models that have run a given scenario is different for each period and scenario, as indicated by the coloured numbers given for each period and scenario at the bottom of the panel. For the same reason, uncertainty across scenarios should not be interpreted from this figure (see Section 10.5.4.6 for uncertainty estimates). (From: *Climate Change 2007: The Physical Science Basis. Working Group I Contribution to the Fourth Assessment Report of the Intergovernmental Panel on Climate Change*, Figure 10.4, Cambridge University Press.)

**Fig. 2.10.** The multi-model mean projections for global mean temperature anomaly (relative to 1980–1999). The numbers in each panel denote the number of models used for each century. The shaded areas denote ±1 standard deviation in range of individual models (Fig. 2.9).

defined as the annual maximum number of consecutive dry days. We see from Fig. 2.13c that the variability in the distribution of dry days increases across the century in a way that is positively correlated with the differences in emission scenarios. Finally, again, we see from Fig. 2.13d that this increased variability in the distribution of dry days is not expected to be uniformly distributed across the globe.

Box 2.2 provides an example of how scenarios and GCMs are used in studies of the biological impacts of climate change. Notice that any attempt

to make statements about the impacts of climate change ought to be made in the context of the scenario and GCM to which they may refer, rather than being sweeping statements about the impacts of climate change per se.

## 2.6 Conclusions

Our job as biologists is to say something meaningful about the likely biological impacts of climatic change. As will become clear throughout the rest

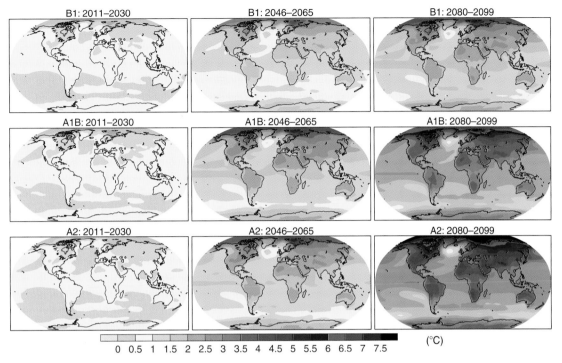

B1: 2011–2030   B1: 2046–2065   B1: 2080–2099

A1B: 2011–2030   A1B: 2046–2065   A1B: 2080–2099

A2: 2011–2030   A2: 2046–2065   A2: 2080–2099

0   0.5   1   1.5   2   2.5   3   3.5   4   4.5   5   5.5   6   6.5   7   7.5   (°C)

Multi-model mean of annual mean surface warming (surface air temperature change, °C) for the scenarios B1 (top), A1B (middle) and A2 (bottom), and three time periods, 2011 to 2030 (left), 2046 to 2065 (middle) and 2080 to 2099 (right). Stippling is omitted for clarity (see text). Anomalies are relative to the average of the period 1980 to 1999. Results for individual models can be seen in the Supplementary Material for this chapter. (From: *Climate Change 2007: The Physical Science Basis. Working Group I Contribution to the Fourth Assessment Report of the Intergovernmental Panel on Climate Change*, Figure 10.8, Cambridge University Press.)

**Fig. 2.11.** The multi-model mean projections for mean temperature anomaly (relative to 1980–1999) at each point in space for three scenarios and at three time periods over the course of this century.

of this book, those impacts are unlikely to be simple, but will depend on factors like how much warming occurs, when during the year the warming occurs, how, when and where precipitation will change, and so on. That is, our conclusions about the biological impacts of climatic change will be context specific. We won't be able to say things like: 'In a warmer world, species X will definitely decrease in abundance'. We won't be able to say things like this because it will turn out that the answer depends on how much warmer, when that warming occurs and, in many cases, how it interacts with changes in precipitation and other climate variables, and other ecological contingencies. So to say something specific and meaningful (as an impact of climatic change) we need to be able to consider specific changes in

climate – not just global annual mean changes, which, let's face it, are ecologically meaningless since no organism ever lives its whole life at the global mean value for any climate variable. So specifics matter. That is why the understanding we tried to convey in this chapter matters so much. If details matter, we need to understand that all climate scenarios, models and projections are not equal and we need to be able to give an accounting of how sensitive our predictions about climate change impacts are to our choice of emissions scenario and general circulation model. A really thorough study would consider multiple combinations of scenarios and GCMs (or RCMs). At a minimum, when reading the results of an impacts study, one has to beware of what 'climate change' the study is considering.

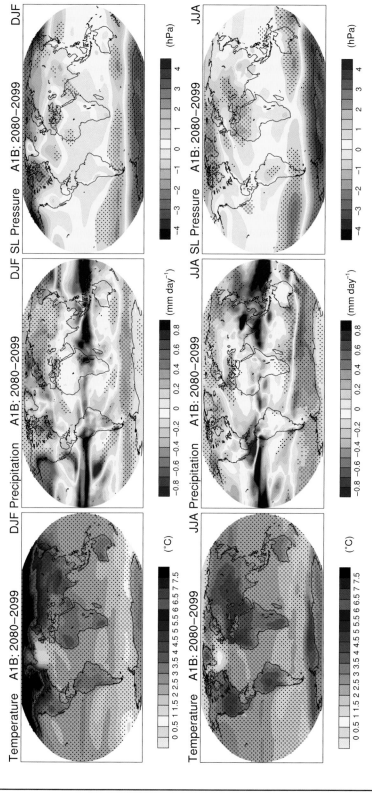

**Fig. 2.12.** The multi-model mean projections for mean temperature anomaly, mean precipitation anomaly and mean sea level pressure anomaly (relative to 1980–1999) and at each point in space for the A1B scenario at the end of this century. DJF denotes the average over the months of December, January and February, while JJA denotes the average over the months of June, July and August. The stippling denotes places where the magnitude of the multi-model mean is greater than the inter-model standard deviation. This is a reasonable indication that the value of the anomaly at that point is likely to be different from zero.

Multi-model mean changes in surface air temperature (°C, left), precipitation (mm day⁻¹, middle) and sea level pressure (hPa, right) for boreal winter (DJF, top) and summer (JJA, bottom). Changes are given for the SRES A1B scenario, for the period 2080 to 2099 relative to 1980 to 1999. Stippling denotes areas where the magnitude of the multi-model ensemble mean exceeds the inter-model standard deviation. Results for individual models can be seen in the Supplementary Material for this chapter. (From: *Climate Change 2007: The Physical Science Basis. Working Group I Contribution to the Fourth Assessment Report of the Intergovernmental Panel on Climate Change*, Figure 10.9, Cambridge University Press.)

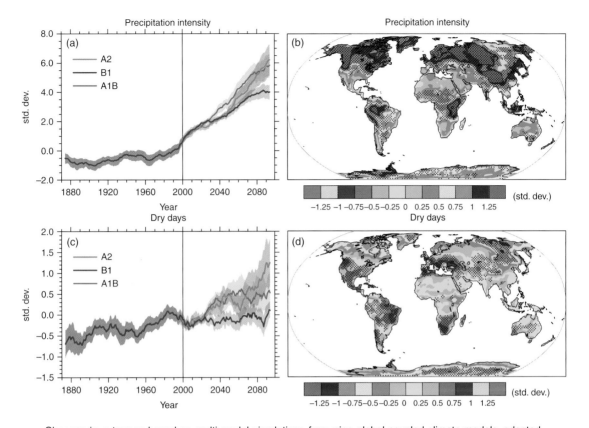

Changes in extremes based on multi-model simulations from nine global coupled climate models, adapted from Tebaldi *et al.* (2006). (a) Globally averaged changes in precipitation intensity (defined as the annual total precipitation divided by the number of wet days) for a low (SRES B1), middle (SRES A1B) and high (SRES A2) scenario. (b) Changes in spatial patterns of simulated precipitation intensity between two 20-year means (2080–2099 minus 1980–1999) for the A1B scenario. (c) Globally averaged changes in dry days (defined as the annual maximum number of consecutive dry days). (d) Changes in spatial patterns of simulated dry days between two 20-year means (2080–2099 minus 1980–1999) for the A1B scenario. Solid lines in (a) and (c) are the 10-year smoothed multi-model ensemble means; the envelope indicates the ensemble mean standard deviation. Stippling in (b) and (d) denotes areas where at least five of the nine models concur in determining that the change is statistically significant. Extreme indices are calculated only over land following Frich *et al.* (2002). Each model's time series was centred on its 1980 to 1999 average and normalised (rescaled) by its standard deviation computed (after de-trending) over the period 1960 to 2099. The models were then aggregated into an ensemble average, both at the global and at the grid-box level. Thus, changes are given in units of standard deviations. (From: *Climate Change 2007: The Physical Science Basis. Working Group I Contribution to the Fourth Assessment Report of the Intergovernmental Panel on Climate Change*, Figure 10.18, Cambridge University Press.)

**Fig. 2.13.** Changes in the distribution of 'precipitation intensity' and in the distribution of 'dry days'. (a) and (c) show the time series for the change in the variability of the distribution of these two metrics, while (b) and (d) show the spatial distribution for just the A1B scenario in the time period 2080–2099. Stippling denotes areas in which five of the nine models used in this analysis agree that the change is statistically significant. The solid lines in (a) and (c) denote the multi-model mean and the shaded areas denote the standard deviation.

## Box 2.2. Potential carbon sequestration in cultivated soils.

Lugato and Berti (2008) conducted a modelling study of the interactions between the choice of SRES scenario, GCM and different management practices on the potential soil carbon sequestration (see Chapters 7, 9 and 11) in cultivated soils in north-east Italy. We chose this study because it was unusually thorough in that it not only considered the impacts of the choice of scenario, but also the impacts of the choice of four GCMs.

Lugato and Berti considered the impacts on soil carbon sequestration of changing farming practice from 'business as usual', to either 'reduced tillage' or 'farmyard manure'. They considered a crop rotation of maize–wheat–maize–soybean. 'Business as usual' and 'reduced tillage' used 240 kg of mineral nitrogen fertilization per hectare, while 'farmyard manure' used mineral-N at 140 kg/ha plus an additional organic-N at 100 kg/ha applied as farmyard manure.

To model the carbon sequestration Lugato and Berti used the Century Model (Parton *et al.*, 1988). The Century Model simulates C, N, P and S dynamics on monthly time intervals. It can be used for either natural or cultivated lands, and it has been used extensively in studies of climate change impacts (see Chapter 3). Table 2.3 shows the change in soil organic carbon (SOC) that results from switching farming practices. For comparison they also looked at the carbon sequestration that could be obtained by converting that land to a grassland reserve (GR). They compared the output from this analysis for two different time periods: the first commitment period from the Kyoto Protocol (2008–2012) and in the more distant future (2080).

Table 2.3 shows that switching 'business as usual' to 'farmyard manure' is better for SOC in the short

**Table 2.3.** Changes in soil carbon sequestration. (Data from Lugato and Berti, 2008.)

| Scenario | Management | Change in soil organic matter (g/cm$^2$) | | | |
|---|---|---|---|---|---|
| | | Had3 | GCM2 | CSIRO2 | PCM |
| *2009–2012* | | | | | |
| A1FI | RT | 113 | 109 | 107 | 103 |
| | FM | *180* | **168** | **167** | **165** |
| | GR | 231 | 260 | 247 | 255 |
| A2 | RT | 113 | 109 | 111 | 103 |
| | FM | *178* | **168** | **174** | **164** |
| | GR | 231 | 260 | 239 | 253 |
| B1 | RT | 115 | 105 | 107 | 101 |
| | FM | *181* | **165** | **171** | **163** |
| | GR | 227 | 254 | 227 | 258 |
| B2 | RT | 112 | 106 | 108 | 102 |
| | FM | *179* | **166** | **171** | **164** |
| | GR | 230 | 255 | 229 | 259 |
| *2008–2080* | | | | | |
| A1FI | RT | **323** | **484** | *614* | **545** |
| | FM | 197 | 391 | 529 | 440 |
| | GR | 1589 | 1346 | 1287 | 1350 |
| A2 | RT | **391** | **509** | *597* | **542** |
| | FM | 266 | 432 | 516 | 451 |
| | GR | 1493 | 1384 | 1278 | 1373 |
| B1 | RT | **420** | *502* | *502* | **475** |
| | FM | 285 | 461 | 461 | 416 |
| | GR | 1553 | 1553 | 1308 | 1399 |
| B2 | RT | **442** | **511** | *546* | **485** |
| | FM | 318 | 459 | 469 | 418 |
| | GR | 1455 | 1296 | 1262 | 1404 |

RT, reduced tillage; FM, farmyard manure; GR, grassland restoration.
A1FI is a combination of the A1C and A1G scenarios; the 'FI' stands for 'fuel intensive'.
Bold figures denote the management practice that maximizes carbon sequestration for each scenario–GCM combination. Bold italic figures denote the GCM–management option that results in the maximum carbon sequestration.

term, but in the long term the result is opposite and 'reduced tillage' is ultimately the best. Furthermore, there is an interaction between management practices and the climate change scenario. The biggest difference between 'reduced tillage' and 'farmyard manure' occurs for the A1FI scenario (a combination of the A1C and A1G scenarios, where the 'FI' refers to 'fuel intensive'). Notice too that the conclusion one can reach from this study can differ by more than twofold depending on one's choice of GCM. Certainly, Lugato and Berti's results argue for the use of both multiple GCM projections for the same scenario, and multiple scenarios for the same GCM, in order to have any idea of how robust a study's conclusions are to these choices (see also Chapter 14, 'Which climate change?', p. 239).

# 3

# Methods for Studying the Impacts of Climatic Change

In this chapter we examine the methods that biologists use to study the impacts of climatic change. These techniques can be divided into three categories: (i) observational; (ii) experimental; and (iii) theoretical/statistical. There is a variety of methodological, statistical and philosophical difficulties associated with studies of the biological impacts of climatic change. It will be important to keep in mind that our critiques are criticisms of the work we examine, not the researchers. There is no one 'right way' to address these questions. All of the methods have drawbacks and shortcomings, and so we do the best we can with what we have to work with. Nevertheless it is important that we are consciously aware of the shortcomings of our work, so that we always remember the limitations of our results (see also Chapter 14).

## 3.1 Observational Methods

Observational methods include a wide range of techniques that share the common feature of using non-manipulative approaches to address questions of climate impacts. Often they take advantage of 'natural experiments' such as occur near volcanic vents that outgas $CO_2$-rich air or using historical data that were initially gathered for other purposes. Alternatively, palaeoecological techniques can be used to reconstruct ecosystem processes over time, to determine how organisms have responded to past climatic changes (see Chapter 1).

### Phenological studies

Phenological studies normally make observations of phenological events such as bud burst, flowering time and emergence, either over a prolonged period of time or at two instances widely separated in time. Any differences that are observed are then related to observed changes in climate over the interval of analysis. The observed patterns of change provide both evidence of climatic change and insights into how species may change in response to future climatic changes. Doi (2008) provides a helpful illustration of this approach. Doi compared the trends in the date of first appearance of dragonflies in Japan with those of temperature over the period 1953 to 2005, and showed that first appearances of the common skimmer (*Orthetrum albistylum speciosum*) occurred, on average, 24 days later in the year at the end of this time period compared with the start. Similarly, Doi showed that temperature increased during this time period by different amounts according to season, but in the range of 1–2°C. Doi then showed that temperature was an important predictor of dragonfly appearance even after other potentially confounding factors were accounted for. Doi's results differ from the results of other studies (e.g. Hassall *et al.*, 2007), and perhaps from our a priori expectations. We might reasonably have expected that warmer temperatures would result in earlier dates of first emergence rather than later. Doi offers several hypotheses for how this could occur, relating to the annual timing of the two generations (bivoltinism) and their spring emergence. Nevertheless, Doi acknowledges that this was a correlative study and cannot address the cause-and-effect relationship that might be driving the correlation.

Doi's (2008) conclusion is worth taking note of. The weakness of observational studies is that they cannot easily establish cause-and-effect relationships. For that we need randomized controlled experiments. However, this does not mean that observational studies are not useful or important. The strength of observational studies is that they usually span a much longer time scale (and often spatial scale) than experimental studies. This strength is significant and, as such, observational studies will always play an important role in climate change impact studies.

## Palaeoecological methods

A wide range of tools are available to reconstruct ecosystem composition, processes and changes. Several of these were discussed in Chapter 1, including the analysis of lake sediments for pollen and aquatic organisms and dendrochronology. Once the ecological parameter of interest has been reconstructed it can be compared with independently derived reconstructions of climate to evaluate how species have responded to past environmental change. Rather than review those techniques here, we examine one new technique that provides insights into ecological processes but not climate, and we look in some detail at how the analysis of sediment cores has yielded insights into ecosystem responses to climatic change.

The fact that the Little Ice Age ended in the late 1800s, and that since then mean temperatures have increased by approximately 1.5°C (see Chapter 1), provides an opportunity to directly observe how organisms have responded to warming over the 20th century. Although ecological monitoring has not been undertaken for most regions of the Earth, other historical documents are available that provide both qualitative and quantitative evidence of ecosystem structure and composition. In terms of qualitative data, historical accounts by early explorers and settlers, survey descriptions and taxation records can all provide baseline descriptions of ecological properties. For example, Gedalof *et al.* (2006) reconstructed stand structure and composition at an oak woodland on south-west Vancouver Island, British Columbia, Canada. They found that there was an interval lasting from 1706 to 1850 during which no presently surviving trees established at that site. Were there trees present, which are now absent having been killed by fire or removed by people? Or does this interval genuinely represent a period without establishment? There were no fire scars present,

but the absence of evidence does not reliably provide evidence of absence. The authors found a description of the site from 1849, written by Hudson's Bay Company explorer and Fellow of the Royal Geographical Society, Walter Colquhoun Grant (1857). Grant described the site as 'a fine open prairie extending nearly across to Becher Bay . . . interspersed with oak trees' (p. 282). Later, in a letter to the Governor of the Island, Grant complained about the indigenous peoples' habit of burning land: 'I have endeavoured in the neighbourhood of Mullacherd to check these fires by giving neither potlache or employment to any Indians so long as a fire was blazing in sight of my house'. These historical accounts provide evidence that in 1850 the site was relatively treeless, and was probably maintained that way by frequent low-severity fires that were intentionally set to promote the growth of certain food crops. Following cessation of burning, trees began establishing at the site.

Quantitative ecological data can be derived from herbaria collections and from historical photographs. Herbaria archive plant samples that are used for a wide range of research purposes. Samples always include, among other data, the date and location of collection. Many herbaria have samples that were collected decades to even centuries ago. These samples can be used to determine community composition at points in the past, the date of key phenological events (see 'Phenological studies', above), and the phenotype and genotype of the plants growing at the time of collection. Using herbaria samples from different times or by comparing them with modern conditions, these attributes can be evaluated for change in response to 20th-century warming. Historical photographs provide another source of quantitative ecological data. Although it is often difficult to identify species from these photographs, plant functional type, vegetation density and geographic location can often be determined with a high degree of accuracy. One particularly good application of this method was undertaken by Eric Higgs and colleagues under the 'Mountain Legacy Project' (http://mountainlegacy.ca/). Starting in 1888, early surveyors and explorers began photographing the Canadian Rocky Mountains for the purpose of developing topographic maps. Beginning in 1997 Higgs and his team began reoccupying the photographic stations used by these early explorers and, using comparable photographic equipment, replicated many of their photographs.

These photographic pairs provide a tool for evaluating landscape-scale changes over the 20th century (Fig. 3.1). In particular, they show evidence of widespread encroachment of forests into both high-elevation meadows and low-elevation grasslands, but also evidence of changes from coniferous to deciduous vegetation in some regions (Rhemtulla *et al.*, 2002). Climate is only one of a suite of processes driving landscape change in this environment, but the Mountain Legacy Project provides compelling evidence that changes have been substantial over the past century, and that ecological surprises are likely.

Sediment cores provide a rich source of palaeo-ecological data: communities of aquatic organisms can often be determined from the analysis and identification of hard body parts; wildfire activity can be detected through changes in magnetic susceptibility and the presence of macroscopic charcoal; littoral vegetation can be reconstructed through the preservation of macrofossils such as foliage and seeds. However, preserved pollen grains provide the most commonly used source of palaeoecological information (see Chapter 1 for a more detailed explanation of data collection and dating techniques). Because most pollen that is preserved is derived from local sources, the grains present at a particular depth in the core provide a snapshot of the vegetation present at a specific point in time.

One way that we can use pollen reconstructions of vegetation is to examine how plant communities have responded to climatic change in the past. For example, we might hypothesize that during the last glaciation plant communities were shifted towards the equator or downslope in response to the cooler climate. We can test this hypothesis by reconstructing community composition at the last glacial maximum and comparing the communities present then with the location of those same communities today. This has been done for many places on the Earth, but one study by Colinvaux *et al.* (1996, 1997) is typical in its findings and provides a good illustration of the technique. In these studies, the authors test the hypotheses that: (i) vegetation zones moved as cohesive units in response to climatic change; and (ii) the tropical lowlands were relatively tree-free – and thus

(a)  (b)

**Fig. 3.1.** (a) The view from the Daisy East 2 station in 1913 by M.P. Bridgland and crew. (b) The same view in 2008, taken by members of the Mountain Legacy Project (http://mountainlegacy.ca/). Note in particular the transition from open- to closed-forest conditions, and the upslope migration of trees.

Chapter 3

two error probabilities are related to each other, such that the only way to simultaneously lower both probabilities is to increase the sample size. And there's the rub in climate change experiments. For reasons that will become clear as we progress through the next sections, we are often limited by our sample size.

To help us understand the significance of the power problem, let us illustrate this with something called a prospective 'power analysis'. To do this we need to introduce the idea of a 'standardized effect size', hereafter called simply the 'effect size'. The effect size is defined as follows:

$$\text{Effect size (ES)} = \frac{\bar{X}_{\text{treatment}} - \bar{X}_{\text{control}}}{s_{\text{pooled}}} \quad (3.1)$$

In words, Eqn 3.1 is just the difference between the mean of the treatment group and the mean of the control group, *expressed in units of standard deviation*. By convention; ES=0.2 is considered a 'small' effect size', ES=0.5 is considered a 'medium' effect size and ES=0.8 is considered a 'large' effect size. We can now go on to define the statistical 'power' of a hypothesis test as the probability of detecting the hypothesized effect size, if it exists. So when planning an experiment, we need to consider how big a difference between our treatment and our control we would like to be able to detect, assuming

such a difference actually exists. Being able to detect an ES of 0.2 means being able to detect quite subtle treatment effects. On the other hand, an ES of 0.8 means that we would be able to detect only quite large differences between treatment and control. A rule of thumb is that a good experiment should have a statistical power of 0.8 for the given effect size we want to be able to detect, if it in fact exists. Now, power is related to sample size as shown in Fig. 3.3. In Fig. 3.3, the sample size refers to the number of replicates in each of the treatment group and the control group. A point to remember from Fig. 3.3 is that, in order to detect a 'large' effect size, you need a sample size of at least 50 replicates, 25 for the treatment and 25 for the control. What will become apparent as we go through the next few sections is that very few climate change impact experiments have anything near that many replicates, and so are only going to be able to detect the very largest differences between treatments and controls. That is, most climate change impact experiments are likely to miss any subtle treatment effects.

### Pseudoreplication

This problem is subtle and can be difficult to grasp at times, but is pervasive in climate change research and so it is important to understand.

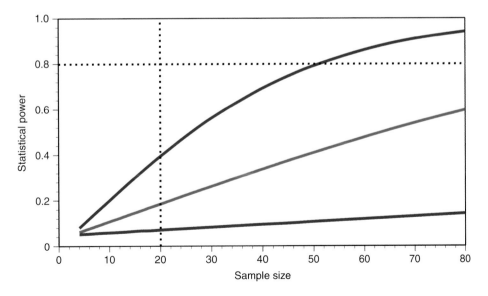

**Fig. 3.3.** The relationship between statistical power and sample size per treatment level. Blue line denotes 'small effect size', red line 'medium effect size' and the green line 'large effect size'.

Pseudoreplication was first introduced to ecologists by Stuart Hurlbert's (1984) landmark paper 'Pseudoreplication and the design of ecological field experiments'. Since that time there have been exchanges in the literature on this issue, most recently by Oksanen (2001) in his paper 'Logic of experiments in ecology: is pseudoreplication a pseudoissue?', in which he answers in the affirmative. The Oksanen paper has been critiqued by Cottenie and De Meester (2003) and by Hurlbert himself (2004). While we think there is no room for serious debate and side clearly with Hurlbert on this issue, readers may wish to decide for themselves by pursuing these and other exchanges (see references in Hurlbert, 2004). We will not review these arguments here.

So what, exactly, is pseudoreplication? Pseudoreplication is a statistical error that arises from failing to appreciate what is the unit of replication for the hypothesis being tested. Let us look at a hypothetical example to try and understand this. Suppose that we are interested in the effects of nitrogen on the concentration of free amino acids in the phloem sap of a tillering grass species. One way to address this question would be to get two plants and give one a high dose of nitrogen and the other a low dose. We could then cut ten tillers from each plant and extract the phloem sap from each tiller. We then analyse these 20 samples of phloem sap and obtain estimates of the concentrations of free amino acids from each sample. The correct way to analyse these data would be to recognize that 'plant' is the unit of replication not tiller. So we have a sample size of two, not 20. Another way to have done that experiment would be to grow 20 plants, ten at high nitrogen and ten at low nitrogen, and then sample a single tiller from each plant. Then we would legitimately have a sample size of 20.

How is the above example relevant to climate change research? Well, as we see below, we are often constrained by the number of, for example, chambers in which to manipulate $CO_2$ or temperature. These chambers can be quite expensive to build and operate and yet the chamber is usually the unit of replication. It is not uncommon to see climate change experiments in which there are only two chambers, say one at ambient $CO_2$ and one at elevated $CO_2$, but in which there are, say, ten plants growing in each chamber. Just as in the hypothetical experiment discussed above, where the proper unit of replication is the plant and the tillers are pseudoreplicates, here the proper unit of

replication is the chamber and the plants are pseudoreplicates. Although it is probably safe to say that everyone recognizes the problem of pseudoreplication, it is still nevertheless tolerated in this field of research, at least in some journals, including some of the best journals. For example, Vestgarden and Austnes (2009) published a paper in *Global Change Biology* on the effects of freeze–thaw cycles on carbon and nitrogen dynamics in soil cores. They had four temperature treatments: constant thaw (5°C); constant freeze (–5°C); fast cycling (–5 to 5°C); and slow cycling (–5 to 5°C). There were five replicates per temperature treatment. The constant thaw treatment was done in a 'cooling room'. The constant freeze treatment was done in a thermostat-regulated freezing chamber, and the two cycling treatments were done in two climate chambers. Here the 'proper' unit of replication is the room/chamber used to manipulate the temperature; the cores are pseudoreplicates. Nevertheless, throughout the analysis the authors treat the cores as if they were genuine replicates.

We mean no disrespect by pointing out this discrepancy. Vestgarden and Austnes are by no means unique in this regard; such analyses are not ubiquitous but neither are they uncommon (see e.g. Box 3.1 for another example). The simple fact is that the authors were unlikely to have access to 20 climate chambers in which to conduct the experiment. We think that, as long as authors are clear about their use of pseudoreplicates, and that readers appreciate the potential problems interpreting such results, then such studies are valuable despite their pseudoreplication. Nevertheless, many researchers will disagree with this assessment and take a very 'right' and 'wrong' view of such experiments. Readers will have to decide for themselves what such experiments mean or do not mean, but most importantly they should be aware of the problem.

## CO₂ manipulations

Figure 3.5 shows some examples of climate chambers that are used to manipulate $CO_2$ concentrations. These same chambers can also be used to control temperature and relative humidity, and of course one has complete control over such variables as soil moisture and fertility (although this is often easier said than done). So the level of control is high. With a sufficient sample size, often a problem with such indoor chambers, such an experiment might be able to detect reasonably small effects of $CO_2$. However, we would be left wondering whether such effects would be detectable against the background of variability in a natural field situation. So the external validity is relatively low.

Indoor chambers vary a lot in how they are constructed, what they are capable of controlling, their size, and how many replicates are available. At one end of the scale are simple growth cabinets (Fig. 3.5c) where it would not be uncommon to have just two of these; or maybe four if you were lucky. These cabinets use artificial light, which is problematic since it is notoriously difficult to mimic natural night. The number of replicates tends to be small because the cabinets are expensive to buy. More towards the middle of the scale are chambers like those seen in Fig. 3.5d. These chambers tend to be cheaper to construct and replication levels tend to be a bit higher. Control of variables like light, temperature and $CO_2$ are more difficult unless one invests in more sophisticated, and expensive, control systems. Finally, at the far end of the scale for indoor chambers would be something like the Ecotron at Silwood Park, in the UK. Here the chambers are large enough to walk into, and there are 16 identical chambers to provide replication. Unfortunately, the Ecotron was very expensive to build and operate. The facility cost approximately US$2.1 million to build and US$200,000 per year to operate, in 2010 inflation-adjusted figures (Lawton, 1998).

Indoor chambers tend to offer very good control of the environmental variables, but they lack external validity and, because of the expense of both the chambers themselves and the indoor floor space, they tend to provide poor replication. This leads to either extremely low statistical power or pseudoreplication. Moving on from indoor chambers, outdoor chambers, particularly open-topped chambers, offer increased external validity and because space in the field is usually cheaper than indoor space, replicate number is usually greater. Figure 3.6 shows some examples of open-topped chambers. Because they are located outside, they are subject to natural variation in sunlight, temperature and moisture, and they have natural soil processes operating. On the other hand, we do tend to see so-called 'chamber effects', such as increased temperature and humidity compared with the surrounding environment. Because they are open at the top, control of $CO_2$ levels is considerably more

difficult than for indoor chambers and requires considerably more $CO_2$ to operate. Also, generally speaking, such chambers are feasible only for studying ecological communities comprising organisms of small stature, such as in grasslands, marshes and so on.

The true 'gold standard' for elevated $CO_2$ experiments is the Free Air $CO_2$ Enrichment (FACE) arena. Figure 3.7 shows some examples of FACE arena used at the Duke experimental forest. FACE systems work by flooding a plot of ground with enough $CO_2$ that the local concentration of $CO_2$ can be raised without chamber walls to help concentrate it. While it sounds like this system would be very variable in the concentration of $CO_2$ it delivers, the control is actually pretty good most of the time. For instance, the Duke forest FACE experiment, over the 11-year period from 1996 to 2007, was within 10% of its $CO_2$ concentration target 60–70% of the time and within 20% of its target 87–93% of the time (http://tinyurl.com/ch3-DUKE-1). More details on a standard design and operation of a FACE site are available online (http://tinyurl.com/ch3-DUKE-2). The FACE system is the gold standard because there are no environmental changes caused by chamber walls, as occurs in the open-topped chambers. FACE experiments therefore have the highest external validity but the lowest level of control of the important climate variables. The drawbacks to FACE experiments are: (i) they are expensive and require long-term funding; (ii) they are technically complex; and (iii) there tend to be few replicates. The Duke FACE experiment for example has cost approximately US$45 million since it started in 1994. This site has eight rings (plots), four are elevated to +200 ppm above current ambient $CO_2$ and the remaining four are left at ambient $CO_2$. This is not an unusual level of replication for a FACE site. There are only 15 FACE sites worldwide, and they vary in the concentrations of $CO_2$ they maintain and in their degrees of replication. Because of the complexity, cost and need for a sustained funding base, FACE sites are not representatively distributed around the world. Figure 3.8 shows the distribution. We see that the sites are located mostly in the USA and Western Europe. They can hardly be said to be representative of the world's ecosystems. Furthermore, most of them focus on a single species of dominant plant, limiting the inferences that can be drawn from their results.

**Box 3.1. Step changes versus natural changes in CO$_2$ concentrations.**

Nearly all experiments on the impacts of elevated CO$_2$ concentrations have one method in common. The experiment is set up using one of the techniques discussed above for manipulating CO$_2$, and then the gas is turned on and the experiment starts. So on day −1 all of the plants, plots of ground, etc. are at ambient CO$_2$; and the next day, day 0, half are at elevated CO$_2$. There has always been a niggling doubt among researchers that perhaps some of the effects observed in experiments and attributed to elevated CO$_2$ are not due to elevated CO$_2$ per se, but to the abrupt manner in which the system was exposed to elevated CO$_2$. Starting in the late 1990s, John Klironomos and his colleagues began a 6-year experiment (Klironomos *et al.*, 2005) in which they compared the effects of an abrupt change in CO$_2$ with a more gradual change in CO$_2$ on the community composition of arbuscular mycorrhizal fungi (AMF). The experiment had three treatments. In one treatment the plants–soil–AMF were continuously exposed to ambient CO$_2$ concentrations (350 ppm). In a second treatment the plants–soil–AMF were switched from ambient CO$_2$ to 550 ppm CO$_2$ in one step-change on day 1 of the experiment. And finally, in the third treatment, the plants–soil–AMF were gradually exposed to increasing concentrations of CO$_2$ from ambient to 550 ppm over a period of 6 years. Since the generation time for these plant–AMF associations is approximately 15 weeks and the plants were replaced every 15 weeks, this experiment represents 21 generations of plant–AMF reproduction. We can see from Fig. 3.4 that an abrupt change in CO$_2$ concentration reduced the AMF species richness compared with the constant ambient treatment. We can also see from Fig. 3.4 that after 21 generations of gradual increases in CO$_2$ concentration the species richness in this treatment was not different from that in the constant ambient treatment.

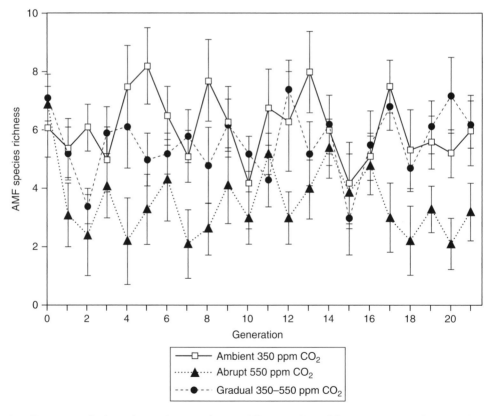

**Fig. 3.4.** Responses of arbuscular mychorrhizal fungi to different ambient CO$_2$ concentrations. (Redrawn from Klironomos *et al.*, 2005.)

This study shows that it is clearly *possible* that abrupt changes in $CO_2$ concentrations can have different impacts from gradual changes in $CO_2$, although it does not show that such changes are inevitable. Klironomos *et al.* just demonstrated that researchers may be right to worry about such artefacts of experimental design. On the other hand, there is little that researchers can do to overcome this limitation of experimental design. Even the 'gradual' treatment in this experiment was raising the $CO_2$ concentration approximately 20 times faster than the 'natural' rate of $CO_2$ accumulation. If researchers hope to be able to provide predictions about the impacts of climate change, then they cannot conduct experiments that approximate the actual rate of $CO_2$ accumulation. Some degree of artificiality will be required, and we cannot know which impacts are artefacts of the abrupt change in $CO_2$ and which are the impacts of $CO_2$ per se. In fact, this study leads us to an important question: what is the potential for species and systems to adapt to climate change? That is, might some of the systems and species we study in these climate change experiments be able to adapt to climatic change, on the time scales at which that change is predicted to happen? We take up this question in more detail in Chapter 8.

Beyond these manipulations, there are some places in the world where nature is doing our experiment for us. These so-called 'CO_2 springs' are places where $CO_2$ escapes from the ground and locally raises the concentration of $CO_2$. Table 3.2 shows the locations and $CO_2$ concentrations for some of the natural springs that have been used for this kind of research. Some of the advantages of natural springs are: (i) they are inexpensive; (ii) the communities have been exposed to elevated $CO_2$ for long periods of time; (iii) they cover relatively large areas; (iv) one can examine responses along a large gradient of $CO_2$ concentrations; and (v) there can be distinctive isotopic signatures of the carbon which can make it possible to conduct stable isotope studies (see also Chapter 1). On the other hand, there are several disadvantages to this approach: (i) $CO_2$ concentrations are not stable, they often fluctuate considerably over all time periods (hourly, daily, weekly, etc.); (ii) choosing an appropriate 'control' or 'reference' site is often problematic; and (iii) the actual purity of the $CO_2$ source itself can be problematic (i.e. the springs do not just add $CO_2$ to the atmosphere; Grace and van Gardingen, 1997).

### Temperature manipulations

Methods for manipulating temperature parallel those for manipulating $CO_2$ and add a few other techniques. Figure 3.9 shows some examples of open-topped and closed chambers used to manipulate temperature. The equivalent of FACE for temperature is achieved using heat lamps; some examples are shown in Figure 3.10. Heat lamps sometimes also emit photosynthetically active radi-

ation; when they do, it must be removed with light filters so as not to confound the treatment effect. More recently, researchers have switched to using ceramic bulbs like those used for keeping snakes, lizards and frogs warm in captivity without affecting their diurnal rhythms.

Sometimes researchers are less interested in the effects of warming on aboveground processes, and are more interested in what happens below ground (see Chapter 9). Again, there is a variety of techniques for manipulating soil temperatures. Soil heating cables have been used for a long time, so have black tarps and cold frames. While these techniques are technically simpler and cheaper than techniques like those shown in Fig. 3.10, they do lack some external validity by not creating the warming via radiant heat (as will happen with global warming). Some researchers have used snow removal as a manipulation (e.g. Groffman *et al.*, 2001), which simulates snow melt and leaves the soil quite vulnerable to freezing. See Shen and Harte (2000) for further discussion on these and other methods.

### Precipitation manipulations

Precipitation manipulations are often designed to impose drought, although, as we have shown in Chapter 2, drought may not always be an appropriate manipulation for the local situation. Drought is usually imposed via one of two primary methods. The first is the use of 'rain shelters', which are shown in Fig. 3.11. The second method is to use water barriers on the soil surface. Such barriers can be made of anything from plastic to more sophisticated semi-permeable membranes. Recently, several experiments have used

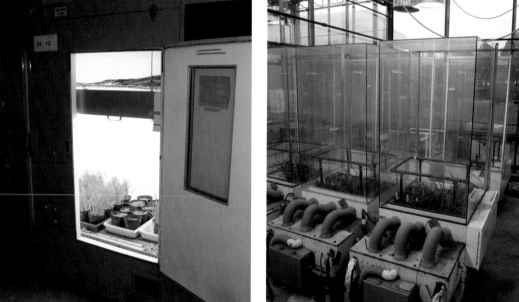

**Fig. 3.5.** Examples of climate chambers used to manipulate $CO_2$ concentrations. (a) The outside of a 'walk-in' climate chamber; (b) the inside of a 'walk-in' climate chamber; (c) a typical 'reach-in' climate chamber; and (d) an example of a chamber used in greenhouses. (Photos: J. Newman, School of Environmental Sciences, University of Guelph.)

(a)

(b)

**Fig. 3.6.** Open-topped chambers used to manipulate $CO_2$ levels. Open-topped chambers can vary quite a bit in size. In (a) the chambers are 1.5 m in diameter, in (b) they are 4-m diameter chambers. Note too, in (b), the chambers are fitted with rain shelters to simultaneously control both $CO_2$ levels and precipitation. Such cross-factoring is uncommon in climate change experiments (with the exception of experiments that cross-factor $CO_2$ and soil nitrogen). (Photo (a): J. Newman, School of Environmental Sciences, University of Guelph. Photo (b): R. Norby, Environmental Sciences Division, Oak Ridge National Laboratory.)

(a)

(b)

**Fig. 3.7.** Free Air $CO_2$ Enrichment (FACE) arena. (a) A forest FACE system run by Duke University; (b) a closer view of the pipes used to deliver the $CO_2$ at the Duke FACE. (Photo (a): J.S. Pippen, Nicholas School of the Environment, Duke University. Photo (b): Duke Photography, Duke University.)

rainwater diversion devices that reduce rainwater in some treatments and divert it to other treatments, so that the effects of both reduced and increased precipitation can be evaluated.

## 3.3 Theoretical and Statistical Methods

Mathematical models and statistical methods play an important role in studying the impacts of climatic change, as we will see throughout the coming chapters. In this section our intention is to give the reader a brief overview of the approaches used and to comment upon their strengths and limitations.

## Bioclimatic envelopes and ecological niche models

The bioclimatic envelope model was briefly introduced in Chapter 1 and is elaborated upon in Chapters 4 and 5. These models relate to the basic ecological concept of the niche. The 'fundamental niche' represents the intersection of all the abiotic factors that limit a species' distribution. Competition, dispersal limitations or other factors may cause the species to occupy a subset of these conditions, termed the 'realized niche'. Figure 3.12 depicts three environmental dimensions that might limit a species' range. Obviously there may be many relevant dimensions and one cannot depict

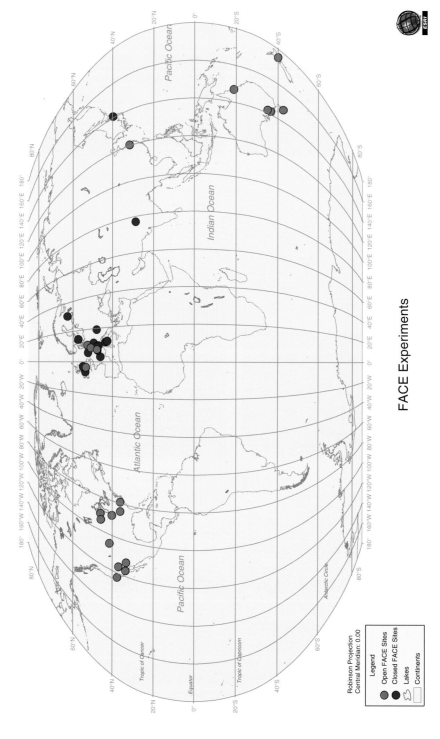

**Fig. 3.8.** The global distribution of FACE sites past and present. (Image: A. Allstadt, Department of Biology, The University at Albany.)

FACE Experiments

Robinson Projection
Central Meridian: 0.00

Legend
● Open FACE Sites
● Closed FACE Sites
〰 Lakes
☐ Continents

this multidimensional space with a picture, but we can imagine such a space, and we can represent that space mathematically. Both bioclimatic envelope models and ecological niche models attempt to describe mathematically the fundamental niche of a species, and then to locate that space on a geographical map as an area of 'suitable climate'. Such models were used before researchers became interested in climate change, but they are now a mainstay of the modelling effort to study climate change impacts – the idea being simply to relocate the area of suitable climate under future climate change projections (see Chapter 2).

The terms 'ecological niche models' and 'bioclimatic envelope models' are sometimes used interchangeably, but sometimes they are differentially used to distinguish how the model is constructed. Bioclimatic envelope models are usually statistical models, in the sense that they use distributional data on the organism of interest and historical climate data and attempt to find a statistical relationship between these two data sets. Ecological niche models on the other hand tend to use information on climate tolerances that are derived from experimental (often laboratory-based) work. For example, suppose that we are studying the hemlock wooly adelgid (*Adelges tsugae*, Fig. 3.13). We might be able to find a statistical relationship between its current distribution and the −25°C February minimum temperature isocline, as did Parker *et al.* (1999). We would use this statistical relationship as part of the bioclimatic envelope model for the species. On the other hand, we might bring eggs into the lab and expose them to different winter temperature regimes and see how these regimes affect overwinter egg survival. We could then build this information into an ecological niche model. Both models effectively use the same information to model potential distribution, but it is generally thought that the ecological niche modelling approach is in some senses 'better', since it relies on the fundamental niche of the organism rather than the realized niche (which is used in the bioclimatic envelope approach).

**Table 3.2.** Natural $CO_2$ springs used in elevated $CO_2$ experiments. (Modified from Grace and van Gardingen, 1997.)

| Country | Region | Source of $CO_2$ | Maximum $CO_2$ concentration (ppm) |
|---|---|---|---|
| Italy | Tuscany | Mineral spring | 1000 |
| | Campania | Mineral spring | 600 |
| USA | Utah | Burning coal seam | 900 |
| | Florida | Mineral spring | 575 |
| Iceland | | Mineral spring | 880 |
| New Zealand | Kamo | Mineral spring | 670 |

(a)

(b)

**Fig. 3.9.** Chambers used to manipulate temperature. (a) An example of an 'ITEX' chamber at the long-term research site on Ellesmere Island in the Canadian High Arctic. ITEX is the International Tundra Experiment (http://tinyurl.com/ch3-itex) and participating researchers have agreed upon a standard design. (b) The British Natural and Environmental Research Council's Arctic Research Station, in Svalbard. (Photo (a): J. Lang, courtesy of Polar Continental Shelf Program, Natural Resources Canada. Photo (b): British NERC Antarctic Survey (www.photo.antarctica.ac.uk).)

**Fig. 3.10.** Chamberless radiant temperature manipulations. (a) An experiment conducted by R. McCulley at the University of Kentucky, USA. (b) An experiment conducted by H.A.L. Henry at the University of Western Ontario, Canada. (Photo (a): J.A. Newman, School of Environmental Sciences, University of Guelph. Photo (b): H.A.L. Henry, Department of Biology, University of Western Ontario.)

**Fig. 3.11.** Examples of rain shelters used for manipulating soil moisture in the field. (Photo (a): M. Richardson, Department of Horticulture, University of Arkansas. Photo (b): B. Shore, Department of Biological Sciences, University of Alberta.)

This view that ecological niche models are better than bioclimatic envelope models is perhaps more philosophical than empirical. The difference between the two modelling approaches boils down to the problem of 'model validation', and both approaches have performed better in different analyses. After constructing either type of model, we would like to 'validate' the model. To do this we compare the results from the model with an independent data set (i.e. using data that were collected independently of those used in the model construction). If there is a reasonable fit between the model and the validation data, then we have some confidence that the model can at

least generate a reasonably accurate prediction of a species' contemporary distribution. For ecological niche models, the distribution data used to validate the model are generally entirely independent of the experiments that generated the niche relationships used to construct the model. With bioclimatic envelope models, distribution data are used to construct the model and to validate the model. This can create a problem of data independence. Modellers attempt to get around this problem by withholding some fraction of the data (typically a third) from the model construction phase to use in the model validation phase. Unfortunately, because all of the distribution data

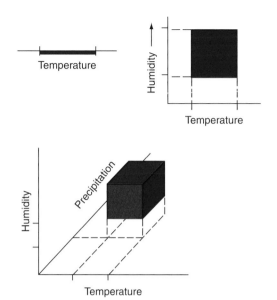

**Fig. 3.12.** Bioclimatic envelope approach.

will be spatially autocorrelated (i.e. non-independent), these holdout data are often not independent of the data used to generate the model.

Additional to the model validation issues, there is a sense that ecological niche models are somehow more 'mechanistic' than bioclimatic envelope models; the latter being entirely statistical. We generally think that mechanistic models are more informative than statistical models, even if they do not always produce better fits to observed data (Thornley and Johnson, 1990). Because bioclimatic envelope models work from an observed correlation between geographic distribution and climate data, to project a correlation between the geographic distribution and the climate data, they seem, in a sense, 'circular'. On the other hand, ecological niche models use one type of data (experiments on climate tolerances) to predict a different type of data (geographical distribution) and so seem like they are doing something more convincing. In selecting the environmental variables to test experimentally,

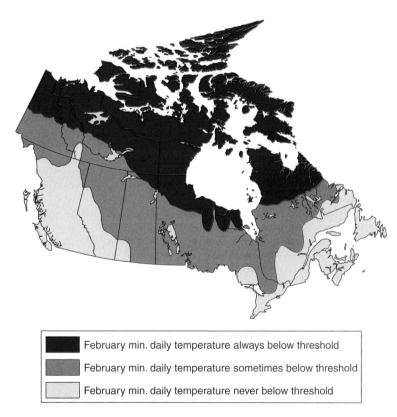

February min. daily temperature always below threshold

February min. daily temperature sometimes below threshold

February min. daily temperature never below threshold

**Fig. 3.13.** The northern boundary of the hemlock wooly adelgid (*Adelges tsugae*) is determined by the −25°C February minimum daily temperature isocline. (Image: J. Newman & A. Mika, School of Environmental Biology, University of Guelph.)

researchers are necessarily choosing to exclude others that may be important to a species' distribution. One example might be helpful in illustrating this problem: whitebark pine (*Pinus albicaulis*) and subalpine fir (*Abies lasiocarpa*) are common associates in high-elevation forests of western North America. At the highest locations, subalpine fir becomes established only in the protection of whitebark pine – which modifies the microclimate at the site of establishment, facilitating the survival of the subalpine fir (Callaway, 1997). Whereas ecological niche models would likely fail to predict establishment of subalpine fir, assuming that the environment would be unfavourable, bioclimatic envelope models would predict their success based on their observed survivorship. Neither modelling approach yet accounts for the sorts of biotic interaction just described.

Regardless of whether or not ecological niche models are better than bioclimatic envelope models, the latter are really the mainstay of climate impact modelling. This is because we have access to current geographical distribution data on far more species than those for which we have good, complete, experimental estimation of climate tolerances. So starting from scratch, it is far easier and far faster to construct a bioclimatic envelope model than an ecological niche model.

Despite the differences in their construction, and perhaps in their inferential power, these two modelling approaches are generally lumped together, and ecological niche models are sometimes referred to generically as bioclimatic envelope models. The reason for this is that they both suffer from many of the same shortcomings and they both produce 'risk maps' that would be indistinguishable if you did not know how the model was produced. An example of such a risk map is shown in Fig. 3.14. The shortcomings of these model approaches are well known. We enumerate them briefly here and refer the interested reader to more thorough critiques of these approaches (Pearson and Dawson, 2003; Hampe, 2004; Segurado and Araujo, 2004; Pearson *et al.*, 2006). Briefly then, these models: (i) rely on current climate tolerances, which may not be the same in the future (i.e. there may be adaptation; see Chapter 8);

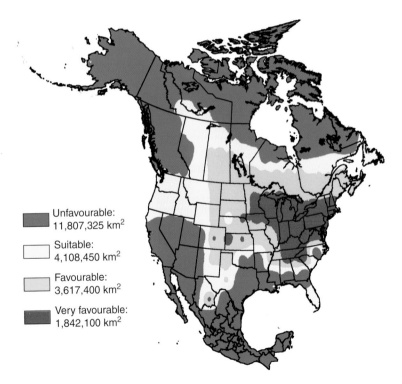

**Fig. 3.14.** Ecological niche model output for the agricultural pest the swede midge (*Contarinia nasturtii*). (From Mika *et al.*, 2008.)

Legend:
Unfavourable: 11,807,325 km$^2$
Suitable: 4,108,450 km$^2$
Favourable: 3,617,400 km$^2$
Very favourable: 1,842,100 km$^2$

(ii) can only show where the climate is suitable for a species, not where it will be found (i.e. take no account of dispersal or biotic interactions); (iii) assume that the species is currently filling its entire fundamental niche; and (iv) assume that the only factor limiting the species' distribution is climate.

In short, knowing where in the future the climate will be suitable for an organism is useful but provides incomplete information for predicting the fate of a species. But just because these methods do not tell us everything about the future distribution of a species, it does not mean they do not tell us anything. Again, this fits into a recurring theme of this chapter; each individual method has strengths and weaknesses. We do not want to rely on a single method to provide strong inferences because of the methodological weaknesses, but as long as we remain cognizant of each method's strengths and weaknesses, they all bring something to the table.

## Physiologically based mechanistic models

We already encountered a crude form of mechanistic models with the ecological niche model discussed in the last section. Defining a 'mechanistic model' is not so easy since one person's mechanism is another person's phenomenon. What we mean by mechanistic models in this section is something different from the ecological niche model. The main difference is that the mechanistic models we have in mind here are dynamic (rather than static, as are the bioclimatic envelope model and the ecological niche model) and generally, but not universally, involve mechanisms at the level of basic physiology rather than whole-organism responses.

One common class of these models might be termed 'ecosystem simulators'. Some well-known examples from the literature would be the Century Model (http://tinyurl.com/ch3-century) for soil organic matter (encountered in Box 2.2), the Hurley Pasture Model (Thornley, 1998; Fig. 3.15) and the Edinburgh Forest Model (Thornley and Cannell, 1996; which we encounter in Chapter 10), just to name a few. These models are often stochiometrically complete, at least for carbon and/or nitrogen. That is, they follow the fate of carbon and/or nitrogen from the time it enters the system until the time it leaves the system, and within the system there is a complete accounting of the element(s). Figure 3.15 shows a schematic of the Hurley Pasture Model. We can see from Fig. 3.15

that the level of detail, particularly process-based detail, is considerable.

If ecological niche models are considered to be in some sense 'better' than bioclimatic envelope models because they are built from data that are independent of the distribution data, then physiologically based mechanistic models are, in the same sense, even better models than the ecological niche models. These physiologically based mechanistic models are based on detailed empirical evidence of the ways in which the environment in general, and climate in particular, dynamically alters these mechanisms. Where ecological niche models usually integrate the effects of climate on the organism over the entire year, based on monthly mean values (but sometimes daily mean values), physiologically based mechanistic models are capable of integrating these effects over much shorter time steps (e.g. 15 min).

These three modelling approaches differ considerably in the time, data and skill required to construct them. For someone who knows what they are doing, and has access to decent distribution data, a bioclimatic envelope model can be constructed and analysed in a couple of weeks or less. An ecological niche model might take the same length of time, but it would be more difficult to obtain data, and thus would be suitable for a smaller subset of species than for a bioclimatic envelope model approach. Starting from scratch, a physiologically based mechanistic model could take a year or more to construct and analyse. Such a model will be feasible only for systems that have been previously well studied.

These are not the only modelling approaches used in the study of climate change impacts, but they are arguably the most common and are ones that we encounter elsewhere in this book. For some alternative modelling approaches, see for example Grimm and Railsback (2005).

## Meta-analyses

A meta-analysis is a formal statistical method for combining the results of multiple experiments to address a specific hypothesis. Meta-analysis is particularly useful for increasing the statistical power of a hypothesis test, and hence can be used to at least partially overcome the problem of typically inadequate replication described earlier in this chapter. So for example, suppose that we had several grassland FACE systems that each had few replicates. Any

(a)

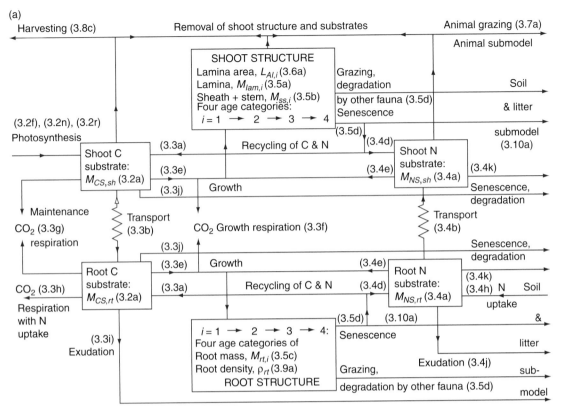

**Fig. 3.15.** Schematics of two of the four submodels that comprise the Hurley Pasture Model. (See Thornley, 1998 for details.)

*Continued*

single experiment in one FACE system will be able to detect only relatively large effect sizes. However, by combining the results from the several FACE experiments, in a formal analysis, we may be able to detect somewhat smaller effect sizes.

It sounds like meta-analysis is a panacea for the problems of weak statistical power discussed earlier. Although meta-analysis certainly helps this problem, it does not eliminate it entirely. First off, unlike some areas of science, climate change biologists are not usually repeating their own experiments or those of others. For example in medicine, you might have available the results of five similar small clinical trials of the effect of a certain therapy. These can be combined using meta-analysis to address the question of whether or not the therapy is effective. In climate change biology, we may be testing the same general hypothesis but in totally different systems. And it is not clear that responses should all be one way or another when we average across the various systems (see Chapter 14).

Equally, meta-analyses are sensitive to the presence or absence of 'negative results' (i.e. ones in which we have failed to reject the null hypothesis) and because of the low statistical power, negative results are not uncommon in climate change research. There is a general publication bias against negative results. If you are a journal editor and you are deciding how to use your precious and expensive journal pages, you could easily be forgiven for not publishing a climate change study with negative results. A negative result might mean that there is genuinely no effect of the treatment (some aspect of climatic change in this case), or that the study lacked sufficient statistical power to detect the effect. For any given study we can never know the answer to this, but we will have more confidence in negative results from studies that were well replicated and thus had a good deal of statistical power (something generally lacking from climate change experiments). So the editor has to make a difficult decision: 'Do I publish this paper even though by

(b)

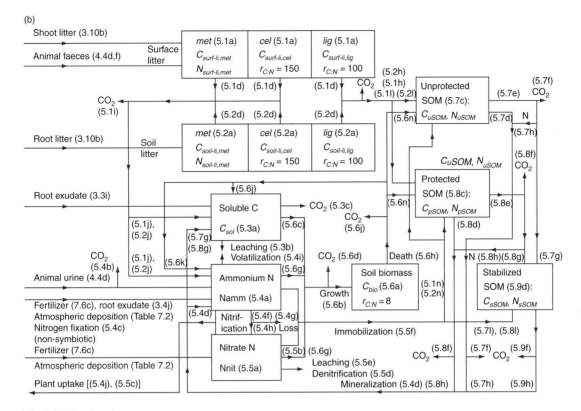

**Fig. 3.15.** Continued.

itself it doesn't tell us very much, but in the long term it may usefully contribute to a meta-analysis?' Perhaps unsurprisingly, the long view is often neglected under space and financial pressure, and negative results are published less often than we would like.

So meta-analysis is a tool that is available to us and it should not be ignored, but equally we should not believe that it will solve all our problems either. Here we briefly review one recent meta-analysis to demonstrate the use of this method, but it comes up time and again in the subsequent chapters. Ainsworth (2008) conducted a meta-analysis of 70 peer-reviewed journal articles on rice growth responses to elevated $CO_2$, published from 1980 to 2007. Averaged across all the studies, elevated $CO_2$ resulted in a 23% increases in yield. However, we can see in Fig. 3.16 that this increase depended strongly on the degree of $CO_2$ elevation and on the method used to elevate the $CO_2$. Ainsworth concludes that rising concentrations of atmospheric $CO_2$ are likely to aid rice production, but that these

positive results are likely to be mitigated by either nitrogen stress or higher temperatures (as are expected with climatic change).

It is worth mentioning that this particular meta-analysis had many results to work with, largely because rice is such an important plant worldwide and has thus been studied extensively. This observation is generally true: economically important species have been disproportionately well studied compared with other species and ecosystems, and this undoubtedly becomes apparent throughout the remainder of this book.

## 3.4 Conclusions

Some important lessons to get from this chapter are, first, that there are many different techniques that can be used to address the same biological questions. In some cases there are methods that are preferable to others, *all other things being equal*, but all other things are very rarely equal. In the real world, there are clearly advantages and disadvantages to all of the

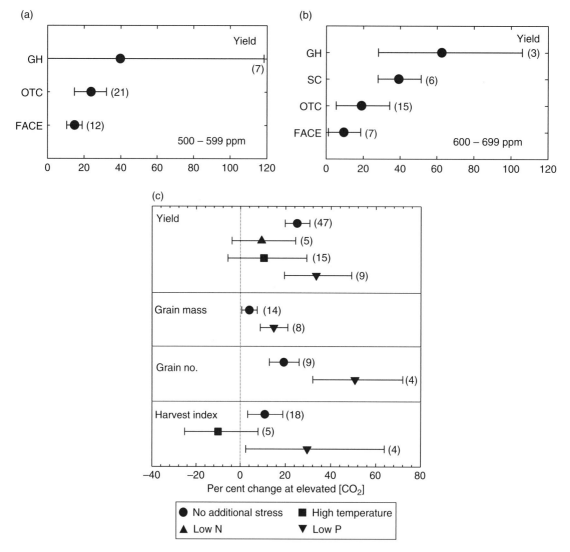

**Fig. 3.16.** Results of a meta-analysis of the impacts of elevated $CO_2$ on rice yield. In (a) and (b) the results are separated by the degree of $CO_2$ elevation and by the method of raising the concentration of $CO_2$. GH, greenhouse; SC, closed sunlit chamber; OTC, open-topped chamber; FACE, free air $CO_2$ enrichment. In (c) the results are separated by nutrient and temperature stress, in addition to the $CO_2$ level.

methods we have discussed in this chapter. The choice of a technique will depend on many factors, not least are time, money and the specific research question of interest. There is a role for each of these techniques in climate change impacts research, but it is important to always bear in mind the limitations of the various techniques. Second, the multiple techniques available for addressing the same question do not always yield consistent results. We caught a glimpse of this in the meta-analysis discussed in the last section. So when we read the research literature we need to appreciate the various trade-offs and compromises necessary to conduct any of this research. We also have to be cognizant of the possible 'technique dependence' in the results of any particular study.

# PART II
# Impacts from Physiology to Evolution

Experimental mesocosm facility of Liverpool University, situated at Ness Botanic Gardens, UK. Forty-eight 3 m³ tanks mimicking ponds and shallow lakes were used to conduct a climate change experiment funded by the EU project EUROLIMPACS. Researchers led by Brian Moss, of Liverpool University in the UK, have been investigating the differential effects of warming and nutrient loading on various components of the food web. Experiments like this one help us to understand the likely impacts of climate change on population dynamics and community structure, as discussed in Chapters 5 and 6. (Photo by H. Feuchtmayr, Centre for Ecology & Hydrology, UK.)

## Overview

Chapters 4 to 8 categorize the biological impacts of climate change by the level of biological organization being considered. We begin with physiological responses to changes in $CO_2$ concentrations, temperature and precipitation. We note that these topics have been studied by plant physiologists, agronomists and ecologists for much longer than we have been concerned with climate change per se. Moving up the hierarchy of biological organization we consider, in turn, impacts on individual populations, ecological communities and ecosystems. We examine research that seeks to address questions about changes in the abundance and distributions of local populations, changes in the composition of biological communities, and changes in the rates of transfer of energy and material (nutrients) within and between ecosystems. We complete this section with Chapter 8, in which we consider perhaps the biggest unknown biological response to climatic change: evolution. All, or nearly all, of the work described in Chapters 4 to 7 is predicated on the assumption that individual organisms will respond to climate in much the same way that they do now. Yet we know that some species are capable of responding to climatic change, via evolutionary adaptation, as fast as the climate itself is changing. This is one area of impact studies in which the theory is further advanced than our empirical evidence.

All of the work discussed in this section relies on the methods we discussed earlier, in Chapter 3, and many make use of specific climate projections from climate models, and for climate change scenarios, discussed in Chapter 2. Finally, some of our research about the responses of species, communities and ecosystems comes from our study of the palaeological record, and hence makes use of the methods and approaches discussed in Chapter 1.

# 4 Physiological Responses

It is fitting that we now begin our considerations of the biological impacts of climate change at the level of physiology. Climatic change is largely, if not entirely, a 'bottom-up' process (but see Barton *et al.*, 2009). The impacts of climatic change are first experienced at the level of the physiology of the individual organism and these impacts combine, amplify, dampen and generally interact as they ripple up food chains, across communities and throughout ecosystems, as we will see in the coming chapters. At the level of physiological responses, it is in some sense a matter of judgement and taste as to what research is 'climate change research' and what is not. Scientists have been studying physiological responses to temperature for many decades, indeed long before we started worrying about climate change. And climate change has become a convenient hook on which to hang just about any work that looks at physiological responses to temperature. We could easily fill whole books with detailed discussions of the impacts of rising temperatures because temperature affects so many physiological processes. Similar comments could be made about responses to drought, and even to some extent elevated $CO_2$ concentrations. Much of this work was not motivated by climate change, and sometimes it is relevant only in a very general sense to the study of the biological impacts of climate change.

Our goal in this chapter is not to survey all of the possible physiological responses, but to introduce the main responses that resonate throughout the other levels of ecological organization that we consider in the subsequent chapters. We start by briefly reviewing photosynthesis and the differences between the mechanisms involved in the uptake of $CO_2$. We then consider how $CO_2$, temperature and drought affect the rate of photosynthesis. This leads us naturally to a discussion of water- and nutrient-use efficiencies. From there we consider how the plant's altered carbon and nitrogen economy impacts growth and reproduction. We then look at the consequences of such physiological changes for changes in the timing of critical life-history events (phenological changes) and finally we look at plant–animal interactions. Plant–animal interactions might more commonly be viewed as population responses (and we discuss them from this perspective in Chapter 5), but in this chapter we are more interested in the aspect of changing plant quality from the herbivore's perspective than we are with changes in population dynamics. Plant quality comprises both nutritional quality from the herbivore's perspective and the plant's investment in chemical defences.

## 4.1 Photosynthesis Review

Virtually all carbon fixed by photosynthesis is ultimately the result of one enzyme, ribulose 1,5-bisphosphate carboxylase oxygenase (Rubisco). Below we review the three main pathways for photosynthesis, but they all share this single enzyme. In general, the three pathways differ in how the *primary* uptake of $CO_2$ (i.e. taking $CO_2$ from the atmosphere) is achieved and these differences are fundamental to understanding some plant responses to climate change.

### $C_3$ photosynthesis

Approximately 95% of all plants are known as $C_3$ plants (Vu, 2005). $C_3$ plants assimilate $CO_2$ directly via the Calvin cycle (Fig. 4.1). Most of Fig. 4.1 is

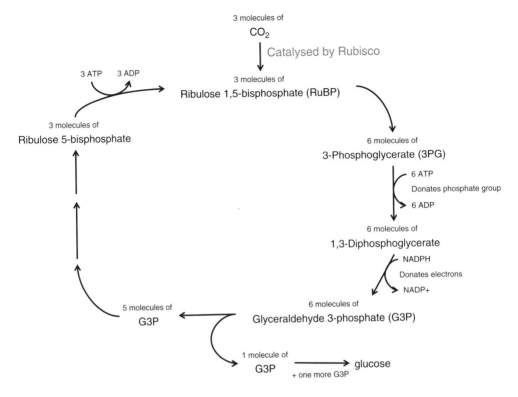

**Fig. 4.1.** The Calvin cycle.

not important for this chapter; what is important is that the uptake of $CO_2$ is catalysed directly via Rubisco. This is important because Rubisco actually catalyses two possible reactions, either the carboxylation of the acceptor molecule (ribulose 1,5-bisphosphate, RuBP) or the oxygenation of RuBP (called 'photorespiration'). Since only the carboxylated molecules can be turned into sugar, this lack of specificity carries both an opportunity cost and an energy cost (in terms of regenerating the RuBP). To see this cost in action, consider Fig. 4.2, which shows the apparent photosynthesis as a function of leaf temperature for white clover. The two lines show apparent photosynthesis for two concentrations of oxygen, 2% and 20% (i.e. approximately current ambient concentrations). So the rate of photosynthesis in $C_3$ plants is very sensitive to the ratio of $CO_2$ to $O_2$ in the atmosphere. Increases in this ratio leads to higher rates of photosynthesis, at least in the short term. Figure 4.3 demonstrates this effect for *Theobroma cacao* seedlings.

### $C_4$ photosynthesis

Approximately 3% of all terrestrial plants use $C_4$ photosynthesis (Sage, 2004). $C_4$ plants come in three different types (NADP-malic enzyme type, NAD-malic enzyme type, or PEP carboxykinase type), but the differences between these three types are not important for our purposes. What is important is that the primary uptake of $CO_2$ is not catalysed by Rubisco but by another enzyme called phosphoenol pyruvate (PEP) carboxylase. The outer cells of the leaf (mesophyll cells) use PEP carboxylase to catalyse the uptake of $CO_2$, converting it to a 4-carbon acid called oxaloacetate (OAA). OAA is then converted into another acid (either malate or aspartate depending on the $C_4$ type) and this acid is transported deeper into the leaf, into cells called 'bundle-sheath cells'. In the bundle-sheath cells, the $CO_2$ is stripped from the 4-carbon acid and it enters the normal $C_3$ pathway, where it is catalysed first by Rubisco (see e.g. Vu, 2005).

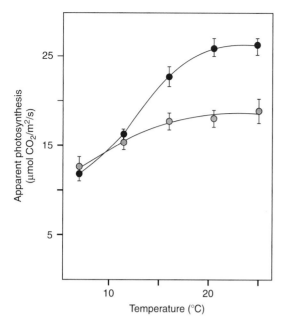

**Fig. 4.2.** Apparent photosynthesis of white clover (*Trifolium repens*) leaves for two oxygen concentrations: 2% (dark symbols) and 20% (light symbols). Photosynthesis is depressed under 20% $O_2$ because of the increased photorespiration. (Redrawn from Schnyder *et al.*, 1984.)

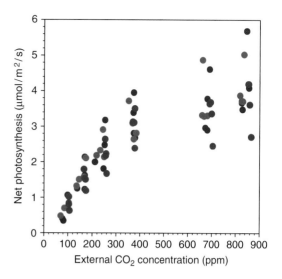

**Fig. 4.3.** An example of how $C_3$ photosynthesis responds to increasing concentrations of $CO_2$ for cacao plants (*Theobroma cacao*) seedlings. The different coloured symbols represent different cultivars and provide some indication of the magnitude of intraspecific variation in photosynthetic responses. (Redrawn from Baligar *et al.*, 2008.)

So all $C_4$ plants also use the Calvin cycle just like $C_3$ plants, they just have this extra step (several steps actually) involving PEP carboxylase. The significance of this 'extra step' is that it concentrates $CO_2$ in the bundle-sheath cells, in the location of Rubisco, minimizing photorespiration. Also, because PEP does not bind to oxygen, the primary uptake of $CO_2$ saturates at a much lower $CO_2$ concentration. These effects can be seen in Fig. 4.4, which depicts a stylized response rate of photosynthesis for $C_3$ and $C_4$ plants as a function of atmospheric $CO_2$. The other consequence of reduced $CO_2$ limitation in $C_4$ plants is that they have more flexibility to close their stomata as necessary to avoid excessive water loss through transpiration (see Section 4.3 below).

$C_4$ photosynthesis seems to have evolved independently at least 48 times in 19 families of angiosperms (Sage, 2004); more recent estimates put this number at 60 (R.F. Sage, personal communication, 2010). Some of these evolutionary events date back as far as 35 million years, but most are fairly recent, occurring within the last 5 million years, and probably arose as a response to arid conditions and low atmospheric $CO_2$ concentrations (Sage, 2004). We also note that there are $C_3$–$C_4$ intermediate plants, but these are few and beyond the scope of this text.

**Crassulacean acid metabolism**

Crassulacean acid metabolism (CAM) is conceptually similar to $C_4$ photosynthesis in that CAM acts as a $CO_2$ concentrating mechanism, and also uses PEP carboxylase to catalyse the primary uptake of $CO_2$. One significant difference between CAM and $C_4$ photosynthesis is that CAM catalyses the uptake of $CO_2$ predominantly at night and stores the $CO_2$ as malate until the next day (Nobel, 1991). CAM is a strategy, common among cacti and succulents, to reduce water loss in extremely arid environments. Although there has been some work done on the responses of CAM plants to climatic change, in the interest of brevity we largely ignore this work (interested readers should see, for example, Nobel, 2000; Poorter and Navas, 2003).

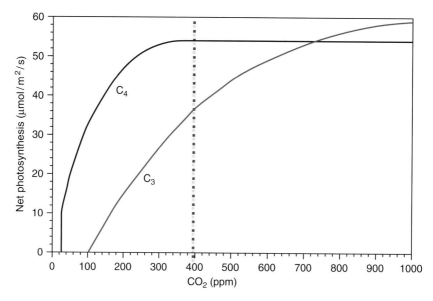

**Fig. 4.4.** Stylized response of $C_3$ and $C_4$ photosynthesis to atmospheric $CO_2$ concentrations. (Redrawn from Kimball *et al.*, 1993.)

## 4.2 Photosynthetic Responses to Climatic Change

So a key point to take from Section 4.1 is that $C_4$ plants are likely to be less sensitive to $CO_2$ concentrations, at least in terms of photosynthetic rates, than $C_3$ plants. Also, $C_3$ photosynthesis should be stimulated by rising $CO_2$ concentrations. As we see below, these predictions are generally supported, but photosynthesis is also affected by temperature and water availability, and these factors often interact to determine photosynthetic rates and hence growth responses.

### CO₂ concentrations

As we mentioned above, plants, particularly those using $C_3$ photosynthesis, respond quite strongly to increases in atmospheric $CO_2$. As shown in Fig. 4.4, both $C_3$ and $C_4$ plants show strong increases in photosynthetic rates as $CO_2$ concentration increases from the minimum concentration, but these rates of increase slow down as $CO_2$ concentration continues to increase. At low $CO_2$ concentrations, both plant types are limited by their abilities to assimilate carbon, and this limitation is eased by increasing the $CO_2$ concentration. However, as this limitation decreases, a new limitation begins to

appear: the rate at which the plants can regenerate RuBP (the carbon receptor molecule in the Calvin cycle – see Fig. 4.1; Long and Bernacchi, 2003). $C_4$ plants are $CO_2$-saturated at approximately 400 ppm $CO_2$, but $C_3$ plants are not $CO_2$-saturated until $CO_2$ concentrations are greater than 1000 ppm. So all else being equal, we might expect that rising atmospheric $CO_2$ will lead to increased rates of photosynthesis in $C_3$ plants, and hence increased biomass production, energy for reproduction and so on. And certainly these effects are very commonly observed in climate change experiments, at least in the short term. This is often referred to as the '$CO_2$-fertilization effect'. On the other hand, we might expect that $C_4$ plants would benefit little from the increased $CO_2$. This is generally true, but because increased $CO_2$ can also lead to higher water-use efficiency in plants (see below), $C_4$ plants may also obtain growth benefits from rising $CO_2$ concentrations, albeit different from the fertilization effect of $CO_2$.

### Temperature

Photosynthesis also responds quite strongly to changes in temperature, in both $C_3$ and $C_4$ plants. Plants will exhibit an optimal temperature at which the net rate of photosynthesis is maximized. This optimal temperature will vary from species to species but, in general, $C_4$ plants have a higher optimal temperature than $C_3$ plants. An example of this trend is depicted in Fig. 4.5, which shows the photosynthetic rates for an extreme contrast: a $C_3$ plant that grows on the cool coastal plains of eastern North America, *Atriplex glabriuscula* (Scotland orache), and a $C_4$ plant from the hot desert of Death Valley in California, *Tidestromia oblongifolia* (honeysweet), as a function of leaf temperature.

Below the optimal temperature photosynthesis is limited by enzymatic reaction rates, so as temperature increases, these reaction rates also increase (Lambers *et al.*, 2008). However, as temperature increases, the solubility of $CO_2$ decreases (relative to $O_2$) as does Rubisco's specificity for $CO_2$, both of which favour oxygenation over carboxylation. Figure 4.6 shows the specificity factor of Rubsico for $CO_2$ over $O_2$ as a function of temperature. We can see that Rubisco's specificity for $CO_2$ declines by a factor of about 3.5 over the range of temperature from 5 to 40°C (Jordan and Ogren, 1984). So for any given concentration of atmospheric $CO_2$, an increase in temperature will result in an increase in the proportion of potential photosynthesis that is lost to photorespiration (e.g. Long, 1991). Eventually, when temperatures become hot enough, photorespiration becomes so large that photosynthesis suffers and begins to decline with temperature.

Not surprisingly, temperature and atmospheric $CO_2$ concentration interact in determining photosynthetic rates. Increased $CO_2$ concentrations shift the optimal temperature for photosynthesis to higher temperatures, as shown in Fig. 4.7. Also apparent in Fig. 4.7 is the idea that the impact of $CO_2$ on photosynthesis will be larger at higher temperatures. Such 'multiple stressors' could interact in ways that make responses at all levels of organization to climatic change difficult to predict (see Chapter 13).

### Acclimatization

There can be significant differences between short-term responses of plants to elevated $CO_2$ and long-term responses. In the short term, nearly all experiments on $C_3$ plants show increased photosynthetic rates, but in the longer term the rate tends to decline, although not necessary as far as the baseline rate. Plants acclimatize to elevated $CO_2$ in many or most environments because of the relationship between the rate of production of carbohydrates (so-called 'source strength') and the rate at which those

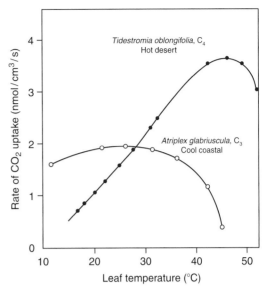

**Fig. 4.5.** Optimal temperatures for photosynthesis differ between $C_3$ and $C_4$ plants. (Redrawn from Berry and Bjorkman, 1980.)

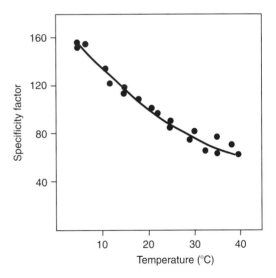

**Fig. 4.6.** The specificity factor indicates the degree to which Rubisco distinguishes between $O_2$ and $CO_2$. (Redrawn from Jordan and Ogren, 1984.)

carbohydrates can be used (so-called 'sink strength'). Under normal environmental conditions, $C_3$ photosynthesis is $CO_2$-limited. Once that limitation is reduced or removed by increasing $CO_2$ concentrations, then photosynthesis becomes limited by something else, eventually by RuBP regeneration.

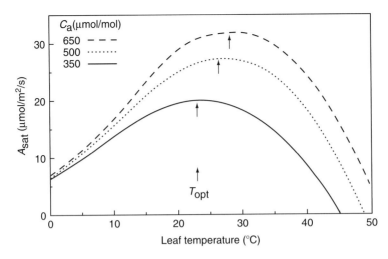

**Fig. 4.7.** $CO_2$ concentration and temperature interact as they affect photosynthesis. Notice the optimal temperature increases as $CO_2$ increases. Notice too that the difference between photosynthesis at elevated and ambient $CO_2$ is greater at higher temperatures. (Redrawn from Long, 1991.)

Naively we might think that if the $CO_2$ limitation on $C_3$ photosynthesis is relaxed, plants will therefore photosynthesize faster and hence accumulate more growth, but this is not the only possible response. Plants might, for instance, regulate their stomata differently such that they fix the same amount of $CO_2$ but are able to do so with less potential water loss (thus increasing their water-use efficiency). Another possibility is for the plant to maintain the baseline rate of photosynthesis not by regulating its stomata, but by reducing investment of nitrogen in Rubsico. This will slow the rate of photosynthesis, but save nitrogen that is then available to relieve shortages elsewhere in the plant (Arp, 1991).

There are many examples and many studies that investigate this 'down-regulation' of photosynthesis in response to elevated $CO_2$. Here we look at just one. Delucia and colleagues (1985) grew cotton plants at 350, 675 and 1000 ppm $CO_2$ for 4 weeks and then estimated the rate of photosynthesis for these plants, comparing them all at each of the same three $CO_2$ concentrations. Their results are shown in Fig. 4.8. We can see that even over 4 weeks, when measured at any given level of ambient $CO_2$, the higher the concentration of $CO_2$ in the growth condition, the lower the rate of photosynthesis and stomatal conductance in the test concentration of $CO_2$. This down-regulation occurred at the same time that the plants accumulated more biomass in higher growth $CO_2$ concentrations (675 ppm = 72% more biomass than

350 ppm, and 1000 ppm = 115% more; Delucia *et al.*, 1985). Delucia *et al.* showed that this down-regulation of photosynthesis correlated with starch accumulation and a reduction in chlorophyll concentration. They suggest that the observed reduction in photosynthetic capacity was due to either feedback inhibition caused by the starch accumulation and/or that the starch accumulation damaged the chlorophyll.

While acclimatization is a potentially complicating factor in understanding the long-term impacts of climate change (Leakey *et al.*, 2009), it is also potentially an artefact of the experimental protocol. Many physiological studies of the impacts of climate change are done on plants grown in pots but, except for the largest pots, plants eventually become root-bound and root growth ceases to become a possible sink for carbohydrates. Carbohydrates then accumulate and lead again to feedback inhibition and down-regulation of photosynthesis (Arp, 1991).

Good long-term data now exist demonstrating that acclimatization does occur, even in field-grown plants, and the degree to which this occurs depends on other environmental factors and differs among plant genotypes. This reduction in photosynthetic capacity at high concentrations of $CO_2$ nevertheless results in rates of photosynthesis at elevated $CO_2$ that are higher than those at ambient $CO_2$ (Leakey *et al.*, 2009). Derived from a meta-analysis of FACE studies (see Chapter 3), Table 4.1 shows that the overall impact on light-saturated photosynthesis is

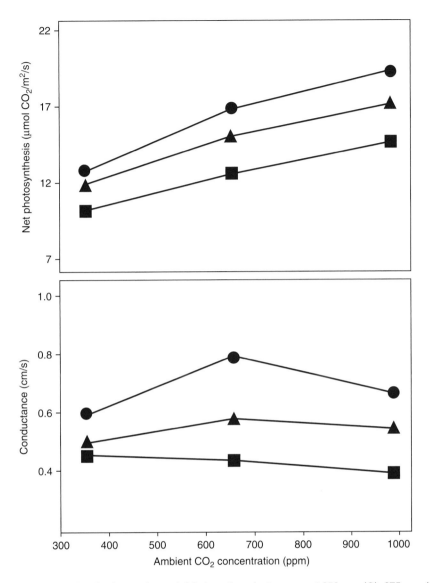

**Fig. 4.8.** Photosynthetic acclimatization to elevated $CO_2$ in cotton plants grown at 350 ppm (●), 675 ppm (▲) and 1000 ppm (■) $CO_2$. Plants grown at higher concentrations of $CO_2$ have lower photosynthetic rates than plants grown at lower concentrations of $CO_2$, when tested in the same concentrations. (Redrawn from Delucia *et al.*, 1985.)

positive (expressed as percentage increase over the baseline $A_{sat}$ at ambient $CO_2$ conditions).

### Drought

Photosynthesis is also limited by drought conditions, but it is controversial whether that limitation is primarily due to stomatal regulation or to metabolic changes, primarily those altering ATP

synthesis (see e.g. Cornic, 2000; Flexas and Medrano, 2002). Since $C_4$ plants have their carbon concentrating mechanism, they may maintain high rates of photosynthesis while reducing their stomatal openings and conserving water through reduced transpiration. We would therefore expect that $C_4$ plants would have a competitive advantage over $C_3$ plants in hot dry climates (Ripley *et al.*, 2007). Paradoxically though, at a regional scale, we

**Table 4.1.** Percentage change in light-saturated photosynthesis under elevated $CO_2$. The 'lower' and 'upper' entries refer to the range of the 95% confidence interval around the estimate of the mean entry. (From Ainsworth and Long, 2005.)

|  | Lower | Mean | Upper |
|---|---|---|---|
| $A_{sat}$ | 35 | 39 | 43 |
| $C_3$ $A_{sat}$ | 30 | 34 | 37 |
| $C_3$ trees | −20 | 47 | 81 |
| $C_3$ shrubs | 7 | 21 | 36 |
| $C_3$ grasses | 30 | 36 | 43 |
| $C_3$ forbs | 3 | 14 | 25 |
| $C_3$ legumes | 12 | 20 | 28 |
| $C_3$ crops | 22 | 36 | 52 |
| $C_4$ $A_{sat}$ | 3 | 11 | 20 |
| >25°C | 26 | 30 | 33 |
| <25°C | 11 | 19 | 28 |
| No stress | 31 | 36 | 40 |
| Low N | 20 | 27 | 35 |

see that the relative abundance of $C_3$ plants compared with $C_4$ plants increases as annual rainfall decreases, at least in grasses (Ellis *et al.*, 1980; Ripley *et al.*, 2007). A possible explanation for this paradox is that the photosynthetic advantage of $C_4$ over $C_3$ plants is diminished in drought conditions. Ripley *et al.* (2007) tested this hypothesis with two subspecies of *Alloteropsis semialata*, one that has the $C_4$ pathway (subspecies *semialata*) and one that does not (subspecies *eckloniana*). They found that, when grown in a common garden with adequate water, the $C_4$ plants had significantly higher rates of photosynthesis than the $C_3$ plants, but this advantage was completely lost during periods of drought and this happened because of greater metabolic impairment, rather than stomatal regulation, in the $C_4$ plants compared with the $C_3$ plants.

The negative impact of drought on photosynthesis can be somewhat ameliorated under higher atmospheric $CO_2$ conditions. As already mentioned, higher $CO_2$ concentrations allow plants the flexibility to regulate their stomata so as to reduce water loss. Also, it is common for plants to respond to higher $CO_2$ concentrations by increasing their ratio of roots to shoots. This change in growth investment probably arises from an increased need for nitrogen uptake to complement the increased carbon uptake, but such an investment also increases the

plant's water-uptake capacity. Figure 4.9 shows an example of observed changes in root:shoot ratios. We now turn to changes in nutrient- and water-use efficiencies.

## 4.3 Nutrient- and Water-use Efficiencies

Increased rates of photosynthesis as a response to increased atmospheric $CO_2$ concentrations are sustainable only if they are matched by increased nitrogen availability or efficiency, otherwise sink limitations lead to acclimatization of photosynthetic capacity and a down-regulation of photosynthetic rates. As mentioned above, it had been thought that plants would not need to invest so heavily in Rubisco to maintain high rates of photosynthesis, and since Rubisco can be a substantial fraction of leaf nitrogen content, this nitrogen could be redistributed and used more optimally (Jacob *et al.*, 1995). Photosynthetic nitrogen-use efficiency (PNUE) can be defined as the net amount of $CO_2$ assimilated per unit of leaf nitrogen. Leakey *et al.* (2009) argue that the evidence from FACE studies suggests that PNUE does indeed increase, by as much as 30% (Peterson *et al.*, 1999), but that this increase is the result of increased carbon assimilation rather than redistribution of leaf nitrogen (Ainsworth and Long, 2005; Ainsworth and Rogers, 2007). Note that most of the experimental work that led to these conclusions was conducted at relatively modest $CO_2$ concentrations, and that as these concentrations increase more substantially by the end of this century, the PNUE could be significantly greater (Leakey *et al.*, 2009). We discuss the implications of low nitrogen (or other nutrient) availability on biological responses to climatic changes further in Chapters 10 and 13.

As we mentioned above, plants control their stomatal openings to regulate the uptake of $CO_2$ and loss of water through transpiration. There is now substantial evidence at the level of the individual leaf that stomatal conductance (the rate at which water vapour exits the leaf) decreases with elevated $CO_2$. It is not always the case that physiological measurements at the leaf level translate into similar results at the level of the whole plant canopy due to effects of microclimatic conditions within and above the canopy. However, in the case of elevated $CO_2$ it does seem that these results on individual leaves do translate into decreased water use at the canopy scale, and this leads to higher

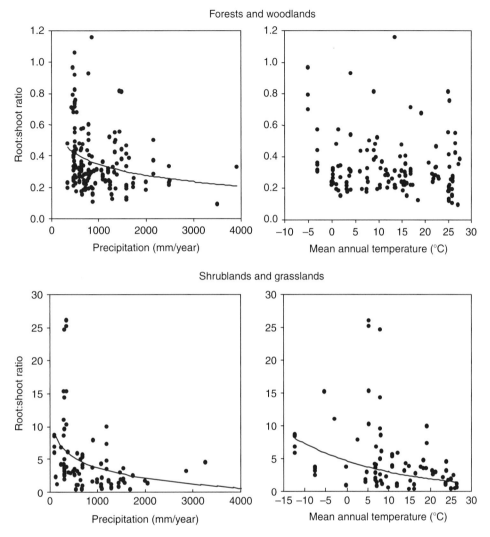

**Fig. 4.9.** Effects of precipitation and temperature on root:shoot ratios in forests and woodlands (top), and grasslands and shrublands (bottom). (Adapted from Mokany *et al.*, 2006.)

levels of soil moisture (Leakey *et al.*, 2009). Water-use efficiency (WUE) can be defined as the net amount of $CO_2$ assimilation per mass of water (sometimes expressed as yield per mass of water). It is debatable whether measurements of WUE made in growth chambers or open-topped chambers are relevant to field measurements, but in a review of the pre-1998 literature, Wand *et al.* (1999) found that $C_4$ plants improved their WUE by 72%. In a more recent meta-analysis of FACE studies, Ainsworth and Long (2005) found that $C_3$ plants increased their WUE by 68%.

## 4.4 Growth Responses

So with numerous caveats, mostly regarding interactions with other environmental variables, elevated $CO_2$ increases photosynthesis and nutrient- and water-use efficiencies. All else being equal, these changes translate into more plant biomass, which can be invested in more vegetative growth, more reproductive growth, more plant defences, differential investment in root versus shoot and so on. What plants do with the extra carbon assimilated has consequences for not just that plant's own fitness, but for its interactions with other species, its impact on herbivores, the plant community composition, decomposition dynamics and hence nutrient cycling, and so on. In this section we look at some examples of these responses, but note that this next step from carbon capture to carbon use often defies simple generalization.

Perennial species in particular face a trade-off between increasing investment in vegetative growth or reproduction. The observed responses are often confounded with the particular part of the growing season in which the experiment takes place, given that many experiments do not follow the treatments through entire growing seasons or lifetimes of the individual plants. In a meta-analysis of the pre-1999 literature, Wand et al. (1999) showed that $C_3$ and $C_4$ wild grasses increased their tiller numbers by about 25% and 10%, respectively. As aboveground biomass increases, so does belowground biomass, and in fact roots tend to increase relatively more than shoots. Figure 4.10 shows an example for the Canada thistle (*Cirsium arvense*; Ziska et al., 2009) and this is generally true for other plants as well: shoot:root ratios generally decrease – or, conversely, root:shoot ratios generally increase – as atmospheric $CO_2$ increases (Bazzaz, 1990; but see Luo et al., 2006).

Under elevated $CO_2$, plants also have more resources to invest in reproduction. In general plants produce more flowers, fruits and seeds, and both individual and total seed mass increase. Crops tend to show the same trends as wild plant species, except that they proportionately allocate even more to fruit and seed mass (Jablonski et al., 2002; see also Fig. 4.12).

In temperate and boreal forests the synergetic effects of recent changes in climate and rising atmospheric $CO_2$ are expected to produce positive responses in tree growth (Bonan, 2008; Salzer et al., 2009). However, accurately predicting the outcomes of atmosphere–forest interactions remains a challenge (Huang et al., 2007). Increased tree mortality in western US temperate forests (Van Mantgem et al., 2009), tree growth decline in European temperate beech forests (Peñuelas et al., 2008) and the loss of temperature sensitivity at the northern boreal treeline (D'Arrigo et al., 2009) are just few examples of unexpected outcomes.

Foreseeing a continuous $CO_2$ fertilization effect, Lamarche et al. (1984) hypothesized that increasing atmospheric $CO_2$ concentrations would positively change the ratio between carboxylation rates and water loss in transpiration, i.e. WUE, promoting tree growth enhancement. Previous studies

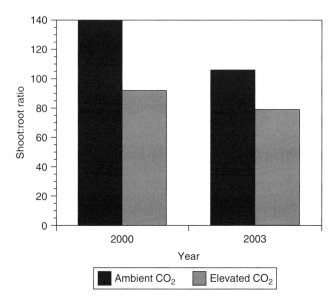

**Fig. 4.10.** Changes in biomass and shoot:root ratio of field-grown Canada thistle (*Cirsium arvense*) with elevated $CO_2$. (Data from Ziska et al., 2009.)

based on tree growth alone aimed to identify this effect under natural conditions, but results have not been conclusive as $CO_2$ fertilization effects can be confounded with those of recent changes in climate, especially in temperature-limited ecosystems (Salzer *et al.*, 2009). Furthermore, warming-induced stresses can simultaneously produce increased WUE and growth decline (Silva *et al.*, 2010). Thus, in order to effectively identify the mechanisms driving changes in forest–atmosphere interactions, information on both WUE and tree growth is required.

## 4.5 Phenological Responses

'Phenology' refers to the timing of significant events in the life history of an organism. Examples from plants include the timing of bud burst in the spring and the switch from vegetative growth to reproduction, flowering and seed set. In animals, examples include the timing of egg laying, moulting, diapause and emergence from overwintering. The timing of such events is often cued by temperature (Cleland *et al.*, 2007). In many ecological applications, for example pest management, managers are really interested in 'physiological time', not actual days, months, etc. Physiological time is often expressed as something called 'degree days'. Degree days accumulate once the temperature rises above the organism's minimum developmental threshold temperature and as long as it stays below their maximum threshold temperature, if there is one (interested readers should see http://tinyurl.com/ch4-degreedays for more details). Many organisms need to accumulate a roughly fixed number of degree days before moving from one life-history stage to the next (see Box 4.1 for a worked example). So obviously as global warming progresses, degree days start accumulating earlier in the season, continue later in the season and accumulate more quickly during the season, meaning that the timing of significant life-history events is likely to change.

Shifts in phenology are some of the most convincing examples of effects of contemporary climatic change (Parmesan and Yohe, 2003; Parmesan, 2007). Schwartz *et al.* (2006) show that, over the last half of the 20th century, the dates of first leaf, first bloom and last spring freeze have occurred approximately 1.2 days earlier, per decade, across nearly all of temperate North America. Many studies of breeding birds have reported an advance in the dates of first egg laying. For example, over the previous half century, the great tits (*Parus major*) breeding in Wytham Wood, near Oxford in the UK, have advanced their date of egg laying by 14 days, a change of more than two standard deviations in their mean date. The date of egg laying is very highly correlated ($r = -0.81$) with the air temperature in the period just prior to egg laying (Charmantier *et al.*, 2008). Other convincing examples exist for butterflies in the USA, Europe and Britain (see e.g. Parmesan, 2007 for references).

One source of phenological information is from observational networks, often staffed by volunteers. Two well-known examples are the British Trust for Ornithology's Breeding Bird Survey (http://tinyurl.com/ch4-BBBS; 1939 to present) and the Kyoto cherry blossoms survey (Aono and Kazui, 2008; 801 CE to present). Recent meta-analyses of phenological data from observational networks have concluded that, globally, spring has been occurring 2.3–5.1 days earlier per decade (Cleland *et al.*, 2007). Generally, data of this sort suggest that the main impact of such contemporary warming has been an earlier spring rather than consistent warming throughout (Cleland *et al.*, 2007).

While much more data exist, spanning far longer periods of time, examining the impacts of temperature on phenology, there is also some evidence that rising $CO_2$ concentrations may affect phenology. These data come from both climate chamber and FACE studies (see Chapter 3), and while far less convincing than the temperature studies, are nevertheless intriguing. In a meta-analysis of FACE studies for crops, the date of first flowering in wheat, rice, sorghum and potatoes tended to occur earlier by a few days. Likewise sorghum and potatoes tended to mature a few days earlier. In general though, crops show little response to elevated $CO_2$ in terms of their phenology (Kimball *et al.*, 2002). In a more recent analysis of 60 studies comprising 111 results on flowering time under elevated $CO_2$, every possible result was found: no change (48 results); accelerated flowering times (41 results); and delayed flowering times (22 results; Springer and Ward, 2007). Recall from Chapter 3 that these 'no change' results might be because there genuinely is no effect of the $CO_2$ concentration, or they might simply represent examples of type II errors; in practice they are probably a mixture of both. Little work has been done on animal phenology

**Box 4.1. Degree-day models of phenology.**

Here we develop an example for the elm leaf beetle (*Pyrrhalta luteola*) in California to illustrate how phenology might change as temperature increases. This beetle accumulates degree days from 1 March onwards. Degree days accrue when the air temperature is between the organism's maximum and minimum temperature thresholds for development, which are worked out experimentally. Table 4.2 shows the necessary number of degree days for the beetle to move from one developmental stage to another, so we can see from Table 4.2 that the beetle needs a total of 1183 degree days to complete two generations.

Now, let us take the 30-year average weather records for, say, Madera County, California (36°57′41″N,

120°3′35″ W); the daily minimum and maximum temperatures are shown in dark blue in Fig. 4.11. So under current climate, the elm leaf beetle completes two generations on or about day 159 (8 June). Suppose that the future climate was projected to be as shown in light blue in Fig. 4.11. We would then expect that it would complete its second generation on or about day 126 (6 May), an advance of approximately 1 month (see Table 4.2 for more details). This advance could have implications for the severity of damage caused to the host plants, the timing of beetle activity relative to the phenology of the tree (white elm, *Ulmus americana*), the beetle's ability to expand its range northward, and so on.

**Table 4.2.** Accumulated degree days necessary to achieve significant life-history events in the life cycle of the elm leaf beetle (*Pyrrhalta luteola*) and the approximate calendar dates on which these degree days fall, based on current climate and future climate for Madera County, California. (From Dahlsten *et al.*, 1993.)

| Developmental stage | Accumulated degree days (°C) | Current | Future |
|---|---|---|---|
| *First generation* | | | |
| Eggs | 283 | 14 April | 24 March |
| First instar larvae | 353 | 21 April | 20 March |
| Second instar larvae | 441 | 29 April | 5 April |
| Third instar larvae | 476 | 2 May | 7 April |
| *Second generation* | | | |
| Eggs | 953 | 1 June | 1 May |
| First instar larvae | 1090 | 7 June | 7 May |
| Second instar larvae | 1142 | 10 June | 9 May |
| Third instar larvae | 1183 | 12 June | 11 May |

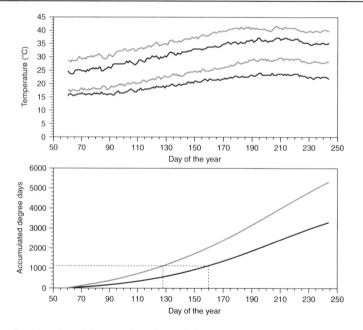

**Fig. 4.11.** The top plot shows the minimum and maximum daily temperatures averaged over the past 30 years for Madera County, California (dark blue) and hypothetical climate change scenario (light blue). The bottom plot shows how degree days accumulate under each condition. Degree day calculations based on http://tinyurl.com/ch4-elmleaf.

and elevated $CO_2$; this is not surprising since few studies follow individual animals through multiple generations, so these studies have usually been restricted to insects.

## 4.6 Changes in Plant Quality and Defences

So in general elevated $CO_2$ increases plant growth, particularly in $C_3$ plants, but plants are not just bigger; they are potentially differently nutritious and defended, from the perspective of their herbivores. In this section we examine how changes that start with photosynthetic rates begin to resonate up the food chain and affect populations, communities and ecosystems.

A suite of biochemical changes occur in plants as a result of the changes in photosynthetic rate. The changes are variable from species to species and environment to environment, but there are some fairly general responses. At a very basic level we can understand plants as taking the raw inputs of carbon and nitrogen and turning those inputs into growth, reproduction and survival (which may include defences). Elevated $CO_2$ and hence increased photosynthesis perturbs this economy. At the simplest level, plants might just shunt much of this extra carbon into storage compounds like sugars and starch, and increases in such so-called 'non-structural carbohydrates' are commonly observed. For example, averaged over 27 species of $C_3$ plants, total non-structural carbohydrates increased by 74% under elevated $CO_2$, by far the largest change for any class of metabolite (Poorter *et al.*, 1997). This increase has the effect of diluting the response of other metabolites. In some sense this is a very real and important effect, and in other cases it is a nuisance for detecting other more subtle responses.

One of the first variables that this dilution effect shows up in is the ratio of carbon to nitrogen (C:N). C:N is almost always observed to increase under elevated $CO_2$. In a meta-analysis of 104 published studies, Luo *et al.* (2006) found that both carbon and nitrogen increased in both shoots and roots, but not surprisingly carbon increased faster than nitrogen, leading to C:N ratios that were 11% greater under elevated $CO_2$. This change tends to be true whether one looks at the whole plant, or just parts like leaves or seeds. For example, the nitrogen concentration of seeds grown under elevated $CO_2$ decreased by as much as 25%, although there were some species where there was no observable decline

in nitrogen concentration (Jablonski *et al.*, 2002). From a plant quality perspective, i.e. from the herbivore's point of view, more important than nitrogen concentration is protein concentration (which is correlated with nitrogen concentration, but lower in magnitude). In 228 experimental observations on barley, rice, wheat, soybean and potato, protein concentrations declined significantly under elevated $CO_2$. For the grains (barley, rice and wheat) and potatoes the decline was about 14%, but was closer to 2% for soybeans (Taub *et al.*, 2008). Similar results have been found in non-crop species.

Reduced nitrogen and/or protein concentrations tend to affect herbivores negatively. In a review of 75 plant–insect herbivore studies under elevated $CO_2$, Stiling and Cornelissen (2007) found, on average, increased relative consumption rates (+17%) and total consumption (+19%), increased development time (+14%), decreased relative growth rate (–8%), decreased energy conversion efficiency (–20%) and decreased pupal weight (–5%). So insects respond to the decline in plant quality by increasing consumption to try and compensate for the decline; but Stiling and Cornelissen (2007) found that on average they are unable to do so, despite the increased consumption, resulting in a decline in relative abundance of 21% (but see Chapter 5 for more on plant–herbivore dynamics).

Increased photosynthetic rates often result in increased C:N ratios, implying that carbon products are in excess of those needed for primary metabolic functions. Several theories of plant defences suggest that this excess of carbon ought to result in increased carbon-based secondary metabolites and decreased nitrogen-based secondary metabolites (Karowe *et al.*, 1997). Ryan *et al.* (2010) reviewed the experimental support for these hypotheses from 102 relevant studies under elevated $CO_2$, with 608 measurements of plant secondary metabolites from 102 species. In general they did not find support for these hypotheses, except perhaps for the phenolics class of compounds. Under elevated $CO_2$, nitrogenous defence compounds increased about as often as they decreased (18% versus 16% of cases). For the carbon-based terpenoids, concentrations decreased about twice as often as they increased (27% versus 11%). Similar results were found for the volatile class of carbon compounds (23% versus 17%). However, in the carbon-based phenolic class, allelochemicals increased seven times more often than they decreased (50% versus 7%).

Recall that this hypothesis is based on the assumption that a decline in plant nitrogen concentrations, commonly seen in elevated $CO_2$ experiments, will result in an excess of carbon products and so an increased investment in defence compounds. When Ryan *et al.* (2010) controlled for nitrogen responses, they indeed found that about 70% of 378 studies observed a decrease in plant nitrogen concentrations with elevated $CO_2$. Of those that found a decrease (262 studies) we would expect that the majority of them would also see an increase in allelochemicals, but this was true in only 45% of the results. Furthermore, such changes do not always achieve defence. In a further subset of 216 studies that investigated changes in nitrogen, allelochemicals and insect performance, in only 48% of the cases where allelochemicals increased did insect performance decrease. More surprisingly, in cases where allelochemicals decreased, 53% of these studies also demonstrated a decrease in insect performance. Together, these results suggest that allelochemicals are only part of the total picture of insect performance on plants grown under elevated $CO_2$, and that simple

generalizations about plant–animal responses are probably unwarranted.

## 4.7  Conclusions

Changing $CO_2$ concentrations directly impact plant growth (summarized in Fig. 4.12) via photosynthesis, particularly for $C_3$ plants, and water- and nutrient-use efficiencies. $CO_2$, at least at these concentrations, has little if any direct impact on animals (see e.g. Fleurat-Lessard, 1990). Photosynthetic rates are often down-regulated, probably as a result of carbohydrate source–sink imbalances, but nevertheless plants tend to accumulate more biomass, which is invested in more vegetative growth, more root growth and more seeds and fruits. Temperature directly influences the metabolism of both plants and animals, and in the case of plants interacts with $CO_2$ concentrations to influence photosynthesis. Temperature, and to some extent $CO_2$ concentrations, influence the phenology of both plants and animals, leading for example to earlier spring leaf-out in temperate climates, earlier flowering dates, faster rates of insect development, and earlier egg laying

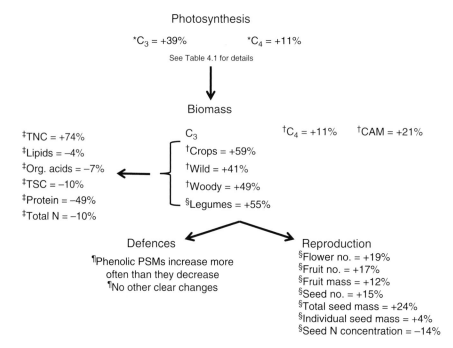

**Fig. 4.12.** Summary of some physiological changes induced by elevated $CO_2$ concentrations. * Denotes data from Ainsworth and Long (2005); [†] denotes data from Poorter and Pérez-Soba (2001); [‡] denotes data from Poorter *et al.* (1997); [§] denotes data from Jablonski *et al.* (2002); [¶] denotes data from Ryan *et al.* (2010). PSMs, plant secondary metabolites.

by birds. Drought, too, affects both plants and animals directly, and again, in the case of plants, interacts with $CO_2$ to influence photosynthesis. All of these effects can influence plant quality from the perspective of herbivores, because nitrogen and hence protein tends to be diluted under elevated $CO_2$. Sometimes herbivores are able to compensate for the decline in plant quality by increasing their consumption and other times they cannot. Even though we might expect carbon-based secondary metabolites to increase and nitrogen-based secondary metabolites to decrease under elevated $CO_2$, there seems to be little experimental support for these hypotheses, except perhaps in the case of phenolics, and this does not always result in increased plant defence.

# 5 Population Responses in Time and Space

In Chapter 4 we examined the impacts of climatic change at the organism and sub-organism levels. In this chapter we examine the impacts of climatic change on the abundance and distribution of groups of organisms, or populations. It would make our job easier if we could find simple generalizations but, as we see in this chapter, like any other sort of environmental change, climatic change is likely to be beneficial to some groups of organisms and detrimental to others. Ideally we would like to discover mechanisms that allow us to distinguish, a priori, which species or populations will benefit and which will not; but we are far from such an ideal end state. So as unsatisfying as it is, in several places in this and other chapters, the best we can do at this point is to recognize that species differ in their responses to climatic change and that we are not, as yet, capable of predicting all of these differences. Nevertheless, as we shall see, some big strides are being made towards this understanding.

## 5.1 What is a Population and How Do We Study its Response?

In this chapter we deal with population-level responses to climatic change. So, to begin, we should define what we mean by a population. We use the fairly loose definition of a population as a group of individuals of the same species that are capable of interacting with each other in a localized area. As we shall see, however, actually defining what we mean by a 'population' of organisms can be tricky. This is because individuals may be difficult to define, particularly in clonally reproducing organisms. In addition, many organisms are mobile and thus the concept of a 'localized area' may not

be very clear. This has given rise to the concept of a 'metapopulation'. A metapopulation is a group of distinct local populations of the same species that are connected via migration (e.g. movement for animals and dispersal for plants). Recent studies (Anderson *et al.*, 2009) have highlighted the importance of considering population structure, such as the spatial structure of metapopulations, in predictions of population responses to climatic change. We examine the consequences of fragmentation of natural populations due to human activities in particular in Chapter 13.

Understanding how climate change impacts natural systems requires studying how populations respond to environmental variation related to such change. This includes responses to changes in elevated $CO_2$, temperature (e.g. air, water or soil) and precipitation (and we have seen many examples of effects of these factors on growth rates in Chapter 4), but also to changes in the variability in these, and other environmental variables such as wind. At a demographic level, increased climatic variability may in turn increase the variability of vital rates (Boyce *et al.*, 2006). We know from general theoretical studies that increased variability in vital rates can lead to decreases in overall population growth rate (Lewontin and Cohen, 1969); therefore, increases in climatic variability could have important consequences for population responses.

In Chapter 3 we examined several different methods used to study climate change effects on biological systems. Studies to investigate population responses to climatic change are undertaken in several ways – through manipulative experiments, observational studies in natural systems and/or mathematical modelling. Experimental studies are often limited to organisms that are easy to track over several generations (e.g. microorganisms and insects) and thus our knowledge of these kinds of effects based on experiments is, to a large degree, taxa-biased. A few observational studies have been able to make use of long-term monitoring of populations over the past century. Macroecology is a relatively new ecological field that encompasses large-scale observed patterns in population distributions and responses and has been particularly useful in this regard. The interested reader should see, for example, Gaston (2000) for further details. Mathematical modelling is another methodological approach. It is frequently used to forecast future changes but it is useful for gaining other sorts of insights too, particularly in cases where experiments and/or observational studies are not possible. We examine evidence of climate change effects on populations from all of these lines of investigation.

## General life-history responses to climatic change

As we mentioned above, populations can respond to climatic change through changes in critical demographic parameters (vital rates) such as fecundity, mortality, immigration and emigration. 'Fecundity' (birth rate) is the term used to refer to the number of offspring produced by an individual (usually female) per unit time. 'Mortality' (death rate) refers to the number of deaths per number of individuals over a given unit of time (e.g. deaths per 1000 individuals per year). Populations of most species are not limited to the confines of a particular area but rather are able to move across space. 'Immigration' and 'emigration' refer to the arrival of new individuals and the departure of individuals from a population, respectively. These aspects of population change are linked to movement and dispersal abilities of organisms and thus affect the spatial distribution and abundance of organisms. These four life-history parameters are essential for understanding basic population dynamics. These models are called BIDE models, as the rate of change of population size is determined by the following conceptual equation:

$$\text{Rate of change} = \text{Birth rate} + \text{Immigration rate} - \text{Death rate} - \text{Emigration rate} \quad (5.1)$$

If individuals are small, or not easily distinguished, population size is often described in terms of biomass. Then the equation is written in terms of changes in biomass; and in the case of consumers, 'birth' refers to acquisition of biomass through food uptake while 'death' refers to biomass loss due not only to death but also starvation. We covered biomass responses in Chapter 4. We have a little more to say about such responses later in this chapter, but we largely restrict our treatment to considerations of clonal growth rather than biomass per se.

Climatic change affects fecundity for many different organisms. Not surprisingly, climatic change has been linked to both increases and declines in reproductive output, mortality and population size (McCarty, 2001), and the implications and the direction of these changes depend on the species, habitat and community interactions. Fecundity-related parameters such as 'fertility' (percentage of eggs that are fertile) and 'recruitment' (the number of new individuals reaching breeding age) have also been affected. As we mentioned above, climatic change will be beneficial to some species and detrimental to others. Below we provide some examples of results from studies of the impacts of warming and rising $CO_2$ concentrations. Like most other areas of climatic change research, it is rare to find studies that combine the impacts of temperature and $CO_2$.

Studies of the impact of warming divide roughly into two sorts. On the one hand, small mobile organisms such as insects and soil fauna, and sessile organisms such as plants, are often subjected to experimental manipulation of temperature (see Chapter 3). Large animals, particularly vertebrates, tend to be studied in 'natural experiments' such as anomalous weather events or in response to long-term trends in temperature. For example, research on caribou (*Rangifer tarandus*) in Greenland suggests a decline in reproductive success with climate warming due to difficulty in meeting energy requirements (Post and Forchhammer, 2008; Vors and Boyce, 2009). Some species of birds, on the other hand, seem to be favoured by warmer climates. Dippers (*Cinclus cinclus*) in southern Norway favour warmer winters, which allow easier access to foraging streams, resulting in subsequent increases in population size (Saether *et al.*, 2000). Clutch size and juvenile survival were higher in warmer springs for pied flycatchers (*Ficedula hypoleuca*; Winkel and Hudde, 1997). In contrast, average clutch size of Arctic-breeding geese (*Chen caerulescens* and *Branta canadensis*) declined during the warming period from 1951 to 1986

(MacInnes *et al.*, 1990). Puffins (*Fratercula arctica*) off the coast of British Columbia, Canada, have shown extreme variation in fecundity in response to anomalously warm sea-surface temperatures associated with inter-annual and decadal variability (see Chapter 1) analogous in their effects to global warming (Gjerdrum *et al.*, 2003). The lesser sandeel (*Ammodytes tobianus*), one of the most abundant fishes in the North Sea, has shown lower recruitment levels in response to warmer sea-surface temperatures (Arnott and Ruxton, 2002). In the Antarctic, changes in sea-ice extent and sea-surface temperatures have had positive impacts on population dynamics of some species of seabird and negative impacts on others (Croxall *et al.*, 2002). Increased temperature generally increases the developmental rate of insects and other poikilotherms and thus can commonly increase reproductive potential (Ayres and Lombardero, 2000), except where those organisms are living close to their physiological upper limit for temperature (see Chapter 4).

Like the impacts of temperature, experimental studies of the impacts of rising $CO_2$ on population dynamics are largely limited to small mobile organisms like insects and to plants. As we discussed in Chapter 3, plants obviously respond directly to elevated $CO_2$ while insects respond indirectly, due to changes in the plants. Thinking specifically about changes in the vital rates, experiments increasing $CO_2$ concentrations over a 3-year period found that loblolly pine trees (*Pinus taeda*) produce three times as many cones and seeds (LaDeau and Clark, 2001; Berryman, 2002). A review of several crop and wild plant species found that elevated $CO_2$ leads to increased seed production (size and/or number) for most crop species, but that wild species on average showed no response in reproductive effort (Jablonski *et al.*, 2002). The impact of elevated $CO_2$ on insects has been more variable than on plants, perhaps because these impacts are filtered through the plants, increasing the total range of responses. Nevertheless, individual studies show all possible responses: increased abundance, decreased abundance and no change (see e.g. Stiling and Cornelissen, 2007). This topic is described in more detail in Box 5.1.

Climate change and how it may contribute to direct mortality and declines in populations depends on how quickly species can respond to change. In regions where climatic changes will be most abrupt

(e.g. in cold regions), declines in populations have already been observed. For example, the decline in polar bear (*Ursus maritimus*) survival during the period between 2001 and 2005 has been attributed to longer annual ice-free periods over the continental shelf, resulting primarily in nutritional stress (Regehr *et al.*, 2009). Amphibian populations, on the other hand, are sensitive in particular to fluctuations in the amount and timing of precipitation. Major declines in populations of frogs in Puerto Rico have been correlated with recent years of low precipitation (Stewart, 1995). Low precipitation may be directly responsible for the extinction of golden toad (*Bufo periglenes*) from the Costa Rican cloud forest, local extinction of harlequin frog (*Atelopus varius*) and drastic declines in populations of several other species (Pounds and Crump, 1994; Pounds *et al.*, 1997, 1999; Still *et al.*, 1999). On the other hand, these extinctions and population declines might be caused by the amphibian chytrid fungus *Batrachochytrium dendrobatidis*, which itself is temperature-limited (Skerratt *et al.*, 2007). Biro *et al.* (2007) found from whole-lake experiments that increased water temperature resulted in increased mortality of fish populations due to indirect trophic effects. Similarly, when climatic change leads to increased temperature and decreased precipitation (drought), it seems likely that most organisms will show increased mortality, at least for aquatic environments. In Chapter 13 we examine some of the effects of interactions of multiple stressors on populations.

Regarding immigration and emigration, it stands to reason that species with the greatest mobility or dispersal ability will have an advantage with respect to rapidly adjusting their distributions with the changing climate. For animals, these are generally species that can fly and/or whose movement is not hindered by habitat fragmentation. Some species have been shown to be incredibly sensitive to climatic warming events. For example, the long-winged morphs of wing-dimorphic bush-crickets (*Metrioptera roeselii*) responded to a climatic warming event by strong immigration into previously unoccupied areas (Hochkirch and Damerau, 2009). This response has also been found for birds (Jiguet *et al.*, 2007). In addition, long-distance dispersal is thought to be important in poleward range extensions (Simmons and Thomas, 2004; Kokko and Lopez-Sepulcre, 2006), but knowledge of these events is lacking owing to the difficulty in tracking organisms over long distances. For plants, dispersal is one of the most poorly understood traits, although it has been made abundantly clear that immigration and emigration effects are critical in understanding plant populations' response to climatic change.

## 5.2 Functional Trait and Within-species Responses to Climatic Change

Population responses may differ depending on both species identity and/or their functional traits. Functional traits are attributes of species that influence their vital rates (survival, growth, reproduction) and ultimate fitness. Some examples in plants include seed size, specific leaf area, height at maturity and lifespan. Some examples in animals include trophic level, degree of mobility and lifespan. Some traits can span several taxonomic groups, such as thermal sensitivity of metabolic rate and dispersal. These traits vary considerably across various groups, suggesting that some functional groups may be more responsive to climatic change than others (Fig. 5.2).

If we want to know why some species respond differently from others, then, it is often useful to examine their functional roles or specific traits related to them. Van Turnhout *et al.* (2010) found that whether or not species of breeding birds have increased rather than decreased in number due to recent climatic changes could be linked to various functional traits of the species. Ground-nesting and late arrival at the breeding grounds in migratory birds tended to be associated with population declines, whereas increasing populations were associated with species that were herbivorous, sedentary, short-distance migrants, herb- and shrub-nesting birds, and large species with a small range. Dalgleish *et al.* (2010) looked at the response of several prairie plants to climatic change variability using population models parameterized from long-term observational studies and found that life-history traits could be used to predict population-level responses to changes. For example, they found that longer-lived plant species were less vulnerable to changes in climatic variability. This seems to be generally true for both plant and animal species (Morris *et al.*, 2008). This finding has interesting implications for the interaction between climatic change and problems associated with short-lived species (e.g. invasive species, agricultural pests, disease vectors). In fact, it has been suggested that some of these 'problems' could possibly be eliminated by increased variability in climate (Morris *et al.*, 2008)!

Even different populations of the same species may not respond in the same way to climatic change, further exacerbating our attempts to understand climate change responses at the population level. For example, a recent study (Balbontìn *et al.*, 2009) demonstrated that barn swallow (*Hirundo rustica*) populations in northern Europe had very different responses in life-history traits from those in southern Europe. Furthermore, some recent studies cast doubt that species will track climatic changes as rapidly as suggested by the majority of previous studies. For example, a recent study (Wilson and Nilsson, 2009) showed that in an Arctic alpine community, overall cover of vegetation did not increase over a 20-year period of warming. Some species increased in abundance and moved up elevation gradients, but others did not.

Finally, it is important to realize that much of our understanding of climate change responses of populations has been based on species of economic importance or charismatic species that are well studied (e.g. birds and butterflies). This is particularly the case with insects (see Box 5.1), but also applies to other taxa.

## 5.3 Complex Population Dynamics in Time: Lags, Cycles and Regime Shifts

As we have seen, owing to changes in basic life-history traits, climatic change will lead to population decline in many species but population size increases in others. However, the climate–population interaction is not always a simple one in which responses over time will be linear and predictable. A large part of population biology focuses its efforts on understanding population changes in time (dynamics). Even without drivers like climate change, this can be complicated by things like time lags, non-linear interactions and sudden, unexpected regime shifts. In this section we elaborate on these dynamics and climate change.

For starters, time lags may exist in how climatic changes affect populations and this can lead to complicated dynamics. A time lag occurs in a response when a variable, say temperature, does not impact a population immediately, but may take a year or two (or more) to result in population-level changes. One recent demonstration of the importance of such lags in natural populations is a study on seabirds (fulmars, *Fulmarus glacialis*;

**Box 5.1.  Are insect herbivore responses to elevated $CO_2$ idiosyncratic?**

A decade ago, in an influential review article, Coviella and Trumble (1999) surveyed the literature as it stood then, looking for generality in the effects of elevated $CO_2$ on insects. They concluded that: 'the limited data currently available suggest that the effect of increased atmospheric $CO_2$ on herbivory will be not only highly species-specific but also specific to each insect–plant system'. This is a situation that ecologists refer to as 'idiosyncratic'. If Coviella and Trumble are indeed correct then this is a worrying situation, since it means that we are unable to predict the impacts of elevated $CO_2$ on any plant–insect pair a priori, and we would have to study every plant–insect pair in order to make any statements about likely impacts. Naively, we might think that a general response of insects would be that they all increase or decrease under elevated $CO_2$. That much we know is not the case, for we certainly see every possible response in the literature: increased abundance, decreased abundance and no change in abundance.

Herbivorous insects are a hyper-diverse group of species, so perhaps this idiosyncrasy is to be expected? Bezemer and Jones (1998) sought to characterize a general response by examining responses within feeding guilds. Their review suggests a bit more of a pattern. They found that leaf chewers were generally able to compensate for reduced plant quality (see Chapter 4) by increasing their consumption and so will probably be able to maintain their populations. On the other hand, they found that leaf miners were not able to compensate for the reduced plant quality and as a result pupal weights were lower, suggesting a long-term reduction in leaf miner abundances. This hypothesis has now been backed up by 9 years of data from a study in Florida showing that six species of leaf miners, on three host plants, were always less abundant under elevated $CO_2$ compared with ambient $CO_2$ (Stiling et al., 1999; Stiling and Cornelissen, 2007). Finally, Bezemer and Jones found that phloem feeders were the only insect group to show a positive effect of $CO_2$ on population growth.

Unfortunately, Bezemer and Jones' conclusions have not generally held up as more experimental results have accumulated. In a meta-analysis of experimental results to 2005,

Stiling and Cornelissen (2007) concluded that there are no differences among feeding guilds. And Newman et al. (2003) surveyed just the aphids (phloem feeders) and found that, again, there was no common response to elevated $CO_2$. In Stiling and Cornelissen's meta-analysis, they concluded that, on average, herbivorous insects will decline under elevated $CO_2$ (by about 20%), but such declines are far from universal.

Where does all this leave us with regard to herbivorous insect responses? It is difficult to say with any certainty, but before we conclude that insect responses are idiosyncratic, there are other explanations for the variable responses that warrant consideration. When comparing the results of experiments, it may be important to keep in mind the variables, and hence mechanisms, that differ across studies. While it may be true that the levels of elevated $CO_2$ are comparable across two studies, a host of other equally important environmental variables may differ, affecting the impact of elevated $CO_2$. For example, soil fertility often directly influences plant quality, but perhaps surprisingly soil fertility is rarely reported in studies of insect responses. Similarly, the temperature at which the $CO_2$ comparison is made can play an important role in understanding the responses. For example, for cereal aphids, Newman (2004) used a physiological model to show that the impacts of 700 ppm $CO_2$ largely offset the impacts of +2°C, resulting in little change in aphid population sizes.

Finally, as hinted at above, it is important to remember that $CO_2$ is only part of the environmental change picture. Newman (2005, 2006) showed for cereal aphids that temperature and precipitation are always important, but $CO_2$ is important only in the low emissions scenarios (because the warming it causes is much more important in the high emissions scenarios). These results are illustrated in Fig. 5.1. Other environmental changes may also mediate the response of insects to $CO_2$ increases. Percy et al. (2002) found that increasing ground-level ozone along with $CO_2$ resulted in much larger increases in aphid populations relative to increasing $CO_2$ alone due to disproportionate effects on aphid predators.

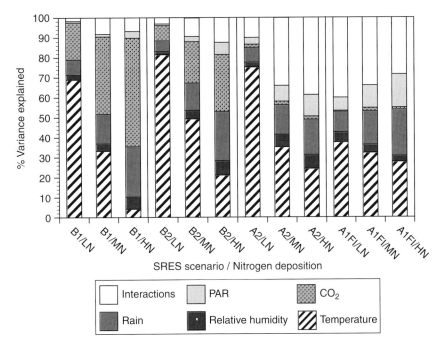

**Fig. 5.1.** The percentage of variance explained in aphid population size by five environmental variables, all of which change under climate change, for four scenarios and three levels of soil nitrogen. Moving from left to right on the x-axis shows results from low emissions (B1) to high emissions (A1FI). As emissions increase, the overall contribution of interactions between the variables increases (white) and the main effect of $CO_2$ (stippled) decreases. PAR stands for 'photosynthetically active radiation'. It is an aspect of climate change that has received very little attention in impact studies, but that GCMs predict will change in the future. (Redrawn from Newman, 2005.)

Thompson and Ollason, 2001), which found that there was a 5-year lag in the effects of ocean climatic changes (temperature and North Atlantic Oscillation) on reproduction of this species. The presence of time lags stresses the importance of examining population dynamics over long time scales, since impacts may not be immediate. In addition, mathematical models have been used to show that time lags can lead to instability in population dynamics, and ultimately greater extinctions, as population levels get very low. They can also affect a population's general ability to recover from perturbations (resilience) particularly in the presence of large amounts of environmental variability (Golinski *et al.*, 2008).

Some population dynamics exhibit particular kinds of temporal trend, such as cycles – rises and falls in population size that occur predictably over time. Perhaps the most famous North American example of population cycles is that of the snowshoe hare that peaks every 10 years. But many other populations exhibit this kind of cycling (including many forest insects). Cycles can occur for many reasons, but one common cause can often be linked to predator–prey interactions. As a prey population expands, there is more food available for predators, increasing predation pressure which then decreases prey abundance. Some climate change biology studies have found shifts in the cyclical dynamics of some populations such as seabirds (Jenouvrier *et al.*, 2005) and voles (Saitoh *et al.*, 2006), suggesting that climate change effects could lead to very important changes in predator–prey relationships. However, there are many theories about what might regulate population fluctuations and an understanding of how these interact with environmental variability, including climate change, is still largely unexplored.

(a)

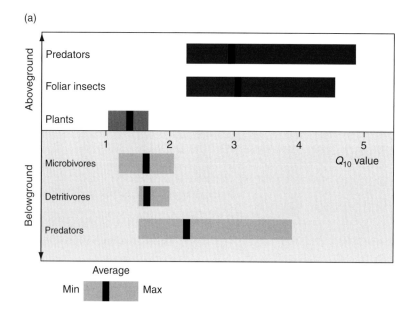

(b)

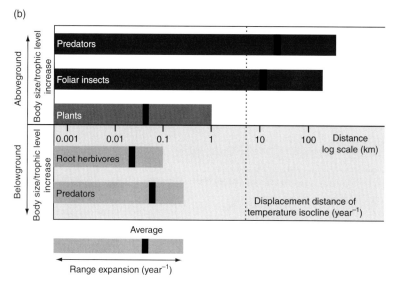

**Fig. 5.2.** (a) Thermal sensitivity of metabolic rate across trophic levels. Here, $Q_{10}$ is a measure of thermal sensitivity for metabolic rate and associated life-history traits for various invertebrate taxa. Horizontal bars show the range of $Q_{10}$ (the increase in trait value from 15 to 25°C divided by 10) for six trophic groups. Vertical black lines give the average $Q_{10}$. Clearly, variation occurs in thermal sensitivity of metabolic rate between trophic levels and within trophic levels. Species occurring in habitats with relatively constant temperatures (e.g. soil layer) have a weaker thermal response in traits compared with aboveground species, which live under more variable temperature regimes. (b) Dispersal rates in a variety of functional groups. Variation can be observed in dispersal rates between and within trophic levels. Horizontal bars give the range and maximum rate (year$^{-1}$, log scale) of dispersal, while the vertical black lines give the average rate of movement. (Figure adapted from Berg *et al.*, 2010.)

Population levels can also fluctuate quite erratically, even under natural conditions, making dynamics already quite difficult to predict even when we do not try to include the effects of climate change. Biologists are working hard to understand these kinds of erratic fluctuations and what might cause them. However, some studies based on model systems have found that increased temperature could lead to even more erratic fluctuations, instability and greater rates of extinction (Zhou et al., 1997). Regime shifts in populations occur when relatively sudden changes occur between contrasting, persistent states of a system. These can occur due to commensurate changes in environmental conditions ('extrinsic' factors), but also to some extent because of non-linear interactions between individuals and/or different populations ('intrinsic' factors). Recent studies stemming from complex systems science are attempting to find ways in which to detect (and thus potentially manage) 'regime shifts' in population dynamics. This work is based on the observation that many complex dynamical systems, ranging from ecosystems to financial markets and the climate, can have 'tipping points'. These are points where a regime shift may occur (see e.g. Anand et al., 2010). One recent study has suggested that a simple measure that reflects a 'slowing down' of changes in population dynamics over time could serve as an early-warning detection tool to regime shifts due to climatic change (Dakos et al., 2008).

## 5.4 Range Shifts and Spatial Distributions

When emigration exceeds immigration and/or mortality exceeds fecundity, species can go locally extinct, and this may lead to changes in a species' spatial distribution or range across either elevations or latitudes. Changes in population range in relation to climate have been of interest to ecologists for many decades and long before the problems associated with contemporary climatic change came to light. This is because climate is an important determinant of spatial distribution and geographic range for many species. Much of the work in this area is based around the 'centre–periphery hypothesis', which predicts that marginal populations are more susceptible to extinctions because they occur in less favourable habitats and at already lower density (Lawton, 1993).

There are three main lines of study in the impacts of climate change on species' distributions. The first uses palaeoecological data (see Chapter 1). For example, based on pollen reconstructions, spruce trees (*Picea* spp.) in North America experienced range expansion following the retreat of the glaciers after the last ice age (the same is true for many other tree species; Fig. 5.3; Williams et al., 2004). The second approach employs observational studies documenting pattern changes over the past century. And the third approach uses bioclimatic envelope or ecological niche models (see Chapter 3) to project future distributions.

### Observed responses over the past century of climatic change

Climatic regimes influence species' distributions through species-specific physiological thresholds of temperature and precipitation tolerance, which we discussed in Chapter 4. With warming trends these 'climate envelopes' become shifted towards the poles or higher altitudes. Generally speaking, species are expected to track the shifting climate and likewise shift their distributions poleward in latitude or upward in elevation (Walther et al., 2002), since it is assumed that most species are temperature-limited in their biology, at least to some degree. In some cases, however (e.g. reef-building corals), range shifts in response to changing temperatures may not occur if latitudinal distributions are also limited by other factors such as sunlight (Hoegh-Guldberg, 1999). Other species' ranges are also strongly linked to available moisture, and changes to the amount and seasonality of precipitation are likely to differ substantially from changes in temperature (see Chapter 2).

While these shifts are a reasonable theoretical prediction, it is difficult to demonstrate the causal relationship between climatic change and observable expansions or retractions, because these occur over broad spatial and temporal scales and often in concert with other important environmental changes that are unrelated to climate (e.g. habitat fragmentation and loss; see also Chapter 13). Nevertheless, several studies have been able to document range shifts in real time. How do range shifts actually occur in response to climatic change? For one thing, range shifts have to do with the ability of species to act quickly in response to yearly climatic variation. Parmesan et al. (1999) suggest that while some migratory species can respond very

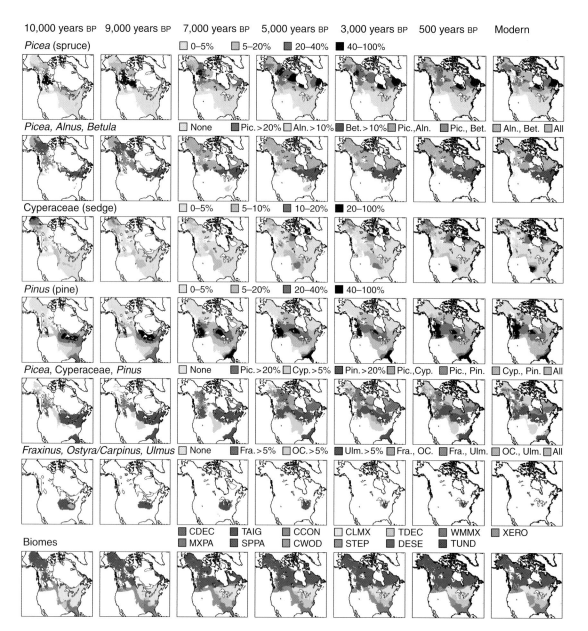

**Fig. 5.3.** Palaeoecological tree species migrations in North America from 10,000 years before present (years BP) to present. Primary colours (red, blue, cyan) indicate regions where only one taxon is present in abundance. Secondary colours (orange, purple, green) indicate associations between pairs of taxa; areas where all three taxa are associated are beige. (Figure from Williams *et al.*, 2004. Animated versions of these maps and others not shown here may be found on the World Wide Web at http://tinyurl.com/ch5-paleo.)

rapidly to changes in climate from year to year (e.g. by altering the timing of their migration), many organisms (both animals and plants) are relatively sedentary and thus responses to warming may be much slower. Capelin (*Mallotus villosus*) is a small pelagic fish that undertakes long-distance migration between feeding, overwintering and spawning locations in the Barents Sea, and its distribution is highly dependent on climate (Huse and Ellingsen, 2008). Due to its sensitivity to climate it has been referred to as a sea 'canary' for marine ecosystem change (Rose, 2005). In response to climatic change capelin is predicted to shift its spawning and adult distribution eastward (Huse and Ellingsen, 2008). A recent study has found that while many migratory species can adjust their behaviour to annual changes in temperature and/or precipitation, the decoupling of climate variables between, for example, breeding and non-breeding grounds can result in mistimed migration and other population responses (Robinson *et al.*, 2009).

Changes in range that are not related to migratory behaviour are due to changes at the population level, in particular in local extinction and colonization at the range boundaries. For example, a northward (poleward) range shift might be expected after net extinction

at the southern boundary or net colonization at the northward boundary. Ranges of species are predicted to either expand or contract depending on particular direct and indirect effects of climatic change (Thomas *et al.*, 2006). When a species is found to shift northward but remains stable at its southern limit, range expansion can occur; otherwise species may undergo range contraction. Figure 5.4 illustrates a northward range expansion associated with a butterfly (*Argynnis paphia*). Otherwise species may undergo range contraction. Large-scale range contraction towards net extinction has been observed for many taxa and will be discussed in more detail in Chapter 12.

Many species are indeed following a general trend of poleward and upward migration. Perhaps the best documented taxa from this perspective are butterflies (e.g. Konvicka *et al.*, 2003; Parmesan, 2006) and birds (e.g. Brommer, 2004; Sekercioglu

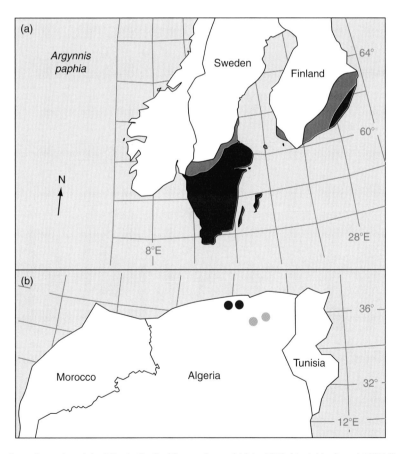

**Fig. 5.4.** The northern boundary (a) of the butterfly (*Argynnis paphia*) in 1970 (dark blue) and 1997 (light blue). The southern boundary (b) shows places where the butterfly was present in 1906–1912 and is either present in the current census (blue) or absent from the current census (orange). (Figure from Parmesan *et al.*, 1999.)

et al., 2008; Zuckerberg et al., 2009). For example, Hitch and Leberg (2007) found compelling evidence linking northward movements of North American birds to global warming. The northern limit of birds with a southern distribution showed a significant shift northward (2.35 km/year). The nature of this shift in species distributions in both North America and Europe is coincident with a period of global warming and so suggests, but does not unequivocally demonstrate, a connection with climatic change. Tracking such changes requires extensive monitoring programmes that are costly and time-consuming. However, climate change studies have benefited from initiatives such as the atlas data gathered through the North American Breeding Birds Surveys (http://tinyurl.com/ch5-NABBS), a large-scale, international avian monitoring programme initiated in 1966 to track the status and trends of North American bird populations.

Plant migration and range shifts over the past century have been difficult to document due to the lack of large-scale and detailed survey data. Forest migration has been identified with increases in biomass at the northern limit (Woodall et al., 2009) and elevation limit (Beckage et al., 2008) of tree

species' ranges and increases in establishment ability of migrating tree species relative to resident species found further north (Ibanez et al., 2008). Vascular plant migration along altitudinal gradients has been observed based on changes in plant species' occurrence along altitudes in the Alps over time. Parolo and Rossi (2008) report a median migration rate of 23.9 m per decade.

Several studies are now emerging to show that a wide array of taxa are experiencing range shifts in response to climatic change (Root et al., 2003; Rodenhouse et al., 2009). For example, a study based on a large national data set showed range shifts in everything from plants to dragonflies to fish to woodlice to mammals. The authors found no clear explanation for differences between taxonomic groups with respect to the extent of shifts (Hickling et al., 2006; Fig. 5.5).

Thus the consensus seems to be that very diverse populations have shown range shifts that correlate with climate warming. However, it is important to keep in mind that most, if not all, of the studies we have mentioned so far deal with temperate or boreal regions. Studies in tropical regions in particular are much rarer for a number of reasons, including the fact that climate change

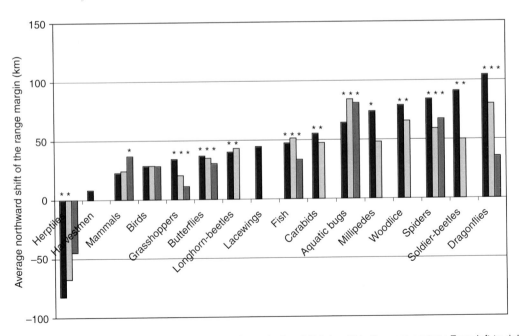

**Fig. 5.5.** Estimated northern range shifts in a variety of taxa in Great Britain within the past century. From left to right, the different shaded bars indicate increasing levels of sampling intensity. (Redrawn from Hickling et al., 2006.)

responses are thought to be particularly strong in northern latitudes. The IPCC found close to 29,000 observational data sets form 75 studies, 89% of which are consistent with directional change in terrestrial biological responses to climatic change (from 1970 to 2004) but only seven of them occurred in the tropics (Figure SPM1.2 in IPCC, 2007). However, studies from the tropics are now emerging that do document range shifts. For example, populations of a number of tropical moth species have shown shifts to higher elevations over the past 42 years in Borneo (Chen *et al.*, 2009). Greater integration and networking of monitoring programmes around the globe will aid climate change studies of population responses in all parts of the world.

Studies of climate change effects on range shifts often examine how populations are responding at their latitudinal or altitudinal extremes. However, recent studies suggest that climate change can also affect the 'spatial core' of distributions; that is, where species find their optimal location for peak abundance. A study to demonstrate this (Lenoir *et al.*, 2008) found that plants in western Europe have shown an upward shift in elevation during the 20th century (Fig. 5.6). Using core information within a species range for several bird species, Devictor *et al.* (2008) found that estimates of species migration rates were much higher than previously calculated using only information from the range extremes.

A recent review has also highlighted the fact that dynamics of populations at the non-expanding range limits could be even more important than previously thought due to the differences in processes occurring at different range limits (Hampe and Petit, 2005; Fig. 5.7). The argument is that the 'leading edge' or colonizing edge is largely controlled by long-distance dispersal events followed by population growth. The poleward migration, both observed and predicted, assumes that environments at this edge will become more suitable with climatic change. However, at the other extreme of a population range, different processes may be occurring. In one scenario, at the 'trailing edge', a population could progressively become extirpated leading to complete shifts in species range. In another scenario, these

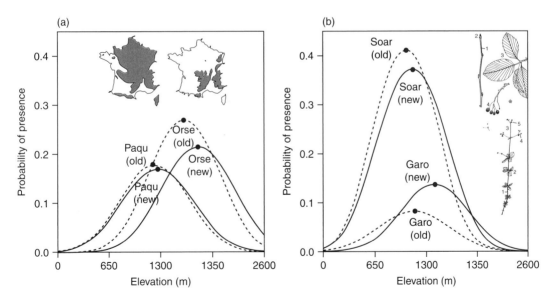

**Fig. 5.6.** Elevation shifts in species optima due to climate change. Shown are four examples of western European plant distributions that have shifted to higher elevations. Response curves depict 1905–1985 (dotted lines) and 1986–2005 (solid lines) for two species according to geographic distribution (a) and life form (b). In (a) a ubiquitously distributed species, *Paris quadrifolia* (paqu) and a mountainous species *Orthila secunda* (orse) are shown. For (b), a grass species *Galium rotundifolium* (garo) and woody species *Sorbus aria* (soar) are shown. The solid circles indicate the species' optimum elevation. (Redrawn from Lenoir *et al.*, 2008.)

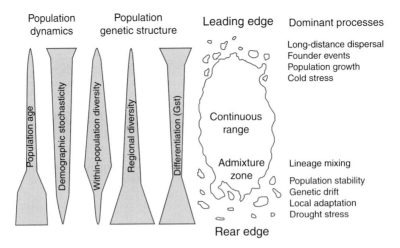

**Fig. 5.7.** Different processes (e.g. long-distance dispersal versus lineage mixing) are thought to be important at different edges of a population's range (e.g. leading edge versus rear edge). The width of the blue bars shown on the left indicates the degree to which certain population features (e.g. population age or demographic stochasticity) might be important at the corresponding position within the range. (Redrawn from Hampe and Petit, 2005.)

populations could remain stable and persist as relict populations from much older periods of time. These populations may persist when environmental conditions are heterogeneous and/or isolation has led to greater genetic diversity, highlighting their importance for biodiversity conservation. We expand on this concept of refugia and its role in conservation in the face of climatic change in Chapter 12.

### Future predicted range shifts

Future range shifts are predicted using several modelling approaches, two of which were discussed in Chapter 3. Predicted range shifts can be used to track whether populations are migrating fast enough by comparing observed trends with those of model predictions.

Species and populations are not randomly distributed, but exhibit distinct ranges in occurrence. These are mostly due to climate-related limitations and this is the basis for the so-called 'bioclimatic envelope models' used to predict climate–species relationships in space and time. Temperature in particular is a major driver of these distributions. However, bioclimatic envelope predictions have their limits as well, since many other factors influence species' distributions (Pearson and Dawson, 2003; Araujo and

Rahbek, 2006). Malcolm *et al.* (2002) provide evidence from coupled GCMs and global vegetation models to suggest that global warming may require migration rates much faster that those observed during postglacial history and that the role of outlier populations, range limits and variation in species' range sizes needs to be further considered. Models are just beginning to incorporate life-history traits, disturbance and habitat availability into predictions of migration rates. In a study using individual-based process modelling (Caplat *et al.*, 2008), rates of migration decreased dramatically when species interactions and disturbance were taken into account.

With increasing rates of change in global temperature, northward expansion of the boreal forest over tundra and a northward migration of temperate forests into current boreal forest ranges are predicted to continue at faster rates than seen in the past (e.g. Bonan, 2008), as illustrated in Fig. 5.8. However, the rates of migration of tree species are still uncertain as dispersal distances and species interactions, and hence migration mechanisms, are still poorly understood (see Box 5.2). A study done in Norway and Finland found that many northern land bird species will be susceptible to major range contractions by 2080 regardless of the severity of the climate change

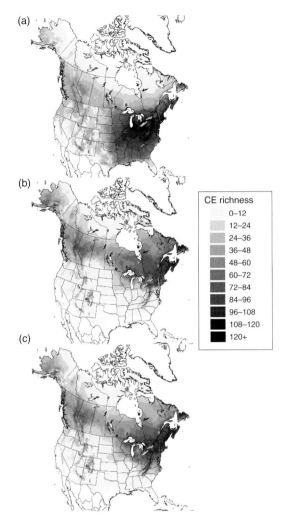

(a)

(b)

| CE richness | |
|---|---|
| | 0–12 |
| | 12–24 |
| | 24–36 |
| | 36–48 |
| | 48–60 |
| | 60–72 |
| | 72–84 |
| | 84–96 |
| | 96–108 |
| | 108–120 |
| | 120+ |

(c)

**Fig. 5.8.** Climate-envelope (CE) richness for 130 North American tree species under (a) current climate conditions; (b) future climate (2071–2100) based on the B2 emissions scenario, in which atmospheric $CO_2$ increases; and (c) future climate (2071–2100) based on the A2 emissions scenario, in which atmospheric $CO_2$ decreases. Maps (b) and (c) are averaged over three GCM outputs. (From McKenney *et al.*, 2007.)

scenario modelled. Over two-thirds of the species considered were susceptible to major range contractions in this geographical region (Virkkala *et al.*, 2008). The authors suggest that one form of mitigation could be to establish connected

reserve networks for these species. We discuss the implications of range shifts for conservation further in Chapter 12.

## 5.5 Clonal Growth Responses to Climatic Change

In temperate zones, it is estimated that 65–70% of vascular plant species are clonal (Klimes *et al.*, 1997). Growth and, importantly, spatial spread in clonal plants occurs by both above- and below-ground production of genetically identical units called 'ramets'. These are typically stolons, rhizomes, roots, bulbs, corms or tubers. The collection of genetically identical ramets is called a 'genet'. For clonal plants, the definition of an individual and population has to be reconsidered and the genet, comprised of many 'individual' ramets, is usually taken to be a population. The migration of clonal plants is also fundamentally different from that of non-clonal plants because it does not typically rely on dispersal by seeds. Studies have shown that ramet production as well as ramet size can increase with increasing atmospheric $CO_2$ levels and temperature (Oechel and Billings, 1992). Table 5.1 shows the diversity of responses that can occur at the population level for clonal plants, with a species of sedge as the example (Callaghan, 1994). Clonal plants also have the important property that genets tend to have very long lifespans and thus are good model organisms to study the effects of contemporary climatic change. For example, the genet lifespan of the evergreen Arctic-alpine cushion plant (*Diapensia laponica*) exceeds 100 years and the clones can be aged from their diameter (Molau, 1996). The species has shown a dramatic decline in biomass in response to global warming, typical of many Arctic tundra species.

## 5.6 Conclusions

Population responses to climatic change are largely of two sorts: (i) changes in population dynamics; and (ii) changes in spatial distribution. Both of these responses may depend strongly on $CO_2$, temperature or precipitation. These changes have been documented: in the palaeorecord; using long-term and large-scale monitoring studies; and for small, short-lived organisms in experimental studies. There are significant taxa and geographical biases

## Box 5.2. How will plant population migration occur?

Plants are generally considered to be sessile organisms, and thus it would seem that the idea of plants migrating through space in response to climatic change is an unlikely phenomenon. However, as we have seen, the palaeoecological record has shown range expansion over hundreds of kilometres, particularly after the most recent retreat of glaciers. How do these plant migrations occur? Although northward migration trends of many plant species are evident from the past, present and predicted for the future, the mechanisms involved with migration are still largely unknown. 'Reid's paradox' is the term given to the inability to fully account for postglacial migrations of tree species. It is named after the Victorian botanist Clement Reid, who could not understand the postglacial spread of oak trees into the UK (Clark *et al.*, 1998). Several other studies (Pearson, 2006) suggest that many species have such small dispersal distances, and thus disperse so slowly, that there is no documented mechanism by which most of these species could have reached their present geographical range since the last glacial maximum. Long-distance dispersal events (e.g. by birds; Johnson and Webb, 1989) have been acknowledged to be important, but still do not completely explain migration patterns. Recent application of molecular techniques based on chloroplast DNA has confirmed that for some species, like American beech (*Fagus grandifolia*) and red maple (*Acer rubrum*), postglacial migration rates may have been slower than those inferred from the fossil pollen record (McLachlan *et al.*, 2005; Fig. 5.9). However, there are some ideas about how this could be the case. Some studies (e.g. Brubaker *et al.*, 2005) suggest that some sparse populations maybe persisted during the last glacial maximum in regions where few, if any, pollen grains have been observed (i.e. out of the bioclimatic envelope); these 'refugia' could have served as points of nucleation for population expansion.

   Some have suggested that a migrating forest species can better establish everywhere in its new range than resident species and it is mainly limited by seed dispersal (Clark *et al.*, 1998). Another hypothesis, heavily based in theoretical work, suggests disturbance as an opportunity for seedlings of

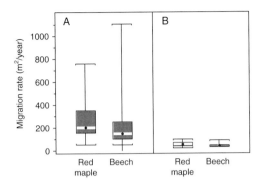

**Fig. 5.9.** Pollen evidence (panel A) versus genetic evidence (panel B) for migration rates for American beech (*Fagus grandifolia*) and red maple (*Acer rubrum*). Postglacial migration rates may have been slower than those inferred from the fossil pollen record. The solid blue dot is the median value. The box indicates the interquartile range. Whiskers indicate the limits of the data. (Redrawn from McLachlan *et al.*, 2005.)

migrating tree species to establish where, in the absence of disturbance, they would be competitively excluded by resident species (see e.g. Moorcroft *et al.*, 2006). Species interactions at the local scale may ultimately determine whether a migrating tree is successful in establishing or not and thus change overall rates of migration (e.g. Caplat *et al.*, 2008). Small-scale disturbances, for example treefall gaps, may play an important role in the shifting composition of species due to climate change (Hart and Grissino-Mayer, 2009), particularly in climatically defined ecotones (Stohlgren *et al.*, 2000). One recent observational study has shown this to be the case in the Great Lakes–St Lawrence forest region in Ontario, Canada, a transitional zone between temperate forest to the south and boreal forest to the north (Leithead *et al.*, 2010). Ongoing studies will help to better understand the mechanisms involved in plant migration and to better help predict species' range shifts and population dynamics responses to climate change.

---

to the literature as a whole. There is a generally well-documented response of species' distributions to move poleward and to high elevations. Population dynamic responses have not so far yielded any such general responses. Clearly some species will benefit from climatic change while others will not. Identifying which species is which has so far not been easy.

**Table 5.1.** Population responses of clonal plants to climatic change. (Adapted from Callaghan, 1994, where more detailed information on source citations is available.)

| Plant parameter | Factor | Response | Source |
|---|---|---|---|
| *Primary environmental effects* | | | |
| Tiller size | $CO_2$ | 11% increase in total weight (under high nutrient conditions) | Oberbauer *et al.* (1986) |
| | | 60% enhancement of photosynthesis (under high nutrient conditions) | Oechel and Billings (1992) |
| | Temperature | Fourfold increase in total dry weight in 12 versus 2°C | Kummerow and Ellis (1984) |
| | | 29% increase in leaf length in warm micro-site | Carlsson and Callaghan (1991b) |
| | | 17% increase in leaf length in warm micro-site | Carlsson and Callaghan (unpublished data) |
| Tiller initiation | $CO_2$ | Sixfold increase in tillering in *Eriophorum vaginatum* | Tissue and Oechel (1987) |
| | Temperature | 13% increase in tiller density in warm microclimate | Carlsson and Callaghan (1991b) |
| Autumn senescence | $CO_2$ | Prolonged photosynthetic activity in autumn in *E. vaginatum* | Tissue and Oechel (1987) |
| Flowering | Temperature | From 0.2 to 0.7 flowering shoots per plant in 9 versus 12°C | Heide (1992) |
| | | 17-fold increase in flowering rate in warm micro-site | Carlsson and Callaghan (unpublished data) |
| Seed set | Temperature | Increase from 0 to ~10 mature seeds/shoot in ~1°C higher average summer temperature | Carlsson and Callaghan (1990a) |
| Seedling recruitment | Temperature | Thermally induced disturbances (e.g. patterned ground, thaw slumps and detachment of the active layer above permafrost) favour seedling recruitment | Lawson (1986); French (1987); Edlund (1989); Jonasson (1992) |
| *Secondary, indirect effects* | | | |
| Tiller survival | Tiller size | Positive correlation between tiller size and survival | Carlsson and Callaghan (1990a) |
| Tiller lifespan | Tillering and flowering | Initiates senescence | Carlsson and Callaghan (1990b) |
| Flowering | Tiller size | Positive correlation between tiller size and flowering probability | Carlsson and Callaghan (1990b) |
| | Tiller type | Phalanx tillers have twice the flowering probability of guerilla tillers | Carlsson and Callaghan (1990b) |
| Seed set | Tiller size and flowering | Positive correlation between tiller size and seed set | Carlsson and Callaghan (unpublished data) |

# 6 Community Composition and Dynamics

In Chapter 5 we discussed how population-level responses to climatic change can shift the range distributions of individual species. Here, we expand this discussion to address the extent to which species respond to climatic change *independently*, or whether species interactions constrain or intensify these responses. We describe the unique challenges of studying community responses to climatic change and provide experimental evidence for community-level responses to warming, altered precipitation and elevated atmospheric $CO_2$. Finally, we explore how climatic change may interact with disturbance to alter the successional pathways of communities, and we consider the complexity of responses across multiple trophic levels.

## 6.1 Changes in the Distribution and Abundance of Coexisting Populations

Ecological communities can be defined simply as assemblages of populations that are spatially and temporally delimited; yet in practice, both the scale and complexity of communities makes them very difficult to monitor and study. A primary achievement of community ecologists has been to identify interactions among species that make community dynamics more complicated to assess and predict than the dynamics of the component populations in isolation. The dynamic nature of communities in response to environmental changes has been of major interest to community ecologists, and the early 20th century produced a spirited debate regarding the mechanism of community responses to such changes. In this debate, Clements (1916) described communities as types of 'super-organisms', with species sharing common evolutionary histories, whereas Gleason (1926) championed an 'individualistic' concept, with coexisting species simply sharing the tolerance of similar environments and existing at the same place and time often by chance. The analysis of species distributions along environmental gradients suggests that Gleason's individualistic concept was the most accurate, because community boundaries are rarely sharp and associations of species are often diffuse. Nevertheless, this debate remains instructive for our discussion of community responses to global climatic change over the next century.

Building on Chapter 5, historical responses of communities to climatic change also provide support for individualistic responses of species, as illustrated by changes in the northern range limits of North American trees following the last glacial maximum (Davis, 1981; see also Section 4.1). These range limits, reconstructed from pollen core data collected from a network of sites across the continent, reveal that species that currently exist together did not advance northward in unison following glaciation (Fig. 6.1). Similar individualistic responses of species have been observed over much shorter time scales for smaller organisms. For example, when microarthropod responses to one year of warming and rainfall manipulations were observed in the sub-Antarctic, springtails (nine species of Collembola) responded differently from mites (22 species of Acari) and variability among individual mite and springtail species was also high (McGeoch *et al.*, 2006). However, as discussed in Chapters 1 and 2, the rapid and sustained climatic change we are experiencing is unprecedented and over the next century it is expected to challenge communities beyond what is observed in the

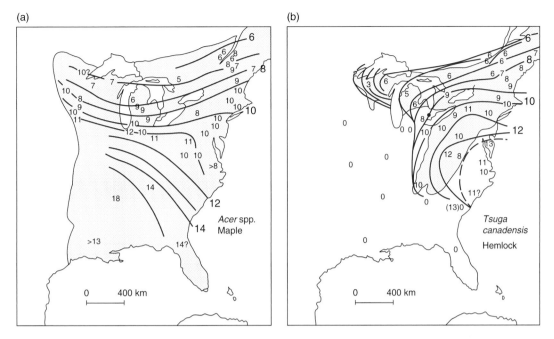

(a)

(b)

**Fig. 6.1.** The northward range expansion of (a) maple (*Acer* spp.) and (b) hemlock (*Conium* spp.) during the Holocene, as reconstructed from the radiocarbon ages of fossil pollen grains recovered in lake sediments. The numbers indicate the number of thousands of years before present to which the range limit corresponded, with the associated lines. (Redrawn from Davis, 1981.)

historical record. Experiments conducted over short time scales cannot adequately evaluate the responses of all organisms to climatic change. Furthermore, despite the diffuse nature of many community interactions, there are many examples of specialized mutualisms and trophic interactions among species, and if species distributions shift in response to climatic change in an individualistic manner, then these important and potentially obligate species interactions may be disrupted.

Changes in species distribution in response to climatic change, discussed at length in Chapter 5, are expected to result in the formation of 'non-analogous communities'. Non-analogous communities are those that are different in species composition from communities we currently observe or have observed in the past (Keith *et al.*, 2009). Nevertheless, at a broader level of observation, it can still be useful to predict how climate will shift the distribution of major biomes, which are defined by the dominant form of vegetation predicted to be reached late in ecological succession under a given set of climate parameters. Higgins and Harte (2006) simulated global biome distribution

under current conditions and a 200-year projection under the A1FI climate scenario modelled by the British climate model (HadCM3; see Chapter 2). Notable responses to the simulated climatic change include a general northward shift of biome types in the northern hemisphere and the expansion of desert in some equatorial regions. However, an important caveat of this simulation is that plant migration was unconstrained. The study goes on to reveal how vastly different biome distributions are predicted when the migrations of grasses, shrubs and trees are differentially constrained to account for differences among these life forms in seed production, dispersal and establishment (see also Chapter 5).

Geographic and physical barriers can constrain species migration, and there are communities at risk of being squeezed out when migration routes are blocked by natural barriers. For both terrestrial and freshwater communities, large bodies of salt water can constrain poleward migration. For example, in a meta-analysis of stream fish in Europe that linked increases in warm-water species to climatic change over the last 25 years, southern species were under-represented at sites located close to the southern border of continental Europe (Daufresne and Boet, 2007). Within continents and large islands, the influence of physical barriers varies widely among community types and areas that function as habitat for one community often function as a barrier for another. Communities at the greatest risk of range contraction are those restricted to specific zones, such as mountain tops and other disconnected areas of high elevation, and exploring changes in the altitudinal zonation of communities in response to climate has been a rich area of study. Mean upslope range expansion over the last 40 years was observed for sub-Antarctic vegetation but, consistent with the individualistic model of community organization, this occurred for less than half of the species and individual range expansion rates varied widely (Le Roux and McGeoch, 2008). Similarly, a 3°C increase in minimum air temperatures over the last century in Yosemite National Park, California, coincided with substantial upward changes in elevation limits for half of the species monitored along a 3000 m elevation gradient, leading to changes in community composition (Moritz et al., 2008). For the latter, although some high-elevation species are threatened because their ranges have contracted, protection of elevation gradients has allowed other species to respond via migration and species diversity has changed little because range expansions have compensated for retractions.

Human land use has created increasingly large barriers to species migration over the last century and will continue to constrain shifts in the distributions of communities in response to climatic change. Higgins (2007) modelled biome distributions for northern South America, but used land-use patterns to constrain the analysis. Currently, species richness (the number of species) in the Guiana Shield is high because this area both receives abundant precipitation and has relatively light anthropogenic land use (Fig. 6.2a). In contrast, species richness is low in much of eastern Brazil, which receives low levels of precipitation and has heavy anthropogenic land use. However, palaeoclimate and modelling studies suggest that the precipitation patterns for the two regions could switch in response to a shift in the location of the Inter-Tropical Convergence Zone. The conditions favourable for species richness in the Guiana Shield would disappear, and although the improved climate in eastern Brazil could theoretically allow species richness increases in Brazil to compensate for species losses in Guiana (Fig. 6.2b),

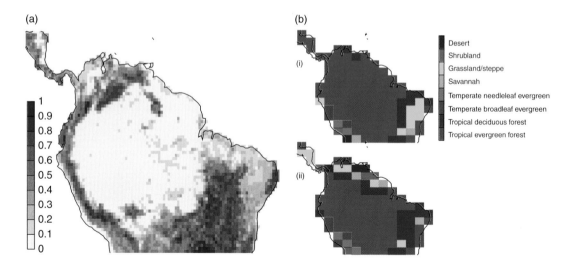

**Fig. 6.2.** (a) The fraction of land used for cropland or pasture in northern South America (adapted by Higgins, 2007 from data in Ramankutty and Foley, 1998; Foley et al., 2005), and (b) modelled biome distributions: (i) under current climate conditions, where rainforest occurs throughout much of the Guiana Shield and drier biomes dominate in eastern Brazil; (ii) with a shift in the location of the Inter-Tropical Convergence Zone, where there is forest loss and a transition to drought deciduous forest in the Guiana Shield region, and the potential (in the absence of human land use) for forest expansion throughout much of eastern Brazil. (From Higgins, 2007.)

Chapter 6

the land-use patterns displayed in Fig. 6.2a would prohibit these increases. Therefore, existing land-use patterns cause the combined species richness projected for the two regions to plummet.

As we suggest above, there are also obligate biotic interactions within communities that could constrain the migration of species. There are many commonly cited examples of specialized plant–pollinator interactions in the ecological literature, such as the sphinx moth (*Macrosilia cluentius*), which possesses a lengthy proboscis to reach into the approximately 30 cm long nectaries of the orchid (*Angraecum sesquipedale*; Wallace, 1876). Similar specializations exist for seed dispersal, as in the case of the tree *Calvaria major* of Mauritius, which reportedly failed to establish new seedlings despite the production of numerous seeds following the extinction of its main seed disperser, the dodo (Temple, 1977). For herbivorous insects, the colonization of new sites could be limited by host plant availability, as with the butterfly *Erynnis propertius* at its northern range limit in North America (Pelini *et al.*, 2009). Of course, there may also be antagonistic species, such as novel predators, competitors or diseases present in newly suitable environments that could constrain migration. The opposite scenario is also plausible and differential migration of species may separate important competitors, allowing formerly rare species to become more abundant.

Overall, community shifts in the context of climatic change can be addressed in the context of ecological 'assembly rules', a concept that is well established in the community ecology literature. These rules are based on the observation that only certain combinations of species exist in nature. While some pairs of species never coexist, either by themselves or as part of a larger combination, some pairs of species that form an unstable combination by themselves may form part of a stable larger combination (Diamond, 1975). Conversely, some combinations that are composed entirely of stable sub-combinations are themselves unstable. A fairly trivial example of an assembly rule is that a consumer cannot arrive in a community before an appropriate producer, although pathways of species establishment, and the number of possible community configurations, become more complicated as further consumers are added to the system (Fig. 6.3). The consequences of assembly rules are that certain sets of species that could be drawn at random from a local species pool may fail to

coexist at some local level, and different sequences of species invasion may lead to different patterns of community composition (Morin, 1999).

Unfortunately, it can be difficult to establish useful sets of assembly rules for ecological communities in nature because of their complexity, which explains why many intentional species introductions and restoration efforts in the past have gone awry. However, the potential importance of assembly rules has been demonstrated in experiments conducted in artificial communities at the microcosm scale. For example, when Warren *et al.* (2003) investigated the potential assembly pathways of protist species in small pools, the numbers of community states and assembly paths were much smaller than theoretically possible but the system had a complex range of assembly behaviours. Many of these paths led to an oscillation between two community types and several communities capable of long-term persistence were unreachable by sequential assembly. One community, dubbed the 'Humpty-Dumpty' community, could not be

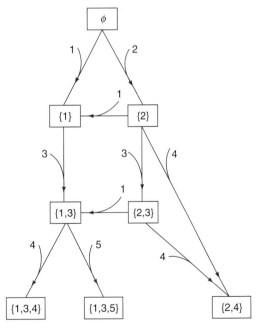

**Fig. 6.3.** Pathways of species establishment, leading to three alternative states. Species 1 and 2 are producers, and species 3, 4 and 5 are consumers with variable prey choices. The arrows represent new species establishments, and the brackets denote the current community composition. (Redrawn from Law and Morton, 1993.)

reassembled from the species it contained. On balance, this study confirms that predicting the assembly behaviours of communities is daunting, even in the absence of climatic change. Nevertheless, useful patterns may emerge regarding the likely community end points.

## 6.2 The Challenges of Studying Climate Change Effects on Communities

Community responses to climate change depend strongly on the temporal and spatial scales on which they operate and at which we study them. Setting aside these scale difficulties for a moment, a second challenge to studying community responses comes once we have gathered the data. Most communities are dominated by one or a few species, with the rest present in relatively low abundances. This uneven distribution presents challenges of interpretation for climate change experiments. We examine both of these challenges in more detail in this section, as well as looking at what mathematical modelling has to offer in terms of studying these responses.

### Temporal and spatial scales

The study of community responses to environmental change is complicated by the wide range of temporal and spatial scales over which the individuals of each species operate. Plants are often central in defining communities, because as primary producers they provide the base food source for animals. They also often serve as the physical matrix occupied by animals and they strongly influence microclimate. While terrestrial plants remain sessile, the individual pollinators, seed dispersers and herbivores they interact with can occupy ranges of metres to thousands of kilometres. Likewise, the lifespans of organisms interacting within a community can range from hours to thousands of years. The aboveground community of course engages in a wide array of interactions with microorganisms and microfauna that live below ground, but we reserve our discussion of climate change effects on soil organisms for Chapter 9.

As discussed in Chapter 3, there are strong spatial constraints on the ecological systems that can be studied using climate change experiments. These spatial constraints have led to a disproportionate number of studies in the global change literature, and more specifically in the community ecology

literature, where the dominant vegetation is of a relatively low stature (e.g. tundra systems, grasslands, algal mesocosms) or the animals of interest are very small or sessile (e.g. microarthropods, zooplankton, intertidal rock-bound organisms). In almost all of these cases what is actually being studied is only a subset of the community, and frequently these observations are restricted solely to the plant community. There is an inherent bottom-up bias in this approach (see also Chapter 4) and, as we discuss below in Section 6.7 on trophic-level effects, experiments targeted at climate change responses at the plant level may exclude important indirect effects on plants.

Temporal scales present a further constraint for climate change experiments on communities. For long-lived species we cannot directly observe responses over an entire life cycle, and for those that reproduce only at a late stage of maturity we cannot track adaptive responses to climate change (see Chapter 8). Furthermore, for processes that occur at the community level, we risk seeing only transient effects, while slower cumulative and equilibrium responses are not revealed. As with spatial scales, misleading results can be obtained when organisms that operate on different temporal scales are studied simultaneously. In a classic study on Darwin's finches (*Geospiza* spp.), populations of the finch community declined sharply in abundance in response to a drought (Fig. 6.4). Although populations of these birds can recover rapidly in response to favourable conditions, they failed to rebound in subsequent years. Following a detailed study of the finch feeding ecology, it was determined that one of the main food sources was seeds of the prickly pear cactus (*Opuntia echios*). Even though these plants exhibited strong recruitment following the drought, they take almost a decade to reach sexual maturity (Boag and Grant, 1984). This concept has been referred to in a broader context as 'multigenerational scale' (Wiens *et al.*, 1986).

Even in the absence of climatic change, community composition is often highly variable over time. A portion of this variability can be explained by inter-annual variation in temperature and precipitation, both of which can exceed the mean changes in these factors predicted over the next century. Nevertheless, after accounting for climatic variability, there can be considerable residual variation in community structure from year to year. A good illustration of this phenomenon is a 20-year exploration of community dynamics in a serpentine

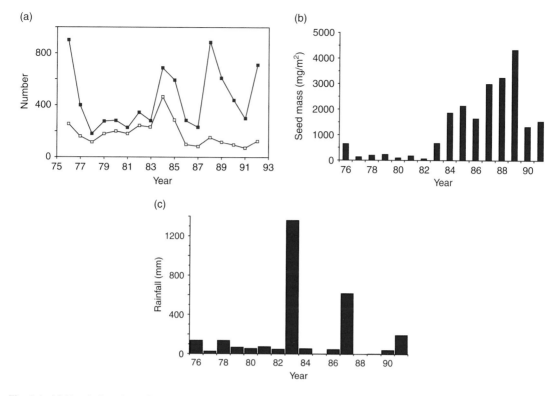

(a)

(b)

(c)

**Fig. 6.4.** (a) Population sizes of Darwin's finches, *Geospiza fortis* (closed symbols) and *Geospiza scandens* (open symbols), on the island of Daphne in the Galápagos; (b) biomass of small seeds on the island at the beginning of the dry season; and (c) annual rainfall on the island. Data cover the period 1976–1991. (Redrawn from Grant and Grant, 1993.)

grassland in northern California. Most of the plant species in the grassland had an annual life cycle, and there was a large array of different responses of individual plant species to inter-annual variation in both rainfall and disturbance by gophers. Meanwhile, some species exhibited consistent declines or increases in abundance over the study period, while others responded in a more idiosyncratic fashion (Hobbs *et al.*, 2007). The results of this type of study help provide a context for shorter-term studies of plant community responses to climatic change.

For the small, short-lived species we tend to use for testing community-level responses to climatic change, experimental results can be influenced very strongly by inter-annual climate variation, particularly if there are interactions between the experimental treatments and year (as there frequently are; see e.g. Section 14.2). For example, the effect of water addition on community composition in a wet year might be less intense than the effect of the same

treatment in a dry year. The solution is clear: community-level field experiments should be run for as long as possible. However, the pace of climatic change is sufficiently fast that control plots often become warming treatments for long-term experiments. Indeed, as we see below, some long-term observational studies of community dynamics in the absence of experimental treatments already provide us with insights into climate change responses, as do space-for-time studies that examine community composition along climate gradients.

### Analysis of community data

Aside from being inherently variable, community data are difficult to analyse because many species are often present at low abundance (these can be subdivided into 'subordinate species', which are stable within the community, and 'transient species', whose composition and abundance are unstable; Fig. 6.5). For example, if a warming experiment

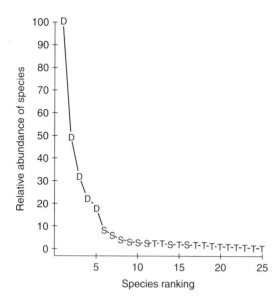

**Fig. 6.5.** A typical rank–abundance diagram of species within a community, featuring abundant dominant species (D), subordinate species (S) and transient species (T). (Redrawn from Grime, 1998.)

conducted in the field contained 48 plots but each of the 20 most rare plant species occurred only in three or fewer plots, there would not be enough statistical power (see Chapter 3) to examine the effects of warming on any of these species. One solution to this problem would be to assemble artificial communities, using equal numbers of each species in a mixed culture, and this is commonly done. However, such a design would reduce the experiment's external validity (see Section 3.2) and inevitably some of the species would not persist over time. For example, when Tilman *et al.* (1996) planted mixed cultures of grasses and forbs in a grassland experiment, plots planted with 24 species declined to an effective species richness of less than ten species over time.

Another way to include rare species in data analyses is to work with a naturally assembled community, but for the purpose of analyses to pool species into growth forms or functional groups (e.g. $C_3$ grasses, $C_4$ grasses, legumes, other forbs), which are represented more ubiquitously across the experimental plots. While this approach is informative on one level, it is not uncommon for species responses within functional groups to be quite divergent (see Chapter 5). Functional versus species-level diversity is discussed further in Chapter 12.

A further alternative is to design climate change experiments that focus solely on the responses of the dominant species in a community or on 'keystone species' (e.g. those, such as certain predators, that exert a highly disproportional effect on community structure relative to their own biomass). The examination of dominant species is particularly effective when attempting to link changes in community structure to ecosystem processes (see Chapter 7). The rationale is that ecosystem controls are proportional to the inputs of individual species to primary production and are therefore determined overwhelmingly by the traits of the dominant species. This concept was described as the mass ratio theory by Grime (1998), although he nevertheless acknowledged that subordinate and transient species may play a role in communities by acting as filters, thereby influencing the recruitment of dominants, particularly following disturbance.

Finally, community responses to climatic change can be assessed based on changes in community metrics such as species diversity, which is a function of the total number of species (richness) and the relative abundances of species (evenness). While this is discussed in length in Chapter 12, we briefly touch on it in this chapter as well. There are widespread concerns that climatic change will decrease biodiversity both regionally and globally, and there have been considerable efforts in recent decades to explore how biodiversity relates to ecosystem function. Based on the results of a grassland field experiment, Tilman and Downing (1994) demonstrated an increase in resilience to drought with increased diversity (Fig. 6.6). The curvilinear relationship between diversity and resilience suggests that declines in diversity bring increasingly severe declines in resilience, although it could equally be interpreted as showing that increasing diversity provides diminishing gains. In a large-scale biodiversity experiment conducted in the same grassland, Tilman *et al.* (1996) described a curvilinear relationship between increases in plant diversity and increased ecosystem productivity. This relationship holds true even when the sampling effect (i.e. the greater probability of including highly productive species in more diverse community assemblages) is taken into account. Although greater community diversity does not increase productivity in all systems, other studies have demonstrated increases in ecosystem predictability, greater nutrient-holding capacity and invasion resistance in response to increased diversity.

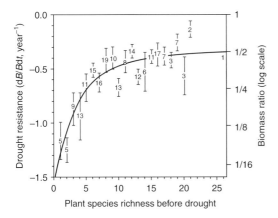

**Fig. 6.6.** Drought resistance of grassland plots and plant species before the severe drought. Biomass ratio (*y*-axis label on right side) indicates the proportional decrease in biomass 2 years after the drought relative to pre-drought biomass. Numbers indicate the number of plots of a given species richness and error bars denote standard error. (Redrawn from Tilman and Downing, 1994.)

### Modelling of community-level responses

The challenges of complexity that plague community-level experiments are extended to community-level modelling and it has been argued that the results of most population manipulations have made very limited contributions to the construction of dynamic models of communities (Abrams, 2001). Therefore, rather than attempting to predict climate change effects on component populations of communities, which requires an understanding of a very complex set of ecological interactions, alternative approaches have advocated making predictions of community-level responses based on knowledge of processes that determine properties such as the distribution of body's sizes in communities (e.g. Jennings and Brander, 2010).

### 6.3   Evidence for Global Warming Effects on Community Composition and Diversity

There has been widespread concern that climatic change will facilitate the spread of invasive species into new communities (e.g. Dukes and Mooney, 1999). Warming is predicted to expand the ranges of cane toads (Sutherst *et al.*, 1996) and fire ants (Morrison *et al.*, 2005), and both of these species have already had severe effects on the novel animal

communities they have encountered. In aquatic microcosms, warming also increased the invasion potential of alpine-lake zooplankton communities by imported species (Holzapfel and Vinebrooke, 2005) and favoured the growth of exotic macrophytes (McKee *et al.*, 2002). Below, we present examples where altered precipitation and disturbance patterns caused by climatic change, and elevated atmospheric $CO_2$, may also favour invasive species. However, we reserve the remainder of our discussion of climate change and invasive species for Chapter 13.

In recent decades, observations of natural communities have provided evidence of climate warming effects on the composition of terrestrial communities. For example, bryophyte and evergreen shrub cover increased in response to 27 years of warming and increased growing season length in High Arctic heath (Hudson and Henry, 2009), and warming combined with increased aridity correlated with decreased plant species richness in a neotropical plant community (Fonty *et al.*, 2009). Likewise, small mammal community composition changed in the northern Great Lakes region of North America over the last 30 years, with species with more southern ranges increasing and those with more northern ranges decreasing (Myers *et al.*, 2009). Such observational studies are limited in their abilities to identify cause and effect by their lack of randomization and controls. We spend the remainder of this section examining lessons learned from experimental manipulations of temperature.

Experiments characterizing warming effects on plant communities have increased dramatically over the last decade and interesting patterns have emerged for the systems that have been most commonly studied. Arctic and alpine communities have received considerable attention, because the former are expected to experience the greatest degree of warming, the latter are constrained by altitude, and all feature small-statured plants that fit under small-scale warming infrastructure (see Chapter 3). Chapin *et al.* (1995) observed that shrub biomass increased and non-vascular plant biomass decreased in warming tents, leading to decreased species richness. A similar pattern of increased shrub dominance with warming has emerged for alpine sites (Harte and Shaw, 1995; Jagerbrand *et al.*, 2009). It appears in both cases there may be positive feedbacks involving shrub-induced changes to snow depth and soil nitrogen mineralization (Wookey *et al.*, 2009; Fig. 6.7). Other warming manipulations in tundra sites have revealed little effect of warming

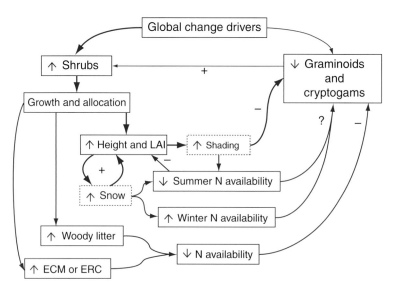

**Fig. 6.7.** Feedbacks resulting in an increase in deciduous shrubs and a decline in both graminoids (grasses and sedges) and cryptogams (mosses and lichens) in response to warming and increased growing season length in Arctic tundra. LAI, leaf area index; ECM, ectomycorrhizas; ERC, ericoid mycorrhizas. (Redrawn from Wookey *et al.*, 2009.)

on plant diversity or community structure (see e.g. Robinson *et al.*, 1998; Jónsdóttir *et al.*, 2005). Nevertheless, a meta-analysis designed to explore warming effects across a range of 11 tundra locations using open-topped chambers documented a switch to dominance by deciduous shrubs, decreased cover of lichens and decreased species diversity and evenness (Walker *et al.*, 2006). The study of animal responses to warming in field experiments has been under-represented, although studies of insect communities in field experiments designed to test plant warming responses have reported reduced species richness (Villalpando *et al.*, 2009).

Nitrogen and water additions in field experiments have indicated that warming often causes shifts in plant competitive ability by altering resource availability (De Valpine and Harte, 2001). Over the next century, atmospheric nitrogen deposition will increase in some regions (see also Chapter 13), and shrub dominance can be reversed back to dominance by graminoids and forbs when warming is combined with nitrogen addition (Klanerud and Totland, 2005). Changes in the competitive hierarchies of plant species with warming have also been demonstrated by more controlled competition experiments that compare the relative biomasses of plants grown in monoculture with those grown in mixed culture (Niu and Wan, 2008).

The success of some plant species in warming experiments has also been attributed to protection from frost damage (De Valpine and Harte, 2001) and, likewise, species adapted to wind-exposed sites with little snow have benefitted from advanced snow melt (Wipf *et al.*, 2009). In addition to affecting plant growth and survival, warming can alter plant recruitment, as in the case of seedlings in a Mediterranean-type shrubland that exhibited decreased species richness in response to warming (Lloret *et al.*, 2004) and of ephemeral species in an arid region of western Australian that displayed changes in germination from the soil seedbank (Ooi *et al.*, 2009).

Shifts in aquatic community structure have also been linked to recent climatic change. The contribution of rising sea temperatures to coral reef bleaching has been discussed extensively (e.g. Hughes *et al.*, 2003), and widespread and dramatic declines in mussel bed community diversity along the Pacific coast of the USA also appear to be related to large-scale processes such as climatic change, rather than local habitat destruction (Smith *et al.*, 2006). There is evidence for poleward shifts in the distributions of marine organisms in recent decades, as in the cases of invertebrate fauna in rocky intertidal zones off California (Barry *et al.*, 1995), groundfish in Icelandic ocean waters

(Stefansdottir *et al.*, 2010) and cetaceans (whales, dolphins and porpoises) north-west of Scotland (MacLeod *et al.*, 2005). The authors of the latter study were particularly concerned that when species respond to changes in local oceanic conditions, they may be moving out of areas specifically designed for their protection (see Chapter 12 for more on conservation).

Warming experiments on aquatic communities have typically addressed shifts in the plankton community in temperature-adjusted mesocosms. One result of these studies has been increased cyanobacterial dominance over diatoms or green algae in lake water (Elliott *et al.*, 2005; Domis *et al.*, 2007). Although there have been concerns that lakes may switch to turbid-water phytoplankton-dominated communities with warming, warming of lake mesocosms has decreased the amount of phytoplankton through shading by floating plants (Feuchtmayr *et al.*, 2009). However, mesocosms designed to simulate warming in alpine ponds revealed a general destabilization of phytoplankton and zooplankton communities, which could reduce the reliability of water quality and food resources for planktivorous fish in shallow coldwater ecosystems (Strecker *et al.*, 2004). In addition, in marine systems warming is expected to lead to plankton community regime shifts (analogous to those at the population level, as discussed in Chapter 5) in pelagic zones, and the formation of algal blooms is strongly temperature-dependent (Aberle *et al.*, 2007).

## 6.4 Evidence for Effects of Altered Precipitation on Community Composition

Many climate change experiments have focused on the effects of altered precipitation *patterns* (i.e. changes in the size and frequency of individual rain events) on plant communities, although changes in total annual precipitation are also expected to occur over the next century. In a mesic grassland, experimentally reducing storm frequency and increasing rainfall quantity per storm increased plant species diversity, independent of changes in total precipitation (Knapp *et al.*, 2002). Similarly, in a salt-marsh plant community, inter-annual variation in precipitation promoted a dynamic community composition and increased species diversity (Callaway and Sabraw, 1994). These responses can be attributed to the 'storage effect' (Chesson, 2000), whereby competitors experience advantages

at different times and can store the gains made during favourable periods. In grasslands, increased winter precipitation may influence the preferential recruitment of grasses over shrubs, succulents (Robertson *et al.*, 2010) and short-lived forbs (Morecroft *et al.*, 2004), despite the presence of summer drought. Community composition may also be affected by interactions between water availability and warming in these systems (Engel *et al.*, 2009). The effects of multiple stressors are discussed in greater detail in Chapter 13.

Tropical cloud forest communities, which receive a large proportion of their moisture from mist and cloud water, may be particularly sensitive to climatic change. Sharp declines in frogs and toads in the highland forests of Costa Rica, along with demographic changes in bird and reptile communities in the area, were linked to recent warming (Pounds *et al.*, 1999). All of these changes were associated with dramatic decreases in dry-season mist frequency and an increase in the altitude of the base of the cloud bank, both of which are correlated with increases in sea-surface temperatures in the equatorial Pacific Ocean since 1976. Similarly, when Nadkarni and Solano (2002) transplanted epiphytes and their arboreal soil from cloud forest trees to trees at lower elevations that are exposed to less cloud water, plants moved to the lower sites had significantly higher leaf mortality, lower leaf production and reduced longevity than control plants moved between trees within the upper site. In addition, after the epiphytes died, seedlings of other tree species grew from the seedbanks within the residual mats of soil, which suggests that climatic change may have negative effects on particular epiphytes to the benefit of species from a previously suppressed seedbank. This example provides another reminder that some species will in fact benefit from climatic change.

Changes in precipitation can also affect aquatic communities by altering water salinity. MacKenzie *et al.* (2007) used regional-scale climate–ocean modelling to predict how climate change will affect the fish community of the Baltic Sea. They suggested that fish diversity will be sensitive to changes in salinity resulting from altered precipitation patterns, with marine species declining and freshwater species likely to expand. Likewise, in a temperate estuary, changes in zooplankton diversity with varying inter-annual precipitation were associated with salinity changes (Primo *et al.*, 2009), and phytoplankton community structure in prairie saline lakes was sensitive to changes in precipitation/evaporation ratios and the resulting changes in salinity (Evans and Prepas, 1996).

## 6.5 Evidence for Effects of Elevated Atmospheric CO$_2$ on Community Composition

Although elevated CO$_2$ tends to stimulate the growth of plants grown individually in a controlled environment (see Chapter 4), the response of communities to elevated CO$_2$ cannot usually be scaled up from individual species because the effects of competition on species responses to CO$_2$ are often unpredictable (Körner, 1995). In many cases, community-level responses appear to be driven more by high-biomass species with weak responses than by low-biomass species with strong responses (Navas, 1998). In addition, the final community composition at a future point in time may depend critically on both the magnitude and the rate of increase of atmospheric CO$_2$ (Ackerly and Bazzaz, 1995).

Temperate grasslands have frequently been used as model systems for examining the effects of elevated CO$_2$ on community structure, not only because the plants are relatively small and exhibit annual turnover of aboveground growth, but because it is difficult to supply tanks of CO$_2$ to more remote communities. In both mesocosm and field experiments conducted in grasslands, species-specific responses to elevated CO$_2$ have been documented, leading to changes in community structure, often but not exclusively in the favour of legumes or other forbs (e.g. Teyssonneyre et al., 2002; Winkler and Herbst, 2004).

The effects of elevated CO$_2$ on community structure frequently interact with other global change factors. For example, elevated CO$_2$ differentially influenced the responses of two birch species to variation in soil moisture (Catovsky and Bazzaz, 1999) and CO$_2$ effects on community composition differed in a wet year versus a dry year in an understorey plant community (Belote et al., 2004). Similarly, plant community responses to CO$_2$ have interacted with temperature in a model grassland (Campbell et al., 1995); with nitrogen addition in both a sphagnum bog (Heijmans et al., 2002) and in a model grassland (Maestre et al., 2005); and with phosphorus in monoliths from a calcareous grassland (Stocklin et al., 1998). A very intriguing result was observed in an annual grassland, where elevated CO$_2$ increased grass biomass on its own, but also suppressed the growth response of grasses

to nitrogen addition, water addition and warming (Shaw et al., 2002). However, the latter response occurred only in the third year of the experiment, and was not evident when the plant community responses over the first 5 years of the experiment were revisited (Dukes et al., 2005). Additional examples and more discussion on how multiple stressors can interact can be found in Chapter 13, and the importance of interactions in general is discussed further in Chapter 14.

## 6.6 Climate Change, Disturbance and Succession

Even under a stable climate, ecological communities do not remain static at a given location (i.e. within biomes). Rather, changes in microclimate and soil properties, often caused by the dominant vegetation, cause communities to progress through successional stages between episodes of disturbance (with the latter defined as a discrete event that removes biomass). An important concept for understanding these dynamics is that of the 'regeneration niche', where the conditions and circumstances required for germination and establishment can be quite different from those experienced and tolerated by mature individuals (Grubb, 1977). A classic example is white pine (Pinus strobus), which is a towering dominant tree in the canopies of northeastern forests in North America, yet it cannot regenerate from seed under the shade of the forest canopy and relies on fire to increase light availability at the forest floor (see also Chapter 10).

Not only is climatic change predicted to alter the frequency and intensity of influential disturbances, such as fire and hurricanes, but the effects of warming, altered precipitation and elevated CO$_2$ on community composition can feed back on disturbance, often to the benefit of invasive species. Sage (1996) suggested a mechanism whereby cheatgrass (Bromus tectorum), an invasive species in the western USA, could increase the quantity of plant litter in response to elevated CO$_2$ and decrease litter quality. The resulting greater fuel load could increase fire frequency, leading to increased invasion by exotic species. An analogous mechanism was proposed by Burgess et al. (1995), who described how increased precipitation could increase fuel loads of invasive buffel grass (Cenchrus ciliaris) litter in the Sonaran desert, leading to an increased fire frequency and increased grass dominance. Elevated CO$_2$ can also directly alter the structure of early successional communities following disturbance, as demonstrated for a model regenerating longleaf pine (*Pinus palustris*)–wiregrass (*Aristida stricta*) community (Davis *et al.*, 2002).

Much like disturbance, increases in extreme climate events such as drought, frost and midwinter melts are anticipated to change the successional trajectories of communities beyond the effects of mean annual changes in climate (Jentsch *et al.*, 2007; Fig. 6.8). Dramatic effects of a severe heat wave in France in 2003 were observed for freshwater mollusc communities, which experienced a sharp decline in diversity and slow recovery (Mouthon and Daufresne, 2006). Similarly, extreme winter warming can severely damage some plant species, as demonstrated in a subarctic site exposed to both ambient conditions and experimentally induced warming over winter (Bokhorst *et al.*, 2009). Extreme summer drought decreased diversity and slowed down post-fire succession in a Mediterranean shrubland (Prieto *et al.*, 2009) and increased invasion by the weed *Rumex obtusifolius* in managed temperate grassland. The succession of pond communities is also very sensitive to drought (Chase, 2007).

Overall, it is clear that changes in disturbance regimes and successional trajectories caused by climatic change will complicate our efforts to both manage and restore natural communities and to predict future changes in these communities.

## 6.7 Multitrophic Responses to Climatic Change

Although a large majority of climate change experiments have focused on subsets of the whole community, and in particular the plant community, there are a growing number of examples of complicated, multitrophic responses to climatic change. Several species of honeycreepers (Drepanidae) endemic to the Hawaiian islands have been driven to extinction over the last century because their forest habitat has been replaced by crop and pasture land, and human introductions of mosquitoes and avian malaria have excluded them from the remaining lowland forests (Fig. 6.9). Landscape analyses of high-elevation forest refuges have revealed that climatic warming is likely to drive several of the remaining species to extinction by facilitating the spread of mosquitoes infected with avian malaria to higher elevations (Benning *et al.*, 2002). There has also been concern that thermodependent bacterial pathogens will cause mass mortality of some species in temperate benthic communities (Bally and Garrabou, 2007). In coastal salt marshes of the south-eastern USA, cordgrass (*Spartina alterniflora*) has recently experienced massive die-off, which has been attributed primarily to drought combined with stress caused by snail grazers that facilitate plant infection by fungal pathogens (Silliman *et al.*, 2005).

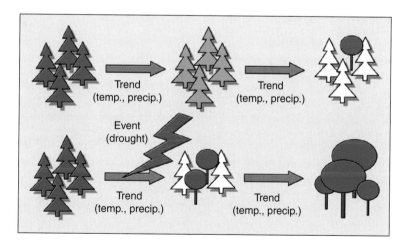

**Fig. 6.8.** Hypothetical successional trajectories of a community exposed to changes in mean values of climate parameters, such as temperature or precipitation (top) and extreme events superimposed on changes in mean values (bottom). Extreme events can bring systems into balance with novel climatic conditions by reducing community inertia, which is otherwise increased by long-lived organisms, competitive balance or clonal reproduction. (From Jentsch *et al.*, 2007.)

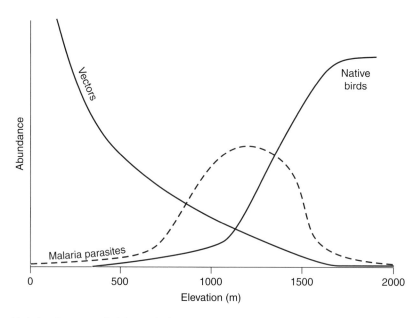

**Fig. 6.9.** Native bird abundances, malarial parasite incidence and mosquito vector levels along an elevation gradient in Hawaii. Mosquitoes are limited to lower elevations by cold temperatures, and the prevalence of malaria is low at low altitudes because the introduced birds present there in great abundance are resistant. (Redrawn from Vanriper *et al.*, 1986.)

Laboratory feeding preference experiments have revealed that elevated $CO_2$ can reduce the nutritional quality of leaves for herbivores (see Chapter 4). However, in community-level studies, the effects of elevated $CO_2$ on herbivory have been mixed. For a model tropical plant community exposed to elevated $CO_2$, leaf quality was not affected under low nutrient conditions (Arnone *et al.*, 1995), and in grassland microcosms, elevated $CO_2$ had no significant impact on herbivore feeding preferences or biomass (Diaz *et al.*, 1998). For herbivory by slugs in an annual grassland, elevated $CO_2$ reduced the numbers of seedlings consumed but there was an interactive effect of elevated $CO_2$ and nitrogen addition that differed among plant species (Cleland *et al.*, 2006).

Earlier in this chapter we discussed how members of a community often operate at highly variable spatial scales. While the plots of climate change field experiments can realistically address responses at the plant community level, the mobility and range of many herbivores are sufficiently high that they are affected only indirectly by the treatments. Furthermore, there is a concern that the infrastructure used to administer the treatments may unintentionally exclude herbivores. For example, in warming experiments conducted in Arctic environments, plastic-walled chambers have typically been used to provide passive

warming because the sites are not located close to convenient sources of electricity (see Chapter 3). Along a forest–tundra ecotone, the shrub *Betula nana* increased in response to warming, but this effect was most pronounced when caribou (*Rangifer tarandus*) were excluded (Olofsson *et al.*, 2009). A similar result was obtained when muskoxen (*Ovibos moschatus*) and caribou herbivory was excluded in Arctic shrub tundra (Post and Pedersen, 2008; Fig. 6.10). These results suggest that herbivores may inhibit the climate-driven shrub expansion on the tundra we discussed earlier, although we now face the difficult task of determining how climatic change and other factors may directly impact herbivore densities in the future (see also Chapters 4 and 5).

Insect herbivores typically have better access than large mammals to experimental plots. In a forest FACE experiment elevated $CO_2$ altered the insect community significantly, with trends towards reduced phloem feeders and increased chewing herbivores (Hillstrom and Lindroth, 2008), but in a different forest it decreased damage by herbivorous insects (Knepp *et al.*, 2005). Insect herbivores also fed preferentially in warmed montane meadow plots that had melted out sooner than the surrounding area, although some species were more abundant in the later-melting plots, which resulted

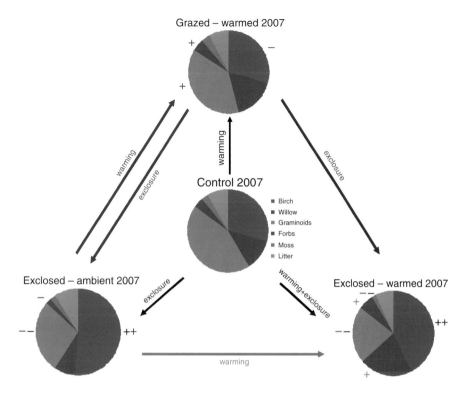

**Fig. 6.10.** Differences in plant community composition between control and treatment plots after 5 years of warming and herbivore exclusion. Arrows indicate treatments and signs (+ or –, colour-coded to treatment arrows) indicate significant (*P*<0.05) differences between pie wedges in the direction of manipulations. (From Post and Pedersen, 2008.)

in an overall shift in community structure in response to warming (Roy *et al.*, 2004). In sub-arctic heath, changes in the abundance of insect herbivores in response to warming were driven by changes in subordinate plant species, rather than dominant species (Richardson *et al.*, 2002). Although free access of herbivores to experimental plots may appear to be desirable, there is a hidden danger that herbivores may simply congregate at unrealistic densities in favourable plots, much like moths around a street lamp, or avoid undesirable plots altogether (Moise and Henry, 2010; see Box 6.1). Thus, herbivore responses may merely be artefacts in many plot-level field experiments.

## 6.8 Conclusions

In this chapter we have presented many examples of community responses to climatic change. However, these examples are restricted to a limited number of study systems and they are biased in many cases

towards studies where climate change effects on communities have been either large or mechanistically interesting (which parallels a similar publication bias in the general climate change biology literature). Although climatic change has the potential to alter community structure dramatically and result in the decline of some communities, other communities may remain relatively intact or even expand considerably. The limitations of climate change experiments in community ecology are daunting, and it may be virtually impossible to experimentally test or model the effects of climatic change on many communities. Nevertheless, the contribution of community ecologists to climate change research will be to identify what communities are most sensitive to climatic change and whether species that play key roles within specific communities are sensitive to climatic change. Ultimately, researchers cannot properly assess population or ecosystem responses to climatic change without accounting for potential changes in community-level processes.

**Box 6.1. Animal communities and climate change.**

Although the infrastructure used in climate change field experiments can restrict the access of influential herbivores or pollinators to plots, Moise and Henry (2010) describe why free access by animals to experimental plots also may result in unrealistic animal densities. Specifically, animals that prefer the local conditions in treated plots may congregate at artificially high densities or those that are repelled by the treatments may avoid them entirely. Therefore, herbivore damage in the plots of climate change experiments may either grossly exaggerate or grossly underestimate the contributions of animals to primary productivity or plant species composition under future environmental conditions. Although behavioural preferences by herbivores may be interpreted as realistic community-level responses to climatic change, in a future world the herbivores would not be privileged with such a choice; rather, they would be forced to tolerate the new conditions or relocate over a long distance to survive.

Herbivore feeding artefacts can explain important changes in plant productivity and species composition in climate change studies and these effects may instead be mistakenly attributed to direct plant responses. Direct plant responses can be isolated through the use of exclosed subplots within the main experimental plots. Measures of animal densities and herbivore damage in climate change field experiments must also be interpreted with caution. These results should be complemented by gradient studies or time series analyses conducted at the landscape scale.

Trophic-level responses to climatic change clearly extend beyond plant–herbivore interactions, yet scaling difficulties make these responses difficult to document and assess. Nevertheless, the results of recent observational studies have been intriguing. For example, in a high-elevation riparian system described by Martin (2007), the abundance of dominant deciduous trees declined, which was correlated with decreased overwinter snowfall. Snow accumulation influenced the overwinter presence of elk, whose browsing altered deciduous tree abundance. The latter decreased preferred bird nesting habitat and increased nest predation rates. The net result was the local extinction and several population declines in some previously common bird species. In an opportunistic aquatic study by Schiel *et al.* (2004), thermal outfall from a power-generating station caused a 3.5°C rise in seawater. After 10 years of this ocean warming, there were significant community-wide changes in 150 species of algae and invertebrates relative to control areas. However, there was no trend towards warmer-water species. Instead, the communities were altered in an apparent cascade of responses to changes in the abundance of several key taxa, including subtidal kelps and intertidal foliose red algae. Overall, the authors concluded that the community shifts were strongly coupled to direct effects of increased temperature on the key taxa and indirect effects operating through ecological interactions.

# 7 Ecosystem Responses

Continuing with our exploration of climate change effects on different levels of biological organization, in this chapter we focus on the ecosystem level. An 'ecosystem' can be broadly defined as a collection of interacting organisms *plus* their abiotic environment. Key ecosystem-level processes include biomass production, organic matter decomposition, carbon sequestration and nutrient cycling (essentially, these processes are manifestations of the interactions among organisms and the abiotic environment). Here, we discuss how these processes will respond to climatic change and elevated atmospheric $CO_2$. We also touch on the issue of ecosystem services and how these may be affected by the changing climate.

Climatic change is driven by the concentration of greenhouse gases in the atmosphere. The human contribution to our changing climate is largely driven by our burning of fossil fuels. Fossil fuels are an example of carbon that has been 'sequestered' for millions of years. Carbon can basically be in three places at any point in time: in the atmosphere; in living organisms; or in the soil and water. In some forms of soil (see Chapter 9) and in some places in the water (e.g. deep in the oceans) carbon enters these forms and places much faster than it leaves them, leading to an accumulation of carbon. This accumulated carbon is 'sequestered' for various lengths of time, from a few years to many centuries. In this chapter we examine the ecosystem processes that lead to carbon sequestration. This first necessitates a brief review of ecosystem processes and then we discuss how climate impacts these processes.

## 7.1 Ecosystems and Carbon

One way to think of an ecosystem is as a collection of pools of organic matter linked by fluxes of energy (with carbon as its 'currency', i.e. energy is stored in organic carbon compounds) and nutrients. For example, the vegetation might comprise one pool, while soil organic matter comprises another pool, and litterfall represents a flux from the vegetation to the soil pool (Fig. 7.1). Environmental conditions influence pool sizes and flux rates. As the environment changes with climatic change, we are interested in how the pools and fluxes will be affected, partly because of feedbacks between organic matter in ecosystems and $CO_2$ in the atmosphere and partly because the size and nature of these pools and fluxes define an ecosystem's functioning, habitat value and capacity to provide ecosystem services.

Carbon sequestration at the ecosystem level depends on the balance of inputs and outputs to and from the ecosystem as a whole. Inputs come from photosynthetic assimilation of $CO_2$ into organic carbon compounds and outputs come from respiration, as the organic matter is used for energy and $CO_2$ is released back to the atmosphere (outputs can also be abiotic in nature, e.g. $CO_2$ released by fire). Before we discuss the factors regulating the overall carbon balance of an ecosystem, let us briefly review the 'ins and outs' of ecosystem carbon cycling in a little more detail: how does carbon enter, move around in and leave an ecosystem? Carbon ($CO_2$) from the atmosphere is fixed by plants during photosynthesis – in other words, it is converted from an inorganic form to organic compounds in

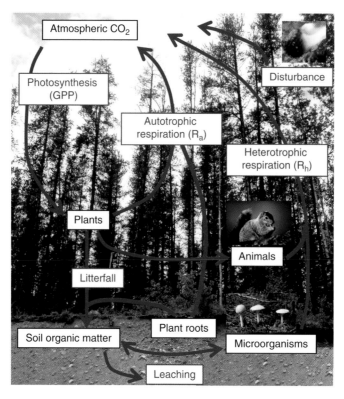

**Fig. 7.1.** The ecosystem carbon cycle.

plants. At the ecosystem level, the total amount of carbon assimilated by vegetation during photosynthesis (over a given time period) is called 'gross primary production' (GPP). Some of that carbon is respired by the plants and thereby returns to the atmosphere as $CO_2$ ('autotrophic respiration', $R_a$). The remainder is available for producing new biomass (some of which will be consumed by herbivores or will be shed as litterfall), supplying rhizosphere organisms with carbohydrates. The amount of carbon left over after autotrophic respiration has been accounted for is called 'net primary production' (NPP). Some of the carbon in NPP will be lost to the atmosphere through 'heterotrophic respiration' ($R_h$), as living or dead plant and animal material is consumed by organisms at higher trophic levels, including the detritivores (in most terrestrial ecosystems, the majority of heterotrophic respiration comes from the soil organisms consuming dead organic matter). The carbon accumulated annually in an ecosystem after both autotrophic

and heterotrophic respiration has been accounted for is called 'net ecosystem production' (NEP). It is possible for NEP to equal zero, if the photosynthetic inputs are equal in magnitude to the respiration outputs. These various terms can be assembled into equations that describe their interrelationships:

$$NPP = GPP - R_a \qquad (7.1)$$

$$NEP = NPP - R_h \qquad (7.2)$$

$$NEP = GPP - (R_a + R_h) \qquad (7.3)$$

There are other ways for carbon to leave an ecosystem besides respiration; these additional outputs may be intermittent or regular occurrences. When carbon losses due to disturbance (such as fire) and hydrological fluxes (such as dissolved carbon leaching through the soil profile into streamwater) occur, the amount of carbon accumulating in an ecosystem is also dependent on these outputs, such that:

$$NEP = GPP - (R_a + R_h + \text{disturbance loss} + \text{leaching loss}) \qquad (7.4)$$

These non-biological carbon outputs are often overlooked in models of climate effects on carbon sequestration (Chapin *et al.*, 2009), but have been receiving more attention lately.

## 7.2 Factors that Regulate Carbon Sequestration

Ecosystem carbon sequestration is controlled by the factors that regulate inputs and outputs of carbon. Here, we examine the controls on NPP, heterotrophic respiration and NEP.

### Net primary productivity

Net primary productivity (NPP) is controlled by factors that influence its components, namely GPP and autotrophic respiration. GPP is controlled by solar radiation, atmospheric $CO_2$ concentration, water, nutrients and temperature. These environmental factors can influence the rates of photosynthesis in a given leaf (see Chapter 4), or can dictate the total number of leaves present to capture solar radiation (usually thought of as the total surface area of leaves or leaf area), or both. In other words, productivity can be increased if plants become more efficient at photosynthesis with a given amount of leaf area or if they increase their leaf area by growing more leaves. If environmental changes are drastic enough (or if changes are moderate but some species in a given plant community exist close enough to their environmental limits; see Chapter 6), the vegetation itself may undergo changes in its composition, thereby changing carbon capture by the ecosystem via changing plant characteristics (e.g. a shift to plant species with greater photosynthetic capacity; Chapter 4). Figure 7.2 shows how the main climatic constraints to NPP vary geographically.

Environmental conditions can also affect the amount of carbon lost from an ecosystem via autotrophic respiration. This process depends, like other physiological processes, on temperature and also on any factor that affects plant growth (more biomass = more respiration). At large spatial scales, plant respiration tends to be roughly a constant proportion of GPP (about half; which in turn means that NPP comprises the other half), but the proportion of GPP lost via plant respiration can be greater under warmer temperatures. Root respiration accounts for the largest proportion of autotrophic respiration in most ecosystems, and often is measured together with respiration of soil heterotrophs. This combination of belowground plant and heterotroph respiration is often called

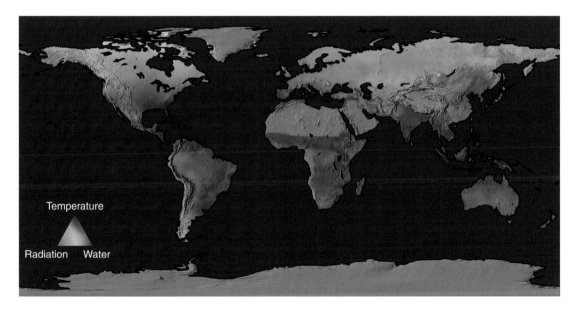

**Fig. 7.2.** Climatic constraints to plant growth across the globe. (From Nemani *et al.*, 2003.)

'soil respiration'. Many field studies of ecosystem carbon dynamics measure soil respiration because it is difficult to separate the two components of $CO_2$ efflux from soils (autotrophic and heterotrophic respiration).

In general we might expect root respiration (autotrophic) and decomposition (heterotrophic) to respond in the same direction to temperature and moisture changes, and even to changes in atmospheric $CO_2$ (elevated $CO_2$ in general stimulates plant growth, see Chapter 4, which in turn stimulates soil microorganisms via increased carbon availability). However, the magnitude of their responses may differ and the balance between them may change. We can also speculate that in some circumstances their responses to a changing environment may differ. For example, imagine a very nitrogen-poor ecosystem where the higher C:N ratios in plant tissues arising from elevated $CO_2$ cause plants to increase their root biomass – and hence root respiration – in search of more nitrogen. At the same time, the lower quality of plant litter in this already nitrogen-poor system may slow decomposition and hence result in lower rates of heterotrophic respiration. So, while it is difficult to measure these two components of soil $CO_2$ efflux separately, it is important to at least consider how they may respond differently to climatic change. It has been suggested that warmer temperatures may in general lead to higher allocation of GPP to plant respiration (Ryan, 1991); however, there is little known about how climatic change will affect the balance between GPP and autotrophic respiration and thus NPP. So much less is known about how autotrophic respiration responds to changing temperature and $CO_2$ that the widely validated notion that plant respiration is equal to about 50% of GPP is often just *assumed* in estimates of ecosystem NPP response to climatic change.

It is more common to measure NPP than GPP in ecosystem studies, because NPP is easier to measure directly (in terrestrial ecosystems it is usually measured as biomass increment plus tissues shed as litterfall; while this method accounts for most of the NPP in many ecosystems, it represents an underestimate because it neglects other, more difficult to measure pathways for NPP, such as root exudates and tissues consumed by herbivores). Also, NPP is often of more interest than GPP since it represents the organic matter, and the energy it stores, that is available to detritivores and herbivores. All of this gives us good

reason to focus on how climate change affects NPP, rather than GPP, in Section 7.3. GPP becomes a more familiar character in Section 7.5 on NEP, where the subject of interest is the balance between gross inputs and outputs of carbon to and from the ecosystem as a whole.

Temperature can affect NPP directly, through effects on rates of physiological processes, and indirectly, through effects on nutrient availability (e.g. warming may increase nitrogen availability by speeding up decomposition and nitrogen mineralization in temperature-limited ecosystems), water availability and species composition (Fig. 7.3; see also Chapter 6). In colder-climate ecosystems, temperature effects may be manifested mainly through the extension of growing season lengths. If a limiting nutrient becomes more available, NPP can increase either because per-leaf rates of photosynthesis are increased or because total leaf area is increased (or both). Temperature also affects water availability due to its influence on evapotranspiration rates. As we have discussed over the past few chapters, in the longer term, shifts in temperature regimes (and precipitation patterns) across the globe will result in shifts in the distribution of organisms, including plants. NPP in a given region will change by virtue of changes in plant functional types and thus ecosystem types, for example from tundra vegetation to boreal forest trees with higher potential NPP rates (Bunn *et al.*, 2005).

$CO_2$ levels can affect NPP directly through increasing carbon availability for photosynthesis (see Chapter 4), but its main effects will likely be mediated through effects on soil water content. Under elevated $CO_2$ plants can conserve water by keeping their stomata more closed, thereby reducing transpiration, while still taking in adequate amounts of $CO_2$ (see Chapter 4). If water is not limiting to plants, they can gain more carbon with the same amount of water use under elevated $CO_2$. Increases in soil moisture due to conserved water can affect nutrient cycling rates and thus nutrient availability, which is a strong determinant of NPP rates. 'Trickle-down' effects of elevated $CO_2$ on NPP will include those arising from the interplay between carbon and nitrogen cycles in ecosystems. (Nitrogen is a limiting nutrient in many ecosystems; climate change effects on the cycling of nitrogen and other nutrients are discussed in the next section and also in Chapter 13.)

For the most part, elevated $CO_2$ and changes in water and nutrient availability brought on by

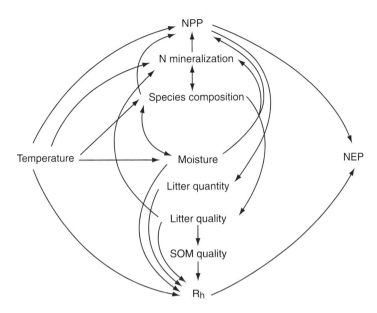

**Fig. 7.3.** Direct and indirect effects of temperature on NPP and other ecosystem processes. SOM, soil organic matter. (Redrawn from Shaver *et al.*, 2000.)

warmer temperatures will likely have the greatest impact on NPP in most ecosystems (compared with direct effects of temperature on physiological processes). In temperate and boreal ecosystems, longer growing seasons will also play an important role.

## Heterotrophic respiration and net ecosystem productivity

Heterotrophic respiration is largely controlled by those factors that affect soil microorganisms, as these are responsible for most of the $R_h$ in terrestrial ecosystems. Most of the energy captured by plants in organic matter flows through the detrital food chain as dead plants and animals are decomposed by bacteria and fungi, releasing mineralized nutrients and $CO_2$ in the process. The decomposer organisms are directly affected by temperature; their metabolic activity increases with temperature up to very high temperatures as long as sufficient moisture is available. In waterlogged environments, warming-induced soil drying can actually accelerate decomposition, and therefore $R_h$, by increasing oxygen availability. So, moisture is also an important factor regulating $R_h$: too much soil water inhibits decomposition due to lack of oxygen, and too little moisture inhibits the diffusion of substances to and from microbial cells (Chapin *et al.*,

2002). The decomposition process is also strongly affected by the quality of the detritus (in most terrestrial ecosystems, mainly dead plant material); that is, the 'palatability' of the plant litter to the decomposers. Generally, litter that is high in nutrients and low in recalcitrant compounds, like lignin, is most palatable to soil organisms and decomposes the most rapidly. The C:N ratio in litter, or the ratio between lignin and nitrogen, is often used as a indicator of litter quality, with high C:N or lignin:N ratios indicating poor-quality litter that is likely to decompose slowly.

The relationship between the biochemical nature of plant litter and the decomposers produces some complicated (and really interesting!) feedbacks. For example, a positive feedback may be triggered if some factor in the environment changes such that decomposition rates (and thus $R_h$) are increased, causing nutrients to be released more rapidly from the dead organic matter, thereby increasing plant productivity and/or the nutrient concentrations in plant tissues. If more litter is produced, and especially if more high-quality litter is produced as a result, decomposition processes may be further stimulated, thus increasing $R_h$ even more. In contrast, a negative feedback may be triggered for example if plant litter quality declines due to nutrient deficiency, causing decomposition rates to be reduced, thereby slowing nutrient release from dead organic matter and exacerbating the lack of nutrient availability. The interactions between multiple stressors (e.g. nutrient deficiency and warming) are discussed further in Chapter 13.

Net ecosystem productivity (NEP) is regulated by the balance between NPP and $R_h$ (or between GPP and the total of $R_a$ plus $R_h$), and so is influenced by the factors that control these processes (Fig. 7.3). Changing environmental conditions may increase NPP, but if those conditions also favour matching increases in $R_h$ there may be no increases in NEP, i.e. the amount of carbon sequestered by the ecosystem as a whole.

In summary, NPP is regulated mainly by the availability of water and nutrients, as mediated by temperature and $CO_2$. With time, changes in these factors (along with growing season length) can result in shifts in species composition and vegetation type, thereby influencing NPP by changing the 'players'. Disturbances such as fire can also affect NPP by disrupting plant life. $R_h$ (mainly from soil organisms – the decomposers) is controlled mainly by temperature, moisture, and plant litter quality and quantity. NEP, which represents an ecosystem's net accumulation of carbon, is regulated by the balance between NPP and $R_h$.

Climate (and atmospheric) change thus has lots of opportunity to affect rates of carbon sequestration, by direct (e.g. rising $CO_2$ levels and temperature) and indirect (e.g. changes in water and nutrient availability) means. Short-term and long-term responses may differ, as plants and decomposer communities acclimatize to the changing conditions, as the balance among limiting factors shifts and changes, and as species replace each other according to geographic shifts in optimal and tolerable environmental conditions.

## 7.3 Impacts on Net Primary Productivity

There is evidence that the warming trend of the past few decades has already produced changes in terrestrial NPP. Globally, a 'greening' trend seems to be occurring, as indicated by increases in terrestrial NPP of about 6% in the final two decades of the 20th century (Nemani *et al.*, 2003). Regionally, changes have been variable, with the largest increases seen in tropical and boreal ecosystems (Fig. 7.4). Increases in tropical NPP have most likely been due primarily to a factor that does not figure prominently in most discussions of climatic change, namely increased solar radiation resulting from reduced cloud cover (Nemani *et al.*, 2003; Fig. 7.2).

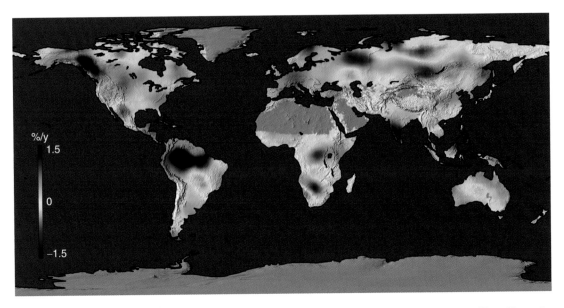

**Fig. 7.4.** Changes in terrestrial NPP (per cent per year) driven by climate change from 1982 to 1999. (From Nemani *et al.*, 2003.)

The >100 ppm post-industrial increase in atmospheric $CO_2$ levels that we have already experienced has likely also contributed to increased plant production in the tropics (Laurance *et al.*, 2004; Wright, 2005) and elsewhere. In northern ecosystems, temperature likely plays a larger role: year-to-year variation in plant production (photosynthetic activity as seen by satellite) in Canadian boreal forest and tundra has been linked to inter-annual temperature patterns, and the current warming trend has resulted in greater productivity in these biomes over the past few decades, as well as a poleward-marching treeline (as discussed in Chapter 5). NPP increases of nearly 20% in China's ecosystems from 1982 to 1999 have been attributed mainly to longer growing seasons and less so to increased rates of NPP within the growing season (Fang *et al.*, 2003). Increased productivity in tundra ecosystems has been linked to both longer growing seasons and greater rates of NPP during the growing season. There are already hints that this 'greening' trend may not last long: evidence from the past few years shows that while circumpolar tundra vegetation continues to ramp up its productivity, many forested areas in North America are showing declines in productivity, possibly because of late summer drought (Goetz *et al.*, 2007).

Some modelling studies that combine GCMs with global vegetation models predict that the positive trend in global terrestrial NPP will continue well into the 21st century (Cao and Woodward, 1998; Alo and Wang, 2008) due to the combination of warming and elevated $CO_2$. Others predict that climatic changes and elevated $CO_2$ will act on global NPP in opposite ways, with negative effects of warming counteracting the positive effects of elevated $CO_2$ (Cramer *et al.*, 2001; see also Chapter 13).

Zooming in from the global scale to regional and local levels, a tangled web of interactions among environmental variables is revealed. First, let us consider what happens with warming alone (without elevated $CO_2$). Modelling studies and field experiments have shown that the response of NPP to warming depends largely on the availability of water. Warming has a drying effect because it increases evapotranspiration, so water-limited ecosystems would need to see increases in precipitation or increased WUE in order to be able to respond positively to warming. In a modelling study of a variety of ecosystem types, Luo *et al.* (2008) found that moderate warming alone had little effect on NPP, except in a boreal conifer forest and a wet

heath – both temperature-limited ecosystems with moisture to spare (all other ecosystem types in the study were in warmer, drier environments). When increased precipitation was combined with warming however, all ecosystems included in the analysis responded with greater NPP. Negative interactions occurred when warming was combined with decreased precipitation (Luo *et al.*, 2008). Many other modelling studies predict similar patterns. Chiang *et al.* (2008) found that the direction and magnitude of NPP response to warming depended on the degree of warming: in US forests, mild warming (about 2°C) and matching increases in precipitation resulted in NPP increases of as much as 25%, while more severe warming (about 5°C) without adequate precipitation increases caused NPP to decline by as much as 60% (Fig. 7.5). Parton *et al.* (2007) used a physiologically based mechanistic model, parameterized with data from a prairie heating and $CO_2$ enrichment experiment, to show varying responses to warming, again depending on water availability for plant growth. Similarly, a cold, dry spruce forest in China modelled by Su *et al.* (2007) increased its NPP by 18% in response to warming as long as precipitation also increased.

Predictions about limitations on NPP in a warmer climate are corroborated when experimental warming studies are carried out in field settings. The results from some of these studies echo the modellers' findings: water deficiencies will prevent NPP from increasing in response to warming. For example, one group of researchers warmed the air by 3°C over plots in a temperate grassland, and found that even though the heated and unheated plots were given the same amount of water, the heated plots dried out faster and had 25% lower NPP (De Boeck *et al.*, 2008). In other cases, nutrient deficiency is the culprit (see also Chapter 13); the tundra warming experiment carried out by Hobbie and Chapin (1998) found no effect of warming on NPP. It turns out that tundra vegetation is not necessarily temperature-limited itself, but its productivity is nitrogen-limited – and nitrogen mineralization is temperature-limited in Arctic ecosystems. The experiment did not heat the soil enough to speed up nitrogen cycling, so the vegetation did not respond to the warming. In a seemingly contrasting example, another tundra ecosystem warming experiment revealed increases in plant biomass production overall (Biasi *et al.*, 2008). In that experiment, soil temperature was raised, stimulating nitrogen cycling and improving the availability

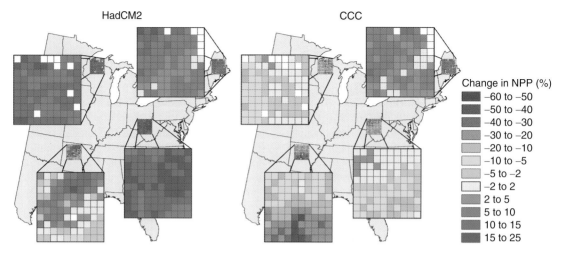

**Fig. 7.5.** Modelled NPP changes in four different US forest regions under doubled $CO_2$, resulting in mild warming (about 2°C) and moderate precipitation increases (as projected by HadCM2) or more severe warming (about 5°C) and small precipitation increases (as projected for the non-SRES scenario, the Canadian Climate Centre scenario). (From Chiang *et al.*, 2008.)

Change in NPP (%)
- −60 to −50
- −50 to −40
- −40 to −30
- −30 to −20
- −20 to −10
- −10 to −5
- −5 to −2
- −2 to 2
- 2 to 5
- 5 to 10
- 10 to 15
- 15 to 25

of this nutrient. The different outcomes of these two studies also point to the importance of species-specific responses (see Chapter 14) and the influence of the dominant species on the overall ecosystem response. In the second study, lichens were the dominant vegetation type and these increased their biomass production significantly, strongly influencing the overall results. In the former study, biomass production of a shrub species increased, while that of other species declined or did not change, leading to no overall change in NPP.

It is possible for NPP to increase without increases in biomass production if the NPP is allocated to rapid turnover pools within the plant, like root exudates. Likewise, it is possible for NPP to increase without increases in long-term carbon storage in plants if the extra NPP is allocated to short-lived tissues (e.g. fine roots instead of woody stems). However, one cannot have increased biomass production without increased NPP. Many studies of the responses of ecosystems to climatic change do not measure NPP completely, but measure plant biomass production instead; reported increases in plant growth or biomass production indicate that NPP is increasing.

Despite a fair amount of variability in the outcomes of individual warming experiments, a meta-analysis of 32 such experiments revealed that, across all of these studies, warming of up to 6°C increased plant productivity by 19% on average (Rustad *et al.*,

2001). This illustrates, however, an issue with much of the research that has been done to date in this field: most of the experiments included in their analysis were performed in temperate, boreal and Arctic ecosystems. So, while there seems to be a general consensus that warming alone can increase NPP, research done in warm, dry climates suggests that NPP will not respond positively to warming (e.g. Peñuelas *et al.*, 2007) and tropical ecosystems have been largely neglected. The jury seems to be out on how these will respond to warming, with some suggestions that, regardless of moisture availability, some tropical ecosystems are close to their upper temperature limits already, beyond which NPP will decline (Doughty and Goulden, 2008). It will be important to sort out climate change impacts on tropical ecosystems, particularly forests, as these contribute more to global NPP than any other ecosystem type!

While there have been numerous field experiments to test the effects of warming, there have also been ecosystem-level experimental manipulations of one of the other major factors involved in climatic change: atmospheric $CO_2$. There has been much speculation, well-founded in the principles of plant physiology, that $CO_2$ will have a fertilizing effect on vegetation (see Chapter 4) and that this may help to counteract any negative warming impacts on plant production. Experimenting with the effects of elevated $CO_2$ has been done on a variety of scales (e.g. individual plants, small plots under chambers), but

Chapter 7

we concentrate on the ecosystem-level free-air $CO_2$ enrichment (FACE) experiments (see Chapter 3 for a description of FACE and how this method compares with others).

On balance, across all of the various FACE experiments around the world (although these are mostly in the temperate zone; Fig. 3.8), plant production has increased under elevated $CO_2$ (usually about 550 ppm; Ainsworth and Long, 2005). Not all plants respond equally: trees have shown the greatest responses (on average, biomass production increases of 28%) while, at the other end of the spectrum, $C_4$ plants (mostly grasses) do not seem to respond very much to elevated $CO_2$. $C_3$ grasses on average increased their biomass production by only 10% in FACE experiments, so we might predict that forests and grasslands will respond very differently to elevated $CO_2$. In other words, there will be no one-size-fits-all response; the response of a given ecosystem to elevated $CO_2$ will depend on the type of vegetation that defines it (see Chapter 4). The type of plants present will not be the only factor regulating ecosystem response to elevated $CO_2$, however: limiting resources will also come into play, as will other stressors such as pollutants (see Chapter 13).

Reich et al. (2006) demonstrated experimentally that nitrogen limitation can constrain the positive response of NPP to elevated $CO_2$, however it took 6 years of treatment (in a temperate grassland) to see this response: for the first 3 years, $CO_2$ stimulated growth equally in both ambient and elevated nitrogen treatments. After 3 years, growth was stimulated more in the elevated nitrogen plots, and progressively less over time in the ambient nitrogen plots. Similarly, a young temperate forest exposed to elevated $CO_2$ showed a positive growth response for 3 years and no response for several years thereafter (Oren et al., 2001; Fig. 7.6a). However, when nitrogen fertilizer was added, tree biomass production in the elevated $CO_2$ treatment once again surpassed the controls (Fig. 7.6b). These examples illustrate how nutrient limitation moderates biomass production increases in any nutrient-limited ecosystem.

While the FACE experiments have provided excellent insights into ecosystem responses to elevated $CO_2$, these experiments have been conducted in a limited number of ecosystem types. Tropical and boreal forests, for example, which each represent large carbon reservoirs, have been ignored so far (Hickler et al., 2008). Also, while several FACE experiments have taken place in temperate forests, young stands have been the focus.

Only a small handful of studies have addressed how mature forests may respond to elevated $CO_2$. These ecosystems have accrued large amounts of carbon, but their rates of NPP are low compared with young, rapidly growing forest. Will mature forests respond to elevated $CO_2$ with increasing NPP? Körner et al. (2005) investigated the response of mature deciduous forest in Europe to elevated $CO_2$ and found no evidence of increased biomass production in a 530 ppm $CO_2$ environment. Similarly, neither elevated $CO_2$ nor warming temperatures, nor the two combined, had any effect on tree stem growth over 4 years in a mature forest in Norway (Rasmussen et al., 2002). So, while there is evidence from FACE sites, all in young forests, that elevated $CO_2$ will increase NPP and biomass production, it is not necessarily the case that the same will be true for mature forests.

As we have emphasized elsewhere in this book, rising temperatures and $CO_2$ levels, as well as other climate change factors such as changing precipitation regimes, are happening simultaneously, not separately – what will be the interactive effects of these factors on NPP? As we learned earlier, while the effects of elevated $CO_2$ on NPP are generally positive, warming can have positive or negative effects depending on the initial conditions in a particular ecosystem. Warming can increase nutrient availability in cold-climate, nutrient-limited ecosystems by speeding up temperature-limited nutrient cycling rates. Warming can also exacerbate water limitations by drying out soils in moisture-limited sites. On the other hand, elevated $CO_2$ levels can ameliorate water stress to a degree and, where nitrogen is limiting, can also allow plants to allocate more carbon to fine roots to aid in nitrogen acquisition (Thornley and Cannell, 1996). So, the ultimate outcome of the two factors in combination depends on the water and nutrient dynamics of a given ecosystem (Thornley and Cannell, 1996). These will determine if the warming and elevated $CO_2$ are acting together to increase NPP, or if warming is counteracting the positive effects of $CO_2$.

Modelling studies tend to show that $CO_2$ and warming enhance each other's effects on ecosystems; greater positive effects are often seen when the two are in combination. For example, Su et al. (2007) modelled climate and elevated $CO_2$ effects on spruce forest in north-western China and found

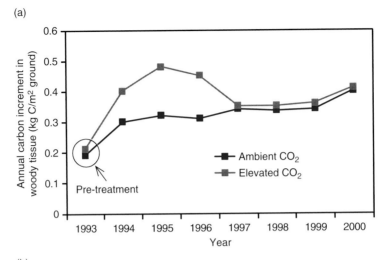

(a)

(b)

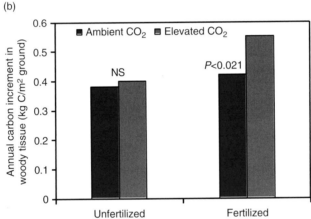

**Fig. 7.6.** The response of tree biomass accumulation (in wood) to elevated $CO_2$ over a 7-year time span (a) and to elevated $CO_2$ plus nitrogen fertilizer (data from 1999 and 2000, after the initial $CO_2$ effect had 'worn off') (b) at a temperate forest FACE site. (Adapted from Oren *et al.*, 2001.)

that when just elevated $CO_2$ was modelled, NPP increased only by a small amount due to moisture and nitrogen limitations. When both warming and $CO_2$ changes were taken into account, NPP increased more dramatically, by 25% or more, because available moisture increased and warming stimulated nitrogen cycling and availability. In another study, NPP in most vegetation types in Alaska increased under various climate change scenarios, using various GCMs (Euskirchen *et al.*, 2009; see Chapter 2). Increases were related to extended growing season length, increased nitrogen uptake and greater light capture due to increased leaf area.

On balance, then, we can likely expect (and are already seeing) positive NPP responses to climate and atmospheric change, at least over the next few decades. However, does increased NPP in response to warmer temperatures and elevated $CO_2$ mean that ecosystems will sequester more carbon? To answer that question, we first have to consider that increases in NPP do not necessarily mean that more carbon is being sequestered by plants if that NPP is allocated to pools with fast turnover rates, like root exudates and fine roots, which are quickly lost from the plant and at the mercy of soil organisms. Even if it turns out that increases in NPP do result in increasing biomass in tissues that remain on the plant for longer periods, we need to consider the other end of the carbon cycle: decomposition and heterotrophic respiration. If heterotrophic respiration keeps pace with NPP, there will be no net carbon sequestration in the ecosystem.

## 7.4 Impacts on Heterotrophic Respiration

The bulk of heterotrophic respiration in ecosystems comes from the decomposers – mainly soil microorganisms consuming dead plant material. So, most of the interest in the effects of climatic change on heterotrophic respiration is focused on organic matter decomposition and the resultant release of $CO_2$ from the soil. Globally, soils represent a huge reservoir of carbon – as much as 2400 Pg C to 2 m depth (Jobbagy and Jackson, 2000). In some ecosystems organic matter can be found many tens of metres below the surface, so this estimate of global soil carbon is actually low. Compare this amount with the carbon currently in the atmosphere (about 750 Pg, and climbing) and it is obvious why there is great interest in the fate of the carbon currently stored in soil. Even a small increase in the net rate of transfer of carbon from soil to the atmosphere could result in a significant increase to the atmospheric pool.

Within any particular ecosystem, carbon is input to soil mainly via plant litterfall (senesced leaves and other plant tissues, including belowground 'litterfall', mainly sloughed fine roots) and exits the soil pool as the litter is decomposed and the carbon is released back to the atmosphere as $CO_2$ (some carbon may also leave an ecosystem by leaching down through the soil into groundwater). Organic carbon builds up in soil if decomposition processes do not keep pace with litter inputs. With climatic change, the soil carbon pool could increase or decrease depending on how the balance between these inputs and outputs is affected. We have already examined the potential effects of climatic change on NPP and, of course, litter production rates are determined by plant production rates. So, in many ecosystems, climatic change will likely result in increased carbon inputs to soils because of increased NPP. But what about carbon losses from soils via heterotrophic respiration?

We know that the main factors regulating decomposition rates (and thus heterotrophic respiration rates) are temperature, moisture and the composition of the microbial community in the soil.

Substrate quantity can also be important – in some ecosystems, the decomposer community is carbon-limited and will respond positively to increased inputs of carbon-rich plant litter. Climatic change has the potential to alter any or all of these regulating factors. For example, warming may increase the activity of the decomposer community or could lead to changes in community composition. Plant litter quality may change due to elevated $CO_2$ or due to shifts in plant community composition promoted by warming or changes in moisture levels (for a comprehensive review on the effects of temperature increases, in particular, on decomposition processes, see Aerts, 2006).

As we explore the results of studies of climate change impacts on heterotrophic respiration, it is important to keep a couple of points in mind. First, heterotrophic respiration is not usually (for a variety of reasons) measured directly in field studies; and second, the heterotrophs that are usually focused on in ecosystem ecology studies are the decomposers. Consequently, we see that the research results discussed below encompass a variety of approaches for estimating the effects of climatic change on the activity of soil heterotrophs, while ignoring other heterotrophic organisms (remember that in most ecosystems, more carbon and energy flows through the detrital food web than through aboveground food webs). For example, some researchers have measured $CO_2$ efflux from soil and made assumptions about the relative contributions of heterotrophic versus root respiration (this has been estimated to be about half each; e.g. Bond-Lamberty et al., 2004), while others have measured the real-time decomposition rate of plant litter, and still others have measured the change in soil carbon content after the application of an experimental treatment.

It is no secret that decomposition rates are positively related to temperature, because microbial activity and hence respiration rates are higher in warm environments than cold ones, and respiration rates generally fluctuate seasonally with temperature. Countless laboratory soil incubations have demonstrated positive relationships between microbial respiration and temperature. In the field, however, warming can lead to drying, the effects of which depend on how wet the environment was in the first place. Drying can slow microbial activity if moisture becomes limited, but drying can also stimulate decomposition in formerly waterlogged environments by increasing oxygen availability.

Ecosystem-level warming experiments have shed some light on how heterotrophic respiration may respond to climatic change. A meta-analysis of 32 such experiments (all located in temperate, boreal and Arctic ecosystems) found that total soil respiration rates increased by 20% across all sites (Rustad et al., 2001). Some of this increase was no doubt attributable to root respiration, but increases in nitrogen mineralization rates indicated that decomposition rates (and thus heterotrophic respiration rates) were also increasing. A more recent tundra warming experiment (Biasi et al., 2008) similarly relied on indirect evidence; this study also found increases in soil $CO_2$ efflux and, since soil organic carbon content was significantly reduced over the warming period, the researchers inferred that increases in heterotrophic respiration must be at least partially responsible for this result.

Other warming experiments have demonstrated that the response to warming will not necessarily be straightforward. In a boreal black spruce (Picea mariana) forest, Bronson et al. (2008) experimentally increased soil temperature and found that soil respiration rose by 25% in the first year; but in the second year there was only an 11% increase over the control. Because the warming treatments reduced fine root biomass, the authors attributed the increases in respiration to decomposition and attributed the decline in response to depletion of the labile organic carbon pool in the soil (i.e. decomposition outpaced plant litter inputs and the microbial community 'ran out' of easily decomposed fresh litter). Others have also observed a decline in the warming response over time. For example, in temperate forest plots subjected to experimental heating, short-term increases in respiration rates were not maintained over the 15-year study period (Bradford et al., 2008). This was attributed to the depletion of soil carbon pools, a reduction in microbial biomass and a change in microbial community composition (to a community with less respiration per unit mass). In other cases, microbial activity may be reduced by warming, even in northern ecosystems, if temperature increases also lead to soil drying (e.g. Allison and Treseder, 2008).

Although alterations of plant litter quality could also have an important impact on decomposition rates and the release of $CO_2$ from soil, there is not much experimental evidence that plant litter quality is directly reduced by warming in the short term. Instead, it is likely that warming-induced shifts in plant community composition, even just in the relative abundance of existing species, will have a larger effect on litter quality than will changes within the existing plant species (leaving aside, for the moment, any $CO_2$-related changes to plant quality). Large differences among plant species in the decomposition rates of litter have been widely observed and, in warming experiments, plant community changes have been seen in as few as 5–10 years (Aerts, 2006). For example, as described in Chapter 6, after 9 years of experimental warming in a tundra ecosystem, Chapin et al. (1995) found that deciduous shrub abundance increased at the expense of mosses, lichens and evergreen shrubs. This shift in the vegetation community was attributed to the increased nitrogen availability that had resulted from a warming-induced rise in decomposition rates, and the increased dominance by the deciduous shrub was in turn causing decomposition rates to speed up further – an example of a positive feedback loop. However, the ultimate effects of plant community changes on decomposition processes will depend on ecosystem-specific conditions. In a montane meadow ecosystem heated experimentally for 7 years, soil moisture levels dropped due to increased temperatures, promoting the increased dominance of shrub vegetation over the original forbs and grasses (Saleska et al., 1999). Initially, soil carbon content dropped, as decomposition outpaced litter inputs from the unproductive shrubs. However, the researchers predicted that in the long run, the low-quality shrub litter would lead to slower decomposition rates and carbon build-up in the soil (a negative feedback loop), in contrast to the findings of Chapin et al. (1995) in the tundra.

When it comes to climatic change-induced carbon losses from soils, all eyes are on northern ecosystems, where the equivalent of a perfect storm may be brewing. First, vast organic carbon stores have built up in cold or frozen soils because of temperature limitations on decomposition. Second, this material is thought to be highly sensitive to warming, so dramatic increases in decomposition rates may be seen with even a small amount of warming. Third, it is at these high latitudes that the greatest magnitude of warming is expected to occur.

Permafrost soils in the northern hemisphere (Fig. 7.7) have been estimated to contain half or more of total global soil carbon stocks, and possibly hold twice as much carbon as is presently in the atmosphere (Schuur et al., 2008). As permafrost thaws due to warmer air temperatures, sequestered

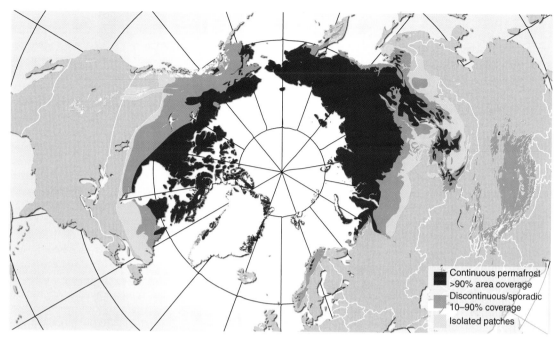

**Fig. 7.7.** Extent of permafrost in the northern hemisphere. Carbon locked up in frozen peat and mineral soil is vulnerable to release as the climate warms. (From http://tinyurl.com/ch7-permafrost (accessed 25 January 2010). Image by: Hugo Ahlenius, UNEP/GRID-Arendal.)

organic matter is made available to soil micro-organisms. High-latitude soils that are not frozen year round can also contain vast amounts of carbon, such as in peatlands, where organic matter accumulates due to a combination of cold temperatures and the anoxia that accompanies waterlogging. In northern peatlands, warming-induced water-level reductions can interact with direct impacts of warming and result in faster peat decomposition. For example, an 8-year whole-ecosystem warming experiment in subarctic peatlands showed that a temperature increase of 1°C during the growing season could increase peat decomposition rates by more than 50% (Dorrepaal *et al.*, 2009). However, while the reduction of waterlogging can be favourable for decomposition rates, too much of a decrease in water level can result in the drying-out of surface peat layers, which inhibits decomposition.

If peatlands remain waterlogged, or if thawed permafrost soil becomes waterlogged due to lack of drainage, anaerobic decomposition processes will result in the release of methane instead of $CO_2$ (for example, thawing permafrost can result in thermokarst – the formation of depressions as the ground subsides; if a depression fills with water the soil underneath is waterlogged, leading to methane production). Anaerobic decomposition is slower than aerobic decomposition, but the resulting methane is a more potent greenhouse gas than $CO_2$.

While it is generally expected that, given adequate moisture, warmer temperatures will increase rates of organic matter decomposition in many ecosystem types, there are still some uncertainties surrounding the sensitivity of decomposition processes to temperature (e.g. Kirschbaum, 2006). However, in cold-climate ecosystems, decomposition is certainly limited by temperature and will speed up with warming in most cases (possible exceptions: waterlogged systems where lack of oxygen may limit decomposition rates despite increases in temperature and systems where dry conditions are exacerbated by warming).

It is important to keep in mind that soil carbon is not a homogeneous entity. Fresh plant litter represents a pool of carbon on the soil surface that is relatively labile (although not all plant litter is equally labile, for example woody material and even conifer needles can be fairly recalcitrant) and can break down relatively quickly depending on environmental conditions. It does not break down completely, however. As litter decomposes, some of the partially decomposed organic matter, along with by-products of decomposition, collects as humus – this substance is full of organic compounds that resist being decomposed (i.e. they are 'recalcitrant'). Humus can become incorporated into the mineral soil (rather than sitting on top, like fresh litter) through the actions of soil organisms like earthworms, and organic carbon compounds can also be carried deeper into the soil profile by water percolating down through the soil, where they may remain for long periods of time. So, soil carbon varies in age from fresh litter (days to months old) to partially decomposed organic matter that may be years to millennia in age. Often a distinction is made between 'fast' and 'slow' pools of carbon in the soil (although of course there is a gradient of ages within any soil, as opposed to distinct pools). It is possible that climatic change may affect the fast and slow carbon pools in different ways; for example, it has been postulated that warming may speed up the decomposition of fresh litter without affecting the release of carbon from older, more recalcitrant soil organic matter. There is some evidence, however, that the decomposition of older soil organic matter is actually more sensitive to temperature than that of fresh, labile materials (e.g. Conant *et al.*, 2008). This means that the carbon stored for centuries or millennia in some soils (which can equal large amounts compared with the fresh litter layer) may be vulnerable to release as global temperatures rise. The jury is still out on whether and how the different soil carbon pools will respond differentially to warming.

We have explored how increasing temperatures may affect heterotrophic respiration, but how might elevated $CO_2$ and interactions among various climate change factors affect this ecosystem process? Elevated $CO_2$ might conceivably affect heterotrophic respiration through changes in plant productivity and litter production, changes in plant litter quality and changes in soil moisture levels. There has been much less research in this area than in the area of temperature effects, and the available evidence suggests a variety of responses. A review of results from both pot-level and field studies found that elevated $CO_2$ resulted in lower decomposition rates due to reductions in plant litter quality (in particular, higher C:N and lignin:N ratios; Ball, 1997). On the other hand, in a temperate forest FACE experiment, 9 years of elevated $CO_2$ had

no effect on plant litter quality or on the decomposition rates of fresh plant litter (Lichter *et al.*, 2008). The FACE researchers did, however, find evidence of increased decomposition rates of organic matter below the surface in the mineral soil, suggested by increased nitrogen release rates from this material under the elevated $CO_2$ treatment compared with ambient. Further evidence that elevated $CO_2$ may increase decomposition rates comes from a review of 47 studies suggesting that the increased plant production brought on by elevated $CO_2$ would stimulate microbial activity through increased litter inputs (i.e. the microbes' food source becomes more abundant; Zak *et al.*, 2000b). We examine these belowground changes in more detail in Chapter 9.

Because ecosystem-level field studies rarely address combinations of climate change factors, we must, so far, rely on computer modelling studies to gain insight into how temperature, precipitation and $CO_2$ changes may interact to influence ecosystem functioning. Luo *et al.* (2008) modelled the interactive effects of temperature, moisture and atmospheric $CO_2$ changes in a variety of ecosystem types and climate zones, and found that heterotrophic respiration rates increased under warming alone, increased precipitation alone and elevated $CO_2$ alone. Greater increases were seen, however, under combinations of warming and elevated $CO_2$, warming and increased precipitation, elevated $CO_2$ and increased precipitation, and all three combined. Drier conditions, due either to reduced total precipitation or summer drought, resulted in lower rates of heterotrophic respiration. The wettest (though not waterlogged) sites used in the study showed the smallest responses: microbial activity was already high in these systems.

## 7.5  Impacts on Net Ecosystem Productivity

Now that we have examined the inputs of carbon to ecosystems (NPP) and the outputs from ecosystems (heterotrophic respiration), we have arrived at the centre-stage question regarding ecosystem interactions with the atmosphere in a changing climate: what will be the result of the new balance of inputs and outputs in any given ecosystem? Will climatic change cause ecosystems to sequester more carbon from the atmosphere (inputs > outputs) or will they become net sources of carbon (inputs < outputs)?

The amount of carbon retained, or *sequestered*, in an ecosystem over a period of time (usually considered on an annual basis) is its NEP. NEP is strictly defined as the net result of all fluxes of carbon in and out of an ecosystem, and thus includes not only NPP and $R_h$ (or GPP, $R_a$ and $R_h$ if you prefer to think of it that way), but also other flux pathways such as those due to leaching, fire and erosion (Randerson *et al.*, 2002). However, it is not uncommon for NEP estimates to be based solely on photosynthetic inputs and respiratory outputs, often estimated indirectly as the changes over time in plant biomass and soil organic matter or soil carbon. A similar concept is net ecosystem exchange (NEE), which refers to $CO_2$ fluxes in and out of ecosystems, and can be measured directly with a technique called 'eddy covariance' (see Box 7.1). NEP, which can also be thought of as the carbon balance of an ecosystem, is thus regulated by those factors that affect inputs and outputs of carbon to an ecosystem and the balance between those inputs and outputs.

We have seen that warmer temperatures can affect both NPP and heterotrophic respiration, often increasing both; if these carbon fluxes both increase, they cancel each other out to some degree. So, warming could conceivably have drastic effects on both productivity and respiration, but little overall effect on NEP. Hobbie and Chapin (1998) found just this in a tundra warming experiment: NEP changed little in the face of warming, even though carbon inputs and outputs both increased significantly with the warming treatment (Fig. 7.9).

Ultimately, climate change effects on carbon balance depend on the relative sensitivity of input and output processes, and on their respective magnitudes. For example, in a given ecosystem, $R_h$ may be more sensitive to warming than is NPP – let us say that warming causes $R_h$ to increase by 200% while NPP only increases by 40%. This would seem to suggest that warming will turn the ecosystem in question into a carbon source, with a negative NEP; but if the magnitude of the NPP inputs are several times higher than that of the $R_h$ outputs in the first place, the $R_h$ can increase significantly without approaching the magnitude of the inputs and the NEP will remain positive though reduced. This is just a hypothetical example. We shall see in some ecosystems that the magnitudes of inputs and outputs of carbon are similar and small changes to one or the other can result in changes to NEP, in some cases even tipping the balance from carbon sink to source.

Ecosystem warming experiments have been carried out in several different types of ecosystem, but most of these studies did not estimate total carbon balance. Tundra warming experiments seem to be the exception, perhaps because the vegetation is of a manageable stature, and because the question of ecosystem carbon balance is especially critical here due to the large amounts of stored carbon in tundra soils and the large temperature increases expected at high latitudes. In a review of a variety of tundra ecosystems subjected to experimental warming, the carbon balance was found to be driven by the response of respiration (total ecosystem respiration, or $R_a + R_h$; Oberbauer *et al.*, 2007). Although the responses varied at different tundra sites, some generalizations could be made. In very wet sites, the increase in respiration was muted due to excessive moisture and increases in GPP meant that NEP increased with warming. In dry sites, NEP declined with warming because the respiration increases were so large (Biasi *et al.*, 2008). Whether or not the increased NEP seen in wet tundra sites would persist under long-term warming remains to be seen; if warming is accompanied by drying in these ecosystems, the boost to NEP may not last long.

While warming experiments are very useful, providing empirical evidence for ecosystem impacts, modelling has the advantage of being able to gauge ecosystem responses to warming in combination with future $CO_2$ levels. While some studies have shown increases in NEP due to warming alone (e.g. Yarie and Billings, 2002), more often it seems that rising $CO_2$ will come to the rescue (at least for as long as the $CO_2$ fertilization effects last), raising NEP in combination with warming, while warming alone leads to declines in NEP. A modelling study of the separate and interactive effects of changing temperature, precipitation and atmospheric $CO_2$ on a suite of ecosystem types found that while rising temperature alone would decrease NEP in a range of ecosystems, elevated $CO_2$ combined with warming increased NEP (Luo *et al.*, 2008; Fig. 7.10). Even in dry environments, NEP can increase with warming if combined with rising $CO_2$ – a modelling study of a semi-arid grassland found that NEP increased by a small amount because of increases in NPP that outweighed increases in respiration (Li *et al.*, 2004). Any warming-induced drying was counteracted by the decreases in plant water losses (via transpiration) caused by the elevated $CO_2$ (see Chapter 4).

**Box 7.1. Sink or source? Assessing ecosystem carbon balance.**

One way to determine an ecosystem's carbon balance is to quantify the various pools of carbon repeatedly over time, to see if they are growing or shrinking (e.g. plant biomass can be measured annually to see if it is increasing thereby accumulating carbon, and soil carbon can be measured periodically to look for increases or decreases in this pool; the net result of changes in all of the pools tells us if the ecosystem is losing or gaining carbon over time). These 'inventory' methods are limited by difficulties in capturing pools in their entirety (e.g. belowground plant biomass is notoriously difficult to measure) and by difficulties in dealing with high spatial variability, especially in soil conditions (Baldocchi, 2003).

The eddy covariance method represents a more direct way of assessing ecosystem carbon balance by measuring the continuous movement of $CO_2$ between land and atmosphere on an ecosystem scale (i.e. net ecosystem exchange, NEE). Photosynthesis results in $CO_2$ transfer from atmosphere to vegetation, while respiration (of both autotrophs and heterotrophs) results in $CO_2$ being released to the atmosphere. The net result of this exchange dictates whether an ecosystem is a net source or sink of carbon (recall from Section 7.1 that $NEP = GPP - (R_a + R_h)$, keeping in mind that there are other potential terms in this equation, as carbon can also be lost via disturbance, leaching, etc.). As air flows over a vegetation canopy it becomes turbulent, resulting in the formation of down-ward and upward eddies. The eddy covariance technique involves measurements taken from 'flux towers'– towers used to position micrometeorological equipment just above the vegetation canopy. In a nutshell, analytical equipment measures the speed of the upward- and downward-moving air plus the concentration of $CO_2$ in these eddies and, via some fairly complicated calculations, estimates can be made of the absolute amount of $CO_2$ moving into and out of the ecosystem. These estimates can be integrated over periods from hours to years, to determine the sink-or-source status of an ecosystem on the time scale of interest and to explore how carbon balance responds to environmental variability (e.g. seasonal changes, directional climatic change, etc.). For example, Dunn *et al.* (2007) used eddy covariance to track $CO_2$ exchange in a 160-year-old black spruce forest in Manitoba, Canada, over the course of 11 years of flux measurements (Fig. 7.8a). They then investigated the relationships between NEE and climate variables, discovering that the ecosytem was a net carbon sink in warmer, wetter years and a carbon source in cooler, drier years (Fig. 7.8b).

There are several different research networks worldwide using flux towers and eddy covariance to assess carbon balance in a variety of ecosystem types (e.g. Fluxnet, Fluxnet Canada, Ameriflux, CarboEurope, etc.). For further reading on this topic, see Baldocchi (2003).

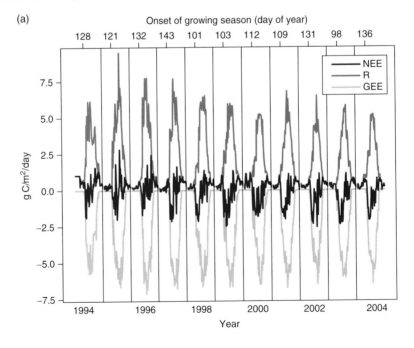

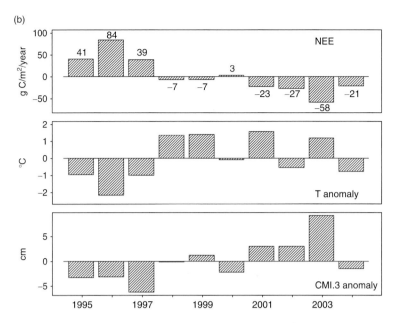

**Fig. 7.8.** (a) Carbon balance (NEE) over 11 years in a boreal black spruce forest, estimated as the net result of photosynthetic $CO_2$ uptake from the atmosphere (denoted here as GEE, or gross ecosystem exchange, equivalent to GPP) and respiration (R) losses to the atmosphere. NEE values <0 indicate net uptake from the atmosphere. (b) Comparisons of annual NEE with yearly anomalies in temperature (T) and moisture (CMI). (From Dunn *et al.*, 2007.)

Lagergren *et al.* (2006) modelled NEP in Swedish forests under a 4°C increase in annual temperature and a doubling of $CO_2$ and found that NEP increased in forests of all age classes (by up to 100% in the coldest, northern forests). Both GPP and total ecosystem respiration ($R_a$ + $R_h$) increased, but GPP increased more (by 43 to 47%, compared with increases in ecosystem respiration of 37 to 41%), leading to the increase in NEP (see also Davi *et al.*, 2006). Otherwise, warming resulted in a decline in NEP in most forest types, with coniferous forest NEP declining by half. These patterns were mainly due to the effects on GPP. With elevated $CO_2$, GPP increased with warming in all forest types over the whole time period. Without elevated $CO_2$, GPP declined in most forest types due to water stress brought on by increased summer temperatures. The increase in GPP in both of these studies was attributed mainly to an extension of the growing season and

to a $CO_2$ fertilization effect, which was modelled as affecting only photosynthesis and not respiration, thereby having a purely positive effect on NEP. Similarly, a modelling study by Harrison *et al.* (2008) found that while changes in climate since the mid-20th century have reduced the ability of terrestrial European ecosystems to sequester carbon, changes in atmospheric $CO_2$ have done the opposite, with the net result being greater NEP. The positive effects of the combination of elevated $CO_2$ and warming on NEP have been detected repeatedly in modelling studies of a variety of ecosystems all over the world. The positive effects of elevated $CO_2$ on NEP have been borne out in a temperate forest FACE experiment, where 4 years of experimentally elevated $CO_2$ resulted in increased NEP. Interestingly, $R_h$ also increased in this experiment, but not enough to offset the increases in tree productivity (Hamilton *et al.*, 2002).

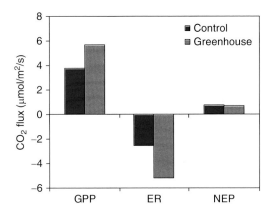

**Fig. 7.9.** The response of tundra ecosystem $CO_2$ fluxes to experimental warming (greenhouse treatment). GPP = gross primary production; ER = ecosystem respiration ($R_a$ + $R_h$); NEP = net ecosystem production. (Redrawn from Hobbie and Chapin, 1998.)

Most global-level models of climate warming predict a net release of $CO_2$ from terrestrial ecosystems due to reductions in NPP and increases in carbon release via decomposition (Luo, 2007). However, when elevated $CO_2$ is incorporated, models predict that most terrestrial ecosystems will likely become stronger carbon sinks because of increased NPP while the $CO_2$ fertilization effect lasts (Cao and Woodward, 1998), but that carbon losses will overpower gains in many ecosystems by the end of the 21st century (Cramer *et al.*, 2001; Schaphoff *et al.*, 2006). After reviewing the available evidence from global-level modelling studies, the IPCC's Fourth Assessment Report (see Chapter 2; Fischlin *et al.*, 2007) concluded that the world's terrestrial biosphere will likely become a net source of $CO_2$ by the end of this century, meaning that for all of the world's terrestrial ecosystems collectively, the loss of carbon from ecosystems via respiration will outweigh the intake of carbon through photosynthesis (Fig. 7.11).

As we have seen, the responses of ecosystem carbon sequestration to climatic change depend on more than just direct effects of temperature on photosynthesis and respiration. Other, indirect, effects such as growing season length, species shifts, nutrient cycling and water availability, also come into play (Luo, 2007). The variability of experimental results stresses the need for more empirical research in this area if we are to understand climate change impacts on ecosystem carbon balance and the resulting feedbacks to the atmosphere. It is important to always keep in mind that climate change-related factors cannot be considered in isolation (even though experiments and modelling studies have often done so due to logistical constraints and/or to gain insight into the relative importance of the main effects of the treatments); it will be the interactions among factors (changes in temperature, $CO_2$, precipitation, etc.) that ultimately dictate how climatic change will impact ecosystems.

Climatic change will also affect ecosystems through changes to natural disturbance regimes; for example, the frequency and intensity of fire in forests, grasslands or tundra ecosystems. Fire, and other disturbance agents such as wind and severe storms, can have strong impacts on ecosystem carbon sequestration, potentially dwarfing any direct impact of increases in temperature and $CO_2$ (e.g. Bond-Lamberty *et al.*, 2007). Thus, to fully understand climate change impacts on regional and global carbon balances, we need to consider how disturbances play out over the landscape (see Chapter 10 for a discussion of climate change–disturbance links in forests and Chapter 13 for a discussion of general interactions with disturbance).

## 7.6 Impacts on Nutrient Cycling

Ecosystem nutrient cycling and carbon cycling are tightly intertwined. Nitrogen is the limiting nutrient in many ecosystems (LeBauer and Treseder, 2008) and as such is often the focus of nutrient cycling studies. Plants take up nutrients from the soil to support various physiological functions and these nutrients become incorporated into organic compounds (nitrogen, for example, is found in proteins, enzymes and photosynthetic pigments). When dead plant tissues are shed from the plant, the decomposers break down this organic material into its mineral components, gaining energy for themselves in the process. $CO_2$ is released alongside nutrients in their inorganic form (e.g. ammonium nitrogen); these inorganic nutrients are once again in a form suitable for plant uptake, although some will be used by the microbes themselves (taken up into the soil's microbial biomass, or 'immobilized'). Limiting nutrients such as nitrogen are cycled tightly between plants and soil. In some ecosystems, nutrients such as nitrogen are in such short supply, and the competition between plants and microbes so fierce, that plants (and their mycorrhizae) will take up small, nutrient-rich organic

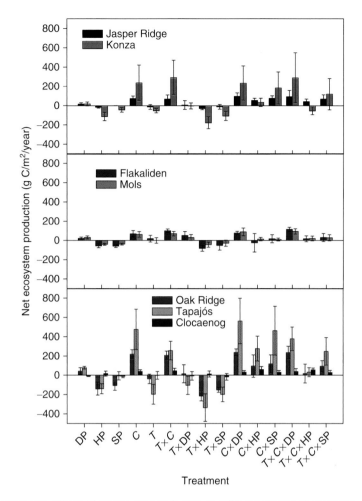

**Fig. 7.10.** Modelled interactive effects of climate change factors on NEP in a variety of ecosystem types. Flakaliden = conifer forest, Sweden; Mols = heathland, Denmark; Jasper Ridge = grassland, California; Konza = tallgrass prairie, Kansas; Oak Ridge = temperate forest, Tennessee; Tapajós = tropical forest, Brazil; Clocaenog = wet heathland, Wales. DP = doubled precipitation; HP = halved precipitation; SP = summer drought; C = elevated $CO_2$; T = elevated temperature. (Redrawn from Luo *et al.*, 2008.)

molecules such as amino acids rather than 'wait' for these to be further broken down into inorganic form. Because nutrients can be limiting to plant growth and microbial activity, the effects of climatic change (and atmospheric change) on ecosystem processes such as carbon sequestration will hinge, in part, on how nutrient cycles are affected and on nutrient availability in general.

There is considerably less information available on climate change impacts on nutrient cycling compared with carbon cycling. Most models, even those that couple climate and atmospheric change to vegetation dynamics, do not yet include detailed

representations of plant–soil interactions and nutrient cycles even though these factors are key to making accurate predictions about climate change impacts (Hungate *et al.*, 2003).

Interestingly, more ecosystem-level research has been done on the effects of elevated $CO_2$ on nutrient cycling (specifically nitrogen) than on the effects of climate warming. Based on a meta-analysis of nitrogen dynamics in the FACE experiments, De Graaff *et al.* (2006) reported that elevated $CO_2$ did not change soil nitrogen concentrations (total soil nitrogen, i.e. including nitrogen in both dead organic matter and microbial biomass) or rates of nitrogen

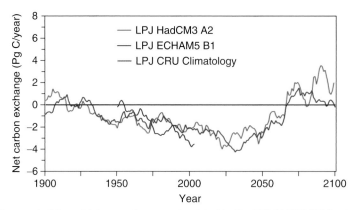

Net carbon exchange of all terrestrial ecosystems as simulated by the DGVM LPJ (Sitch *et al.*, 2003; Gerten *et al.*, 2004 – negative values mean a carbon sink, positive values carbon losses to the atmosphere). Past century data are based on observations and climate model data were normalised to be in accord with these observations for the 1961–1990 data (CRU-PIK). Transient future projections are for the SRES A2 and B1 emissions scenarios (Nakićenović *et al.*, 2000), forcing the climate models HadCM3 and ECHAM5, respectively (cf. Lucht *et al.*, 2006; Schaphoff *et al.*, 2006). In contrast to previous global projections (Prentice *et al.*, 2001 – Figure 3.10), the world's ecosystems sink service saturates earlier (about 2030) and the terrestrial biosphere tends to become a carbon source earlier (about 2070) and more consistently, corroborating other projections of increased forcing from biogenic terrestrial sources (e.g., Cox *et al.*, 2000, 2004; White *et al.*, 2000a; Lucht *et al.*, 2006; Schaphoff *et al.*, 2006; Scholze *et al.*, 2006; see Figure 4.3 for maps on underlying ecosystem changes). Note that these projections assume an effective CO2-fertilisation (see Section 4.4.1). (From: *Climate Change 2007: Impacts, Adaptation and Vulnerability. Working Group II Contribution to the Fourth Assessment Report of the Intergovernmental Panel on Climate Change*, Figure 4.2. Cambridge University Press.)

**Fig. 7.11.** Net carbon balance of all terrestrial ecosystems combined, projected to the end of the century by models that incorporate future emissions scenarios, climate forcing, vegetation dynamics and $CO_2$ fertilization. Net uptake is indicated by negative values; net release by positive values.

mineralization, but that rates of nitrogen immobilization (in microbial biomass) increased, indicating that microbial biomass growth was stimulated by elevated $CO_2$. This was likely because of the increase in the decomposers' 'food supply', as increased plant growth led not only to increased leaf litterfall, but also increased belowground 'litterfall' (fine root sloughage) and root exudates. Elevated $CO_2$ also increased the C:N ratio of soils, suggesting an eventual decrease in nitrogen availability, as organic material with high C:N is slow to decompose. This could represent a negative feedback on plant growth under elevated $CO_2$: positive plant growth response increases carbon input to soils, which stimulates microbial growth and nitrogen immobilization, reducing nutrient availability to plants and thus slowing the plant growth response.

A related sort of feedback loop is central to the idea of progressive nitrogen limitation (PNL), proposed by Luo *et al.* (2004) as a potential result of long-term exposure of ecosystems to elevated $CO_2$. If elevated $CO_2$ promotes increased NPP, then plant demand for nitrogen will go up to support increased growth and nitrogen may become increasingly locked up in living and dead organic matter, especially if decomposition rates do not keep pace with increased growth rates. The situation can become exacerbated if diminishing nitrogen availability causes plants to increase the C:N ratios of their tissues, effectively creating more biomass per unit of nitrogen, but reducing the decomposability of plant tissues, slowing nitrogen mineralization rates in the soil (Fig. 7.12).

Although initial levels of available nitrogen will constrain an ecosystem's plant growth response to elevated $CO_2$, PNL can occur regardless of initial nitrogen status – it is an issue of nitrogen *distribution* within the ecosystem, rather than the absolute amount of nitrogen present. Regardless of how nitrogen-rich an ecosystem may be to begin with,

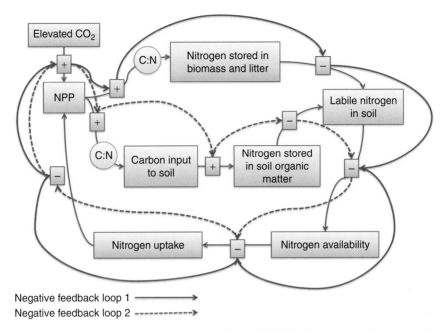

**Fig. 7.12.** Feedback processes involved with progressive nitrogen limitation in ecosystems in response to elevated atmospheric $CO_2$. C:N ratio is flexible and alters the rates of flow. (Adapted from Luo *et al.*, 2004.)

if increased plant growth causes more and more nitrogen to be locked up in plant biomass and soil organic matter, and no new nitrogen inputs occur, over time nitrogen can become progressively limiting to plant growth. So, the processes that allow for enhanced carbon sequestration in the beginning (increased plant growth, increased carbon inputs to soils) can in the end slow themselves down through their effects on nitrogen availability.

But how will elevated $CO_2$ interact with warming to affect nutrient cycling (nitrogen in particular)? There are no easy answers to this question, due to the complexity of the plant–soil interactions that define ecosystem nutrient cycling.

Based on the discussion of climate change impacts on decomposition in Section 7.2 we already have an idea of how nutrient cycling processes may respond to warming, because heterotrophic respiration and the release of $CO_2$ from organic matter goes hand in hand with nutrient mineralization. In other words, as long as moisture conditions are favourable, it is likely that rates of nutrient release from organic matter will increase in many ecosystems as the climate warms (particularly in colder-climate ecosystems where decomposition is currently temperature-limited). Indeed, a

meta-analysis of warming experiments across a variety of ecosystem types (from forest to grassland to tundra) found that net nitrogen mineralization rates (i.e. what's left over after microbial immobilization has been accounted for, so it represents what's available for plant uptake) increased on average by 46% (Rustad *et al.*, 2001). Increased net nitrogen mineralization could have positive effects on plant growth in nitrogen-limited ecosystems via increased nitrogen uptake. And, because mineral nitrogen availability can also constrain microbial activity in some ecosystems, enhanced gross mineralization may further accelerate decomposition rates (Mack *et al.*, 2004).

Climatic change-driven shifts in species composition can also impact nutrient cycling; for example, the encroachment into some tundra ecosystems of shrubby vegetation with nitrogen-rich litter can contribute to faster decomposition rates (Weintraub and Schimel, 2005). Hobbie and Chapin (1998) found that the shift to greater dominance by shrubby vegetation in Alaskan tussock tundra subjected to experimental warming was mediated by changes in the distribution of nitrogen among species: nitrogen uptake by one shrub species in particular increased at the expense of others.

In addition to nitrogen, climate change might also affect other nutrients. Chronic metabolic stress and biomass decline due to ongoing calcium loss are already evident in Canadian aquatic biota (Cairns and Yan, 2009). In some forest ecosystems, calcium and phosphorus may now be as important as nitrogen in limiting tree growth rates (Bigelow and Canham, 2007) and the interactions between these nutrients and climate change may limit growth in temperate and boreal forests, where nutrients are particularly limited (e.g. St Clair *et al.*, 2008). A recent study in fact has shown widespread decline in growth of several tree species in this region and can attribute these declines to drought in only some of the sites, in many other sites nutrient limitation has been suggested (Silva *et al.*, 2010).

## 7.7 Conclusions

In this chapter we have seen that there is the potential for ecosystems to respond in a variety of ways to climate and atmospheric change. Ultimately, the effects of these changes on ecosystem processes will be specific to each ecosystem and will depend on initial conditions in the ecosystem. In particular, the existing environmental factors that limit biomass production, decomposition and nutrient cycling will determine how these ecosystem processes respond to changing environmental conditions. For example, we can expect warming to increase decomposition rates in cold-climate ecosystems while these rates may be reduced in warm-climate ecosystems where drier conditions result from increasing temperatures. We have also learned that observed and modelled ecosystem responses to climatic change are often the result of indirect effects and interacting factors, rather than direct effects. Consider that increases in NPP with warming are less likely to be due to direct temperature effects on plant productivity than to effects on growing season length, plant community composition or nutrient availability (via increased decomposition rates). And recall that while plant productivity in many ecosystems will likely respond positively to the combination of rising temperature and elevated $CO_2$, the same cannot necessarily be said about the response to warming alone. As ecosystem processes are impacted by climatic change, so will be the services that these processes support, such as food production, climate regulation, etc. In some parts of the world these services may be enhanced, while in others they may decline, impacting the well-being of humans and other organisms.

# 8  Evolutionary Responses to Climatic Change

Throughout most of this book we have tacitly assumed that biological responses to climatic change will be of three sorts: (i) accommodation; (ii) migration; or (iii) extinction. We are not alone in doing so; the vast majority of climate change biology is focused on these three possible outcomes. There is, however, another possible response, in some senses the great-unknown response: evolutionary adaptation. In this chapter we examine what we do know about the potential evolutionary responses of species to climatic change.

Throughout this chapter we refer to 'climate' and 'environment' interchangeably. Clearly climate is only a part of the total biotic and abiotic conditions that comprise a population's environment, but as we are considering rapid and large climatic change we can think of the population as encountering a new environment. In this way it may be helpful to think of rapid and large climatic change as the species 'invading a new environment', where we substitute time for space. With climatic change, the population remains where it is in space and the environment changes around it. With biological invasions the climate remains constant and the population moves in space (see Chapter 12). In either case we may think of the population as encountering a largely new environment in an ecologically short period of time. Indeed, to push the analogy even further, in both biological invasions and rapid climatic change, both populations may well have gone through a severe population bottleneck, resulting in a sample of genetic variation that may be much less than that which existed before the change in time or space.

As we just mentioned, evolutionary responses to climatic change have received far less attention than most areas of biological impacts research (Bell and Collins, 2008). This is perhaps because predicting the course of evolution in detail is probably not possible, but also because there seems to be a tacit assumption that evolution is not fast enough to respond to climatic change (Gomulkiewicz and Holt, 1995). The lip service regularly paid to evolution often begins and ends with statements to the effect that: large populations with very short generation times may show adaptive responses to climatic change on the time scale at which it occurs. The implication is that viruses, bacteria, algae and perhaps some insects are likely to show adaptation, but most other species are not. Although we acknowledge that such adaptive responses are possible, they tend to be ignored in the literature. Our aim in this chapter is to sketch some of the work that has been done in this area, drawing at times from related work on adaptive responses to other rapid changes. We largely conclude that such research is a promising and interesting line of work, one that certainly needs a thorough accounting if we are to ever meaningfully predict the impacts of climatic change, but that to date has fallen far short of our needs.

## 8.1  Phenotypic Plasticity and Ecological Fitting

The ability of one genotype to express different phenotypes in different environments is referred to as 'phenotypic plasticity' (see e.g. Bell, 2008). The relationship between a character value (e.g. height) for a given genotype and a particular environmental gradient is sometimes called a 'reaction norm'. For example, when a species undergoes selection in an environment of a certain temperature, it does not mean that any adaptation is confined to that temperature (Bell, 2008). Phenotypic plasticity is what allows a species to invade a new habitat, or to tolerate significant changes in its current habitat, because plasticity allows individuals of a given genotype to maintain a positive fitness value in its new environment. Plasticity means that a population does not have to await a favourable mutation in order to persist or even to flourish as the climate changes (see Section 8.3 on 'adaptive rescue'). Furthermore, phenotypic plasticity itself is often under genetic control, and thus it is a trait under selection pressure (West-Eberhard, 1989; Reusch and Wood, 2007). So phenotypic plasticity may be thought of as the raw material on which natural selection *initially* acts as the climate changes (Stearns, 1989; Thompson, 1991; Jump and Peñuelas, 2005; see also Section 8.6).

Many climate-related traits have a strong genetic component yet they respond to changes in climate variables. For example, in the south-west Yukon in Canada, the average spring temperature has increased by about 2°C for the period 1975 to 2001. During that time, the mean lifetime parturition date for the Yukon red squirrel (*Tamiasciurus hudsonicus*) has advanced from 8 May for females born in 1989 to 20 April for females born in 1999. That is 18 days in 10 years, or about 6 days per generation. Of this variation, 3.7 days per generation was the phenotypically plastic response to the increased abundance of food (correlated with the temperature) and 0.8 days per generation was the genetic component of the response (Fig. 8.1). This indicates that the response of these squirrels to rapid climatic change was a function not only of significant phenotypic plasticity but also of a measurable change in gene frequency over just four generations in response to strong selection pressure (Réale *et al.*, 2003).

Phenotype plasticity allows organisms to fit into their environments, habitats and biological communities. Imagine that we observe a species throughout its known range and we are principally interested in its response to two climate variables, temperature and precipitation. We depict this situation in Fig. 8.2. Here the light blue rectangle denotes the combinations of temperature and precipitation in which we find the species persisting – that is, it has positive fitness in these geographical regions. This is the bioclimatic envelope that we discussed in Chapter 3. Now suppose that the species is actually able to obtain positive fitness outside the regions in which it is found, but it is not found in those regions, perhaps because of dispersal limitations (depicted in Fig. 8.2b;

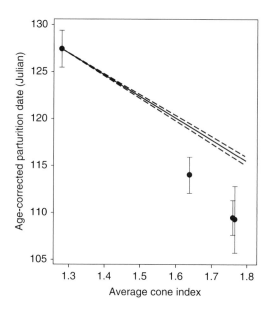

**Fig. 8.1.** Changes in the mean lifetime parturition date for the Yukon red squirrel (*Tamiasciurus hudsonicus*) across generations. The closed symbols represent the mean observed parturition date and the error bars the standard error of the mean, as a function of the average cone abundance (which itself is a function of temperature). The solid line shows the predicted within-generation change in parturition date, with the standard error shown by the dashed lines. This change represents the phenotypically plastic component of change in response to cone abundance. (Redrawn from Réale *et al.*, 2003.)

see also Chapter 5). Agosta and Klemens (2008) refer to this 'extra' space as the 'sloppy fitness space'. This sloppy fitness may arise through phenotypic plasticity, phylogenic conservatism (i.e. because of shared ancestry) or through correlated trait evolution (see below), although Agosta and Klemens argue that much of it will be due to plasticity and that it seems likely that all organisms possess some degree of potential fitness in combinations of climate variables different from those in which they evolved (see e.g. Jinks and Connolly, 1973). Now suppose that the climate rapidly becomes hotter and drier, leading to the combinations shown in the blue square in Fig. 8.2c, which are seemingly outside the species' bioclimatic envelope. The existence of the sloppy fitness space, in this example, means that the species can obtain positive fitness in this new environment, even though it did not apparently evolve or previously exist within this climate space.

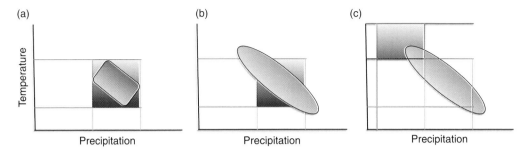

**Fig. 8.2.** Ecological fitting. (a) The observed range of two climate variables in the habitat in which this hypothetical species is found. (b) The area in this 'climate space' in which the species is able to obtain positive fitness values. The difference between the two areas of light blue is what Agosta and Klemens (2008) refer to as 'sloppy fitness space' and is part of the organism's phenotypic plasticity. (c) The case where the climate becomes warmer and drier, showing how sloppy fitness space allows this organism to fit into its new climate even though it evolved elsewhere. (Adapted from Agosta and Klemens, 2008.)

The scenario depicted in Fig. 8.2 has been termed 'ecological fitting' (Janzen, 1985); although used in a narrower sense the idea is the same as 'ecological sorting'. The idea comes into play when we consider why we find certain species (here characterized by genotypes) in certain environments. One answer to this question, which has been and still probably is the dominant view, is that species (genotypes) are found where they are because they have evolved (or perhaps co-evolved with other organisms) adaptations that make them well suited to that environment. Ecological fitting is an alternative view that a species (genotype) may be where it is because it is able to fit into that environment even though it evolved elsewhere and under different environmental conditions. The problem of course is that without knowledge of the species' evolutionary history, or other experiments, one cannot distinguish these two hypotheses as the end result looks the same (Agosta and Klemens, 2008). While this scenario seems rather hypothetical, remember that for most of the past 2.9 million years extensive regions at mid- and high latitudes were glaciated and the climate system was fundamentally different in terms of atmospheric circulation patterns (see Chapter 1). The specific combinations of temperature, precipitation and seasonal variability that characterize the modern climate may not have existed anywhere on Earth over much of species' evolutionary history.

We know that ecological fitting is a real phenomenon from the large number of successful biological invasions, particularly those that are anthropogenic in origin. In many of those cases, the invading species will have no shared evolutionary history with the invaded biological community, nor in many cases will it be moving within its current range of climatic conditions (Agosta and Klemens, 2008; see also Chapter 12).

Just as invading species are sometimes able to 'fit into' their new environment despite their previous lack of experience in that environment, some species will be able to fit into their new climate despite their previous lack of history with that climate. They will largely do this through phenotypic plasticity. After the environmental change sorts through the available phenotypic plasticity, it starts to exert a selection pressure. The selection pressure needs genetic variation on which to act, and we take up this issue in the next section.

## 8.2 Genetic Variation

For a quantitative trait (e.g. date of flowering) the variation in the value of that trait within a population is the result of the amount of genetic variation in the population, the environmental variation, and sometimes the interaction between genetic and environmental variation. Genetic variation comes in three types: (i) additive; (ii) dominance; and (iii) interactive. With additive genetic variance each locus contributes additively to the expression of the quantitative trait. Dominance genetic variation occurs when a dominant allele masks the effects of recessive alleles in the heterozygous form. Interaction genetic variance occurs when genes interact to influence the expression of the quantitative trait. The amount of additive genetic variance (rather than the total amount of genetic variance) present in a population for a given trait indicates the potential for natural selection to act upon that trait. And since natural selection acts upon additive genetic variance, if the trait contributes to the organism's fitness, we expect selection will rapidly act to reduce that variation (Bradshaw, 1991). Indeed, traits that have clear and direct impacts on fitness often have minimal additive genetic variation remaining in a population (see e.g. Hartl and Clark, 1997), although such variation may continue to persist across populations if those populations are differently locally adapted. So if we see substantial amounts of additive variance in a population, it must mean either that variation is currently selectively neutral or that balancing selection is operating to maintain a genetic polymorphism (Bradshaw, 1991).

When rapid climatic change occurs, a population finds itself in a new environment that will likely exert different selection pressures compared with the old environment. If these selection pressures are outside the species' reaction norm then that species may still persist in the new environment via evolutionary adaptation. For adaptation to occur there must be additive genetic variance. Such variance may exist within the population in the form of hidden or cryptic variation, it may arrive in the new environment via gene flow from other populations, or it may arise in the population via mutation. We examine these possibilities below.

Bradshaw (1991) described the apparent wealth of molecular variation in many populations as 'junk variation' because it does not seem to be under current selective pressure. While this variation may currently be selectively neutral, it might be beneficial in the future, as the environment changes. On the other hand, just because this variation exists does not mean that adaptation to a new climate will occur. It still may be the case that the additive genetic variation *needed to adapt to a new climate* has been exhausted by the selection pressures of the old environment. So the amount of genetic variation currently present in a population is not necessarily a good indication of a population's 'evolutionary potential' (Bradshaw, 1991). Nevertheless, there are certainly examples where such variation does exist within a population and that variation occurs in a trait clearly related to fitness in the current and perhaps future environment. Kelly *et al.* (2003) provide an excellent example of this in their study of the variation in temperature-related traits in silver birch trees (*Betula pendula*). They found, within a single population, a mix of genotypes that correlate with the average temperature in the year of seedling establishment. They found that there are basically two genotypes, those that establish in cool years and those that establish in warm years, and this mix is maintained through time, perhaps by inter-annual climate variation. In any case, the presence of this variation means that variation exists on which natural selection may act as the climate warms. Of course not all populations, and certainly not all species, will have this sort of variation currently present within them. And as Jump and Peñuelas (2005) point out, just because many populations do exhibit substantial genetic variation in climate-related traits, this does not mean that adaptation will occur or that it will occur at a rate necessary to keep pace with the changing climate.

As the climate changes, some populations will obtain the genetic variation needed to adapt from other populations where the necessary traits exist. For example, genes may regularly flow from southern populations into northern populations through, for example, seed dispersal. It may be the case that currently those 'southern genotypes' are at a selective disadvantage and so are rapidly removed from the northern populations. But when the climate changes, those southern genotypes may provide the variation necessary for selection to act and for local adaptation to occur.

If appropriate additive genetic variation does not already exist within a population, and it cannot arrive from other populations, then the only remaining possibility is that it arises via mutation. It now seems pretty clear that mutations occur at a basic underlying rate (Bradshaw, 1991). This basic mutation rate is quite slow, on the order of one in every 100 generations (Maynard-Smith, 1989). However, this figure is misleading in that it implies that populations with even modestly long generation times have no hope of 'adaptive rescue' (see Section 8.3). It is misleading because the mutations to which the rate refers tend to be selectively neutral, because deleterious mutations are rapidly lost from populations, and advantageous mutations are rapidly fixed within populations, in as few as ten to 20 generations (Bradshaw, 1991). There are plenty of examples of beneficial mutations arising over reasonably short time scales. This is particularly true in very large populations, both because there are more individuals in whom the beneficial mutation can occur and because if it does occur, it will tend to increase deterministically within the population rather than possibly being lost through genetic drift. We discuss this issue in more detail in the next section.

## 8.3  Adaptive Rescue

When the environment inhabited by a species changes more than can be accommodated by phenotypic plasticity, and dispersal is not possible, the species must adapt or go (locally) extinct. When is adaptive rescue possible? A fairly simplistic notion is that as long as the change is not 'too large', and that the existing amount of genetic variance is 'large enough', a population can adapt. In this section we consider these notions in more detail, and sketch out what theory can tell us about the potential for natural selection to restore local adaptiveness to a population before it goes extinct.

As we pointed out above, evolution over the course of a few dozens of generations is dominated by the process of sorting the variation already present in a population. On these time scales, mutation can usually be ignored, and migration (gene flow) is often ignored by assumption (although in reality it may be quite an important source of rescue; see Chapter 5). However, evolution over the course of hundreds or thousands of generations relies almost solely on new variation arising, regardless of how much variation was present to start with (Bell, 2008). Predicting the exact course of an evolutionary response to climatic change will usually be impossible since it will depend on which mutations arise and in which order they arise. Nevertheless, there are some general features of adaptation that are independent of the particular genes (Bell, 2008) and we consider some of these features in this section.

Most theoretical considerations of adaptive or evolutionary rescue begin by removing the possibility of simple sorting by assuming that the population is locally adapted and genetically fixed, so that adaptation must occur via mutation. One of the essential questions for adaptive rescue is whether or not a beneficial mutation will actually spread even if it arises. Naively this seems silly; of course such a mutation will spread and eventually become fixed. However, when a beneficial mutation first arises, it will be extremely rare in the population and can easily be lost through stochastic events (e.g. the individual carrying the mutation happens to die before it reproduces). Interestingly, it is not actually the frequency of the mutation in the population that matters here, but rather the number of individuals who carry the mutation. Once the number of individuals bearing the mutation is large enough (a few hundred), the chance of stochastic loss of the mutation becomes minimal (Bell, 2008).

In one of the simplest and most straightforward considerations of the problem, it can be shown that the probability of a single beneficial mutation becoming fixed in a population is approximately:

$$\text{Probability} = \frac{2s}{1 - e^{-2sN}} \qquad (8.1)$$

where $s$ is the selection coefficient and $N$ is the population size. For $N$ even modestly large, $e^{-2sN}$ rapidly goes to zero and Eqn 8.1 goes to $2s$ (see Bell, 2008: 74).

So the probability is approximately two times the selection coefficient. Except for quite large selection coefficients, the probability of a single mutation becoming fixed is fairly small indeed (Bell, 2008). The upshot of Eqn 8.1, and other treatments that lead to similar results (Fig. 8.3), is that a beneficial mutation may need to occur many times before the numbers of individuals possessing it allows it to increase deterministically in the population. The time course involved has two phases: the first being the waiting time until the founder individual appears (i.e. not necessarily the first to carry the mutation, but the individual who gives raise to the successful population) and the second being the passage time from the appearance of the founder individual until the mutation becomes fixed in the population. The waiting time depends critically on the mutation supply rate ($\theta$) and the stochastic process just described. The expected waiting time is then $t = (2s\theta)^{-1}$. The passage time depends mainly on the strength of selection, $s$. Figure 8.4 illustrates this process.

Let us consider a slightly different model of adaptive rescue called the 'extreme-value model'. Here we imagine that the environment changes and the current wild-type genotype lead to an intrinsic rate of growth of $r_0 < 0$ and the population begins to decline. If it is not rescued by a new beneficial mutation then it will decline until it is locally extinct. Among the many mutants that might arise at any given time, most will be deleterious but a few will be beneficial, leading to an intrinsic growth rate greater than zero and hence rescue. The number of these so-called 'rescue mutations' that arise before the population goes extinct depends on the number of individuals who live during this period of decline. Skipping the mathematics (interested readers should see Bell and Collins, 2008), we see the general conclusion of this model in Fig. 8.5. The probability of rescue depends on how severe the impact of climatic

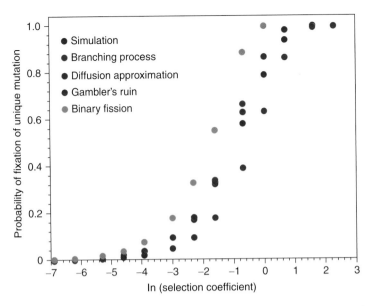

**Fig. 8.3.** The probability of a beneficial mutation becoming fixed in the population. The plot shows results for several different ways of modelling the process, but notice that they all lead to roughly the same conclusion. (Redrawn from Bell, 2008.)

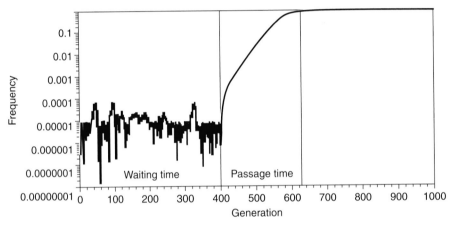

**Fig. 8.4.** A simulation of the waiting and passage times for the fixation of a beneficial mutation. (Redrawn from Bell, 2008.)

change is and on the mutation supply rate. As shown in Fig. 8.5, we would need a mutation supply rate of at least one per generation, a rate very unlikely to be achieved in small populations. The mutation supply rate, together with the degree of adaption of the current population, yields the supply rate of beneficial alleles. This rate is not well known in most populations although it is usually thought to be quite low. The evidence for this assumption is fairly weak and mixed (see Bell and Collins, 2008).

Well that's a sketch of the theory; what about the practice? If climate change will really be so dra-

matic and so rapid, populations may need to evolve very rapidly to keep pace with such change. In the next section we consider the evidence for rapid evolution. If nothing else, this should cause us to take seriously the possibility that evolution can happen on the time scale on which we are considering climatic change (i.e. a few hundred years).

## 8.4 Rapid Evolution

We define 'rapid evolution', as others have (Whitney and Gabler, 2008), as 'genetic change occurring

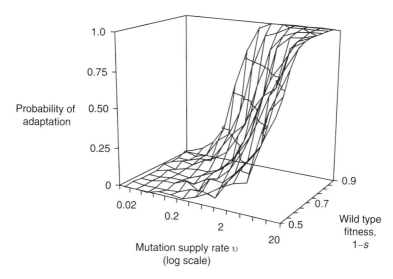

**Fig. 8.5.** The probability of adaptive rescue as a function of the severity of the climate change, denoted by the reduction in fitness of the current wild type and the mutation supply rate. (Redrawn from Bell and Collins, 2008.)

rapidly enough to have a measurable impact on simultaneous ecological change' (Hairston *et al.*, 2005). As we pointed out in the introduction, one piece of 'common wisdom' about species' responses to climatic change is that evolution is too slow relative to the pace of climatic change to allow adaptation. When pushed, researchers will generally acknowledge that for species with generation times measured in hours, days or up to a few weeks, and for populations that are large (and thus have a higher chance of experiencing an adaptive mutation), evolution may occur on the same time scale as climatic change (see for example Box 8.1). We can readily point to numerous examples of the rapid evolution of insecticide resistance in insects or antibiotic resistance in bacteria for confirmation of this view. Nevertheless, there are some excellent examples of rapid evolution in plants and mammals that are worth considering, and we do so in this section. We see that rapid evolution is a possibility even for these longer-lived species, but we should not adopt a naive view that therefore everything will be all right. As we point out later in this chapter, there are clearly limits to adaptive responses and, for every species that displays rapid evolution in response to a particular selection pressure, there are others, exposed to the same selection regimes, which fail to adapt (Bradshaw, 1991).

The 'resistance treadmill' is well known among agronomists. Since the advent of chemical pesticides,

insects have been evolving resistance. Indeed, this effect has been so ubiquitous that resistance to a pesticide seems an inevitable outcome (hence the resistance treadmill). This 'inevitability' is seen by some as a good reason not to develop genetically modified organisms, such as Bt crops (which use the genes from the soil bacterium *Bacillus thuringiensis* to produce constitutive insecticidal proteins), because the insects will simply evolve resistance to them; indeed, we have already seen the evolution of such Bt resistance in the diamondback moth (*Plutella xylostella*) – this has happened within 15 years and despite significant resistance management (see e.g. Stewart, 2004, for more discussion). The general area of insect resistance has been thoroughly treated elsewhere (see e.g. Wood and Bishop, 1981), so we concentrate below on other examples of rapid evolution.

Sites contaminated with heavy metals, usually associated with mining or smelting operations, often have a depauperate flora and fauna compared with nearby but uncontaminated sites (e.g. Shaw, 1989). Nevertheless, except in the most extreme metal concentrations, contaminated sites are not usually devoid of plant life. Several species will develop a tolerance for the metal, and these are well-known examples of evolution in action (see e.g. Macnair, 1981). Furthermore, such *in situ* adaptation can occur very rapidly. For example, in a set of sites in England contaminated with copper, the populations of *Agrostis stolonifera* in all sites

Marine algae play an important role in the global carbon cycle. They fix large quantities of carbon, they grow very rapidly and they die very quickly. If they are not eaten before they die, they sink to the bottom of the ocean where the carbon is effectively sequestered. For the most part, the flow of nutrients and carbon between the upper layers of the oceans and the lower layers is one-way: what goes down tends not to return for very long periods of time. This

so-called 'biological carbon pump' is a globally important sink. It has been estimated that, without this sink, pre-industrial atmospheric CO$_2$ concentrations would have been approximately 180 ppm higher than observed (Sarmiento and Toggweiler, 1984). Indeed, there have been suggestions and serious experiments aimed at exploiting this pump in a form of biogeoengineering to help counteract climatic change. Marine algae are thought to be iron-limited

(a)

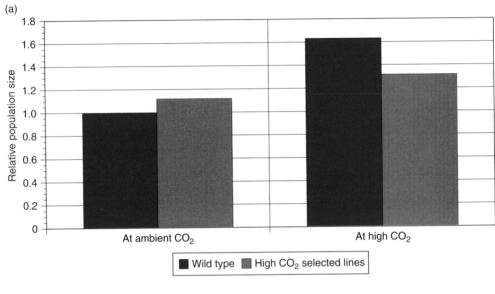

(b)

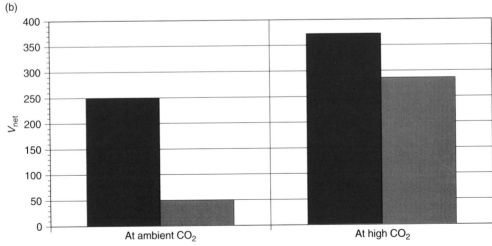

**Fig. 8.6.** Response of *Chlamydomonas reinhardtii* after 1000 generations in either ambient or 1050 ppm CO$_2$. (a) The relative maximum population size and (b) the net carbon uptake ($V_{net}$) relative to the population size estimated from the chlorophyll (Chl) concentration, expressed in μmol/mg Chl/h. (Drawn from data in Collins *et al.*, 2006.)

and it has been suggested that by seeding the oceans with large quantities of iron, we could stimulate algal growth, which would accelerate the biological carbon pump and sequester larger quantities of carbon (see e.g. Schiermeier, 2009).

In general, many algal species are insensitive to $CO_2$ because they possess an inducible carbon concentrating mechanism, similar in principle to $C_4$ photosynthesis except that it can be turned on and off as dictated by the degree of carbon limitation. So while $CO_2$ can have a fertilizing effect, particularly on $C_3$ plants, it may have little effect on many species of algae because they may just turn off their carbon concentrating mechanism and maintain roughly the same rates of photosynthesis in elevated $CO_2$ as they do in ambient $CO_2$.

Collins and Bell (2004) conducted what at first glance appears to be yet another experiment to determine the biological response to elevated $CO_2$ (see e.g. Chapters 3, 4 and 5). They subjected cultures of the freshwater alga *Chlamydomonas reinhardtii* to ambient and 1050 ppm $CO_2$. Perhaps not surprisingly, the algal maximum population size was increased by 63% and the alga fixed about 48% more carbon in elevated $CO_2$. So far this sounds like any of several other studies we discussed earlier in this book, and these results represent the alga's short-term response to elevated $CO_2$. The difference is that Collins and Bell maintained their cultures under ambient and elevated $CO_2$ for about 7 months, which represents about 1000 generations of selection (actually, they slowly increased the $CO_2$ concentration over the first 600 generations and then maintained 1050 ppm for the final 400 generations). After this period of time the results were very different, as we can see in Fig. 8.6. We see that both

population growth rate and net $CO_2$ assimilation were reduced in the high-$CO_2$ selected lines measured at high $CO_2$, compared with the control lines measured at high $CO_2$; i.e. the longer-term response was much reduced compared with the short-term response. We might have expected to see the development of specific adaptations that enabled the alga to exploit the high-$CO_2$ world. In general, Collins and Bell (Collins and Bell, 2004; Collins *et al.*, 2006) failed to find any specific adaptation to elevated $CO_2$. The selected lines did not grow faster nor did they attain larger population sizes than the wild type. Furthermore, these selection lines had very poor fitness when tested at ambient $CO_2$; apparently the high-$CO_2$ selected lines had lost their inducible carbon concentrating mechanism after 1000 generations. Collins and Bell suggest that this may have been caused by an accumulation of conditionally neutral mutations affecting the carbon concentrating mechanism. Perhaps this response is to be expected (Reusch and Wood, 2007); when $CO_2$ concentrations increase it is no longer necessary for algae to maintain their metabolically expensive carbon concentrating mechanism and because there is no selection pressure to do so, mutations (now selectively neutral) accumulate and the mechanism stops working.

Collins *et al.* (2006) then went on to compare the evolved responses they observed with those of algae found around naturally occurring high-$CO_2$ springs (see Chapter 3) near Sienna, Italy and Radenci, Slovenia. They found similar responses in the $CO_2$-spring populations and they suggest that some loss of function in carbon uptake, or perhaps metabolism, is common among algae exposed to high $CO_2$ for long periods of time.

showed marked copper tolerance and some of these sites were not more than 4 years post-contamination (Bradshaw, 1991). It now seems clear from genetic evidence that the variance necessary to evolve copper tolerance is usually already present in uncontaminated populations and selection rapidly acts on that variation. The variance is always very small, but present, of the order of one in 1000 individuals within the population (Bradshaw, 1991).

Another interesting and well-studied example comes from the Park Grass experiment in Rothamsted, England. Established in 1856, this experiment continues to follow the plant species diversity in plots under different fertilizer treatments. In 1903 the plots were split in half and one half began receiving a liming treatment that rapidly increased the pH of the soil. This created a strong selection force on the plants and many

species were extirpated from the limed halves. Some species, however, continued to persist in both halves of the plots. The perennial grass species *Anthoxanthum odoratum* has developed, in less than 70 years of selection, different responses to soil pH and differences in morphology between the contrasting halves of the plots, and this despite what must be significant gene flow between the two halves (Snaydon, 1970; Bradshaw, 1991; reviewed by Silvertown *et al.*, 2006).

These are but two examples of rapid evolution in plants; there are many others (see e.g. Endler, 1986), including many examples of crop weeds that develop herbicide resistance analogous to the insects we considered above. Whitney and Gabler (2008) reviewed 82 cases from 38 species of invading plant and animal species and found evidence of rapid evolution in all of them, across many traits. In particular, they examined 15 cases of rapid evolution for traits directly associated with climatic and environmental tolerance, in birds, copepods and a variety of plants. In about half of those cases the evolutionary change occurred in less than 150 years. The wood frog (*Rana sylvatica*) was able to evolve its thermal tolerance and preference as well as its temperature-specific development rate in less than 40 years in response to changed temperatures (see Skelly *et al.*, 2007 for references and future discussion). And one last example, the stream-dwelling beetle *Gyretes sinuatus* showed a nearly 10% increase in body size over 60 years in response to a 3°C change in air temperature (Babin-Fenske *et al.*, 2008).

At this point we might be tempted to say that this is all well and good for grasses and insects, but what about long-lived large-bodied organisms, like mammals and trees? Oceanic islands have held a special place in the development of both ecological and evolutionary theory. We often see rapid radiations of species on islands, as well as large morphological changes in island compared with mainland populations of the same species. These changes may derive from a lack of natural enemies, competition or from abundant resources. Island evolution theory suggests that the smaller and more remote the island, the stronger will be the isolation effect and hence the more rapid can be the rate of evolution (Millien, 2006). One relatively well-known island pattern is that small mammals on islands tend to increase in body size while large mammals decrease in body size. Thus we have seen gigantism, with some extinct rodents (e.g. *Amblyrhiza inundata*) weighing as much as 200 kg (Biknevicius *et al.*, 1993), and dwarfism in large mammals such as the Sicilian elephant (*Elephas falconeri*) weighing in at a mere 100 kg (Roth, 1992). Furthermore, such changes are not restricted to oceanic islands; they also occur in habitat fragments that may act as islands (Hendry and Kinnison, 1999; Schmidt and Jensen, 2003). Dramatic body size changes in island mammals appear to be the result of adaptive evolution rather than simply a founder's effect (Pergams and Ashley, 2001), and they can occur very rapidly. So this analysis further suggests that given rapid and dramatic environmental change such as climatic change, most contemporary mammal species have the capacity to increase their rate of morphological change by up to a factor of three within a few decades (Millien, 2006; Fig. 8.7).

Carroll *et al.* (2007) summarized the situation as follows:

> Ecologically significant evolutionary change, occurring over tens of generations or fewer, is now widely documented in nature. These findings counter the long-standing assumption

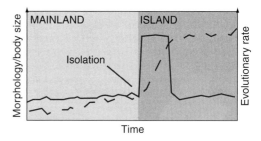

**Fig. 8.7.** Schematic illustrating how the rate of evolution for morphological characters on mammals changes after isolation on oceanic islands (solid line) and how the corresponding values of the trait change (dashed line). Notice that eventually the rate of evolution on the island returns to the background rate seen on the mainland. The same changes are likely to occur when habitats are fragmented on the mainland. (Redrawn from Millien, 2006.)

> that ecological and evolutionary processes occur on different time-scales, and thus that the study of ecological processes can safely assume evolutionary stasis.

Clearly climate change impact studies are remiss in not at least considering the likely evolutionary consequences for the population(s) of interest.

Many of the examples in this section are descriptive studies or accidental/natural experiments, but the real gold standard for estimating the potential for rapid adaptation is experimental evolution studies. In the following section we examine the admittedly few examples of experimental evolution on traits related to climatic change.

## 8.5 Experimental Evolution

Reusch and Wood (2007) reviewed 16 examples of experimental evolution, on a variety of organisms, for traits related to climatic change. These studies and their conclusions are summarized in Table 8.1. It seems clear from these studies that, minimally, many climate-related traits are capable of responding to selection, although this by no means indicates that all such traits are capable of such responses nor that all species will show such responses.

Experimental evolution studies suffer from the same experimental design problems that experimental ecological studies of climatic change suffer from, but two particularly important aspects are: (i) abrupt change in the environment (see also Box 3.1); and (ii) concentration on main effects. With the exception of the *Chlamydomonas* study (see Box 8.1), all of the studies in Table 8.1 cover fewer than 150 generations,

**Table 8.1.** Review of 16 experimental evolution studies on traits related to climatic change. (Adapted from Reusch and Wood, 2007; see original for references.)

| Study type, no. of generations | Species | Selection regime | Evolutionary response | Conclusion |
|---|---|---|---|---|
| *Abiotic stress* | | | | |
| Laboratory, 150 | *Drosophila melanogaster* | Natural selection, mean temperature | (+) Heat-shock survival; (+) threshold temperature acclimatization | Mean temperature correlated with tolerance to heat shock |
| Laboratory, 120 | *D. melanogaster* | Natural selection, water vapour | (+) Water retention; (+) water storage | Two or three mechanisms responded to selection |
| Laboratory, 32 | *D. melanogaster* | Artificial selection, upper and lower knockdown temperatures | Knockdown temperatures (+) in + lines and (–) in – lines | For selected + lines the variation in knockdown temperature was exhausted in 30 generations |
| Laboratory, 26 | *D. melanogaster* | Artificial selection, desiccation | (+) Female survival | Tolerance to water loss and rate of water loss both evolved |
| Laboratory, 10 | *D. melanogaster* | Artificial selection, desiccation | (+) Desiccation resistance; (+) heat-shock resistance | Desiccation and heat-shock resistance were correlated traits (see Section 8.6) |
| Field, 2 | Barley (*Hordeum vulgare*) | Artificial selection, drought tolerance | (+) Grain yield under stress | Rapid evolution of drought tolerance with up to 57% more grain |
| Field, 2–8 | Maize (*Zea mays*) | Artificial selection, drought tolerance | (+) Biomass; (+) N-accumulation under drought | Drought tolerance and N-uptake and metabolic efficiency were correlated traits (see Section 8.6) |
| Greenhouse, 3 | Wild mustard (*Sinapis arvensis*) | Natural selection, water supply | (–) Flowering time; (+) seedling growth rate | Evolution of stress avoidance (via flowering time), correlated trait responses accelerate adaptation (see Section 8.6) |
| *Photoperiodic responses and phenology* | | | | |
| Laboratory, 2 | Blackcap warbler (*Sylvia atricapilla*) | Artificial selection, autumn migration date | (+) Migration date | Rapid evolution possible, trait mean shifted to 1-week delay, high heritability |
| Greenhouse, 9 | Sea beet (*Beta vulgaris maritii*) | Artificial selection, lower critical day length for flowering | (–) Critical day length | Rapid evolution, population mean reduced from 13.5 to 11h |
| *Adaptation to increased $CO_2$* | | | | |
| Field, 1–6 | *Sanguisorba minor* | Natural selection, increased $CO_2$ concentration | (+) Leaves; no change in reproductive traits | Enhanced carbon sequestration via litter production |
| Greenhouse, 5 | Mouse-ear cress (*Arabidopsis thaliana*) | Natural selection, increased $CO_2$ concentration | (+) Biomass; no change in reproductive traits | Adaptation to increased $CO_2$ may enhance reproductive rate |
| Laboratory, 1000 | *Chlamydomonas rheinhardtii* | Natural selection, increased $CO_2$ concentration | No change in growth or photosynthetic rates | Reduced function in carbon accumulating mechanism (see Box 8.1) |
| *Increased temperature mean and variability and increased $CO_2$* | | | | |
| Greenhouse, 7 | *Brassica juncea* | Natural selection, combination of temperature mean and variability, and $CO_2$ | No change in vegetative or reproductive traits; (+) biomass | In-breedings prevented adaptive evolution |

*Continued*

**Table 8.1.** Continued.

| Study type, no. of generations | Species | Selection regime | Evolutionary response | Conclusion |
|---|---|---|---|---|
| *Dispersal* | | | | |
| Laboratory, 25 | Cricket (*Gryllus firmus*) | Artificial selection, wing morphology | (+) Wing length; (+) fatty acid metabolism | Correlated responses of several physiological traits (see Section 8.6) |
| Laboratory, 5 | Flour beetle (*Tribolium castaneum*) | Migration between culture vessels | (+) Rate of exchange between culture vessels | Rapid evolution for migrational behaviour possible |

and most cover fewer than 25 generations. These studies begin by suddenly exposing the populations to the selection pressure and following the responses. However, as we pointed out in Chapters 1 and 3, climatic change may be 'rapid' in comparison with historical climatic change, but it is nevertheless gradual from the perspective of the evolving population. The populations will experience a continuous range of intermediate climatic conditions and this can lead to different conclusions compared with what happens when climate changes abruptly (Bell and Collins, 2008). Also, in most selection experiments, it is common to impose a stressful condition but to keep all other aspects of the environment benign and this has caused some to question the value of these experiments (Harshman and Hoffmann, 2000; Reusch and Wood, 2007).

One interesting feature of Table 8.1 is that within the 16 examples, four of them involved correlated trait responses. Such responses can speed or retard adaptation to climatic change and they are probably fairly common in nature. We take up this issue in the next section.

## 8.6  Correlated Genetic Traits

Up until now we have been considering whether populations can adapt in some climate-related trait fast enough to track the changing climate. The answer to this question depends in part upon the extent to which the trait in question is independent of other climate-related traits, and the direction and strength of the selection imposed by the changed climate. In particular, if we have pleiotropic effects (a single gene affects multiple traits) or linkage effects (genes located near each other on the same chromosome), then selection may not be able to act independently upon multiple traits (Hellmann and

Pineda-Krch, 2007). These genetic correlations can either speed up or slow down the effects of selection. In this section we consider this problem in more detail.

Let us consider our hypothetical species from Fig. 8.2. Imagine now that instead of the climate variables, we consider the associated traits that might be selected by these climate variables. We depict this situation in Fig. 8.8. The ellipse denotes the variation within the population in these two traits, drought tolerance and temperature tolerance. In both Fig. 8.8a and b the two traits are correlated, but in Fig. 8.8a they are negatively correlated and in Fig. 8.8b they are positively correlated. We imagine that the centre of the ellipse sits on the current peak in the fitness landscape (sensu Wright, 1931; reviewed by Bell, 2008) denoted by the point $d_1^*, t_1^*$. Now imagine that the climate in the particular region changes rapidly getting both hotter and drier, and over time results in a new fitness peak $F^* = (d_2^*, t_2^*)$. This peak in the fitness landscape is denoted by the concentric contour intervals. The direction of selection is denoted by the vector $\beta$. The vector $g_{max}$ denotes the 'line of genetic least resistance', meaning that trait changes along this line are easiest to accomplish because the two traits are maximally correlated along that selection line. The population will move towards the new fitness peak, but in a direction that is biased towards $g_{max}$. This direction is denoted by the vector $\delta$. The curves are stylized trajectories through the fitness landscape, with the dots representing the location of the population every 20 generations. Comparing Fig. 8.8a and b should provide a sense that the closer the direction of the selection gradient $\beta$ is to the path of genetic least resistance $g_{max}$, the more direct and more rapid can be the adaptation (Hellmann and Pineda-Krch, 2007).

Chapter 8

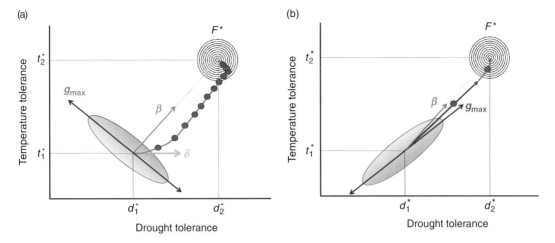

**Fig. 8.8.** Climate change-induced selection on correlated traits. See text for explanation. (Adapted from Hellmann and Pineda-Krch, 2007.)

The fitness difference caused by the population not being located on the fitness peak in trait space is called the 'lag load'. As climatic change shifts the fitness peak in trait space, if the species is able to closely track the change then the lag load may remain small and the species may once again become locally adapted. However, if the lag load is large, then fitness may be reduced to the point at which the population size declines leading to either a population bottleneck or extinction (Hellmann and Pineda-Krch, 2007; see Chapters 5 and 12).

## 8.7 Conclusions

Individuals routinely experience climatic variation from day to day, month to month and year to year, and for longer-lived species like trees, decade to decade and century to century. For most individuals, phenotypic plasticity will be sufficient to allow them to cope with such changes. However, it is predicted that the pace of the coming climatic change will be too rapid, and the change itself too large, to be accommodated solely by phenotypic plasticity. Barring gene flow from populations adapted to warmer climates, species will have to adapt to climatic change through natural selection acting on beneficial mutations.

In this chapter we have considered three main situations: (i) when phenotypic plasticity and ecological fitting are sufficient to cope with large changes in the climate; (ii) general conclusions around adaptive or evolutionary rescue; and (iii) rapid evolutionary responses to climatic change. These lines of thought offer differing views on what we can and should expect from evolution (Gienapp *et al.*, 2008). Examples of ecological fitting and rapid evolution would seem to suggest that species are largely able to adapt to their new climates because they often possess sufficient genetic variation on which selection can act. More probably, what these examples demonstrate is that such adaptation *can* occur, not that it *will* occur. Remember that for every successful example of, say, ecological fitting in the form of species invasions, many more examples exist where a species has failed to successfully invade a new habitat. And for every example of a species successfully adapting to heavy metal contamination, or to a new pesticide, there will be many more counter examples where such adaptation does not occur (Bradshaw, 1991). We also saw that for evolutionary rescue to occur before the population goes extinct, we need a relatively high mutation supply rate and relatively small changes in fitness that result from the changing climate. The smaller the first or the larger the second, the less likely is adaptive rescue. And finally we saw that correlated trait responses can either help or hinder adaptation to climatic change.

We end this chapter with a brief consideration of the plight of the sea turtle, and other species that show environmentally determined sex ratios (Box 8.2). Such species would seem to be prima facie candidates for extinction in the face of climatic change.

## Box 8.2. Environmentally dependent sex determination.

The sea turtle is an example of a species that has environmentally dependent sex determination. In sea turtles it is the temperature experienced by the eggs in the middle third of their incubation that influences the sex of the resulting offspring (although a variety of other factors are also involved; Carthy *et al.*, 2003). The so-called 'pivotal temperature', above which sex ratio is female-biased and below which the sex ratio is male-biased, tends to be well conserved across sea turtles at around 29°C (Mrosovsky, 1994). Several populations of loggerhead turtles (*Caretta caretta*) in Florida, USA are known to be highly skewed towards females, while others further north are closer to a 1:1 sex ratio. Hawkes *et al.* (2007) studied two of these latter populations in North Carolina, USA. They modelled the distribution of mean nest temperatures (Fig. 8.9) and then applied projected changes in air temperature to estimate the effect of climatic change on sex ratio in these populations (Fig. 8.10). Their analysis indicates that, barring any adaptation, these populations are also likely to become seriously female-biased, leading to the fear that the species might well go extinct as a result of climatic change.

To avoid extinction, the loggerhead turtle could alter its pivotal temperature, nest on beaches further north (i.e. migrate), choose different microhabitat locations within the same beaches, or change the time of egg laying so that the mean temperature experienced by the eggs is reduced (i.e. changed phenology). The four options are not mutually exclusive, and changes could occur via phenotypic plasticity or via adaptive mutation (Hawkes *et al.*, 2007). Before we consider any of these options, it is worth reminding ourselves of what leads to the evolution of skewed sex ratios.

In many species, the genetic mechanics of mating give rise to equal sex ratios and we need not look for an evolutionary explanation. However, in systems like the sea turtle, where the conditions of incubation determine gender, we need to know what selective pressures are brought to bear on the sex ratio.

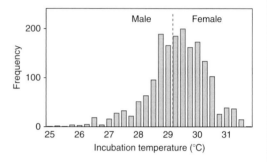

**Fig. 8.9.** Frequency histogram of nest temperatures experienced by loggerhead sea turtles nesting at Bald Head Island, North Carolina. The pivotal temperature is 29.2°C; nest temperatures above the pivotal temperature result in clutches that are biased towards females and temperatures below the pivotal temperature result in clutches biased towards males. (Redrawn from Hawkes *et al.*, 2007.)

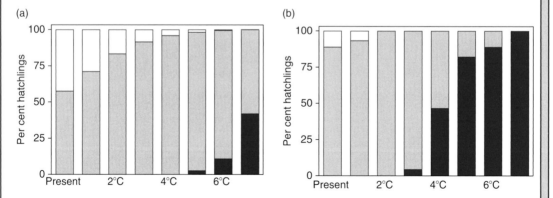

**Fig. 8.10.** Sex ratio at hatching of loggerhead sea turtles under current and future temperatures for (a) Bald Head Island, North Carolina and (b) Cape Canaveral, Florida. The open bars denote males, the light blue bars denote females and the dark blue bars denote the proportion killed by lethal temperatures. (Redrawn from Hawkes *et al.*, 2007.)

Theory indicates that unbalanced sex ratios should rapidly come back into balance because, in a population with a skewed ratio, a mutation that leads to a bias towards the minority sex will be favoured by natural selection since it will produce more grand-offspring (Bell, 2008). So the outcome should be a 1:1 sex ratio, as long as males and females are equally costly to produce, otherwise the sex ratio will be skewed towards the less costly sex. There are examples of temperature-dependent sex determination where sex ratios are maintained at 1:1 despite different temperatures. An example is Conover et al. (1992), who established laboratory colonies of the fish the Atlantic silverside (Menidia menidia) captured from South Carolina, USA, New York, USA and Nova Scotia, Canada. In the lab, each population was raised at a constant temperature of either 17°C or 28°C. Regardless of the origin of the selected population or of the temperature regime of selection, and despite the initial skew to the sex ratio, the minority sex increased in frequency and within a few generations regained a 1:1 sex ratio.

If the environment affects male and female fitness differently then there may be selective pressure for a sex ratio other than 1:1, and we might expect that a mechanism for the control of the sex ratio, like environmentally determined sex, to arise (Trivers and Willard, 1973; Bull, 1981). In sea turtles it need not be the case that the incubation temperature affects male and female egg survival differently, only that incubation temperature is somehow correlated with different fitness for the sexes later on in life. The problem, as noted by Mrosovsky (1994), is that it is difficult to see how this could be the case. Sea turtles mature quite slowly, and it is not clear why the temperature in the middle third of the incubation period would have much of anything to do with fitness later in life. Nevertheless, sea turtles are difficult creatures to study and perhaps this relationship may yet be elucidated.

If it is not the case that altered sex ratios arise either because there is a difference in the cost of producing one sex or because there are differences in the fitness of the two sexes in different environments, then another possibility is that other assumptions of the 1:1 sex ratio result are not true. For example, if siblings influence each other's fitness, positively or negatively, then skewed sex ratios are also possible. Since some species of sea turtles have high nest site fidelity within a mating season, and some have high site fidelity for their natal beach, such sibling interactions also seem possible.

The bottom line seems to be that we do not know what will happen to sea turtles and other species with strong environmentally determined sexes in the face of climatic change. Like other species, the same four possible responses are available: (i) accommodate the changing climate via phenotypic plasticity; (ii) move to other habitats; (iii) adapt in situ; or (iv) go extinct.

# PART III
# Applications

Aerial view of T-FACE (temperature free-air controlled enhancement) plots of wheat at Maricopa, Arizona, March 2009, in the Hot Serial Cereal Experiment: 'Cereal' because it was on wheat, 'Serial' because the wheat was planted serially about every 6 weeks for 2 years, and 'Hot' because on six of the planting dates infrared heaters were installed in hexagonal arrays to warm the wheat. The different planting dates caused the various shades of green. The white dots in hexagonal patterns are infrared heaters used to warm the plots. The triangular shapes are the cables and posts used to deploy the heaters around the peripheries of the plots. The bare plots within the brown strips are due to the fact that the warmed plots matured sooner and were already harvested before the un-warmed plots on the day the photo was taken. The plots were established in 2007 by Bruce Kimball, Jeffrey White and Gerard Wall (USDA-ARS) and Michael Ottman (University of Arizona). Experiments like this one help us to understand the impacts of climate change on agriculture, as discussed in Chapter 11. (Photo by T. Clarke, USDA, Agricultural Research Service.)

# Overview

In Part II we examined the basic biological considerations necessary for thinking about the impacts of climatic change. In the course of our examination we encountered a variety of applications, but in this section of the book we take a more deliberate look at how we might apply this basic scientific knowledge to questions of more direct concern to humans. Soils, and the decomposition and nutrient cycling activities that occur within them, are the foundation for all terrestrial ecosystem functions. In particular, soils are directly relevant to agricultural and silvicultural production. In Chapter 9 we consider how climatic change may impact soil organisms and the functions that they perform. We then move on to consider forest productivity (Chapter 10) and agricultural production (Chapter 11) in greater detail. In Chapter 12 we take up the issue of biodiversity resources and changes that we might expect to result from climatic change. In these chapters we not only consider the impacts, but where available we also consider research directed at adapting to the anticipated impacts. For example, in Chapter 10 we consider the scope for plant breeders to introduce desirable traits to crop genomes, for coping with, or indeed for exploiting, climatic change. In Chapter 12 we consider the controversial proposal to use so-called 'assisted colonization' to aid in the conservation of biodiversity in the face of climatic change.

Throughout these chapters we draw upon the ideas we discussed in Part I, and we make use of the results of research described in Part II.

# 9    Responses by Soil Organisms

In our discussion of community-level responses to climatic change in Chapter 6, we postponed our discussion of belowground community dynamics to the current chapter. In Chapter 7 we examined in some detail the likely impacts of climatic change on ecosystem processes, many of which depend strongly on soil processes. In that chapter, our focus was on the consequences of climatic change on soil *functioning*. In this chapter, our primary focus is on the consequences of climatic change on the soil communities themselves. Secondarily we note the functional consequences of these community changes, and so we repeat some of the insights from Chapter 7. Where we do so, we use different examples and focus on different facets of such process-level impacts.

Although the aboveground and belowground components of communities are inextricably linked, the study of soil community responses to climatic change brings a unique set of opportunities and challenges. With respect to opportunities, many of the organisms are very small and present at incredibly high densities in relatively small volumes of soil (see Table 9.1), which allows community-level data to be obtained from soil samples. Many of these organisms also exhibit rapid rates of turnover and community composition can shift very rapidly in response to changes in the soil environment. Substantial adaptation of soil microorganisms can even be observed on the time scale of many climate change experiments (see Chapter 8).

The small size of most soil organisms also presents a challenge for researchers. Direct counts of soil organisms can be painstaking and laborious, and they often reveal little information

about species composition or function. The soil matrix – a mix of minerals, organic matter and water films – is also difficult to work with (as compared with water, for example) because it exhibits a high level of aggregation, layering and heterogeneity.

The trophic structure of the belowground community mirrors that of the aboveground community, with the main difference being that much more of the energy flow below ground is derived from dead organic material, rather than directly from the consumption of live primary producers. Primary producers (e.g. plant roots and cyanobacteria) are consumed to some extent in soil, but the bulk of the energy flows through decomposers (bacterial and fungi) and detritivores (soil fauna) that directly consume dead organic material. There are then microbivores (consumers of microbes), such as protozoa and nematodes, and multiple levels of meso- and macrofauna that form a pyramid of consumers above these organisms. Soil food webs are also highly circular, with fungi and microorganisms often consuming protozoa and arthropods. Associations of microorganisms with plant roots are common, and many, such as those between roots and mycorrhizal fungi, nitrogen-fixing bacteria and other endophytic bacteria, are mutualistic. Although we can do a fairly good job of describing soil food webs qualitatively, quantitative data describing soil community dynamics are scarce relative to the study of aboveground organisms. Nevertheless, in many cases the responses of soil food webs to climatic change are likely to strongly influence carbon and nutrient flow at the ecosystem level.

**Table 9.1.** Relative numbers and biomass (in grams fresh weight) of soil organisms per square metre commonly found in the surface 15 cm of soil. (Adapted from Brady and Weil, 2004.)

| Organism | Number | Biomass |
|---|---|---|
| Bacteria | $10^{13}$–$10^{14}$ | 40–500 |
| Actinomycetes | $10^{12}$–$10^{13}$ | 40–500 |
| Fungi | $10^{10}$–$10^{11}$ | 100–1500 |
| Algae | $10^{9}$–$10^{10}$ | 1–50 |
| Protozoa | $10^{9}$–$10^{10}$ | 2–20 |
| Nematodes | $10^{6}$–$10^{7}$ | 1–15 |
| Mites | $10^{3}$–$10^{6}$ | 0.5–1.5 |
| Collembola | $10^{3}$–$10^{6}$ | 0.5–1.5 |
| Earthworms | 10–$10^{3}$ | 10–150 |
| Other fauna | $10^{2}$–$10^{4}$ | 1–10 |

## 9.1 Responses of the Detrital System

We can think about the responses of the detrital system along taxonomic lines. The major players are microbes, protozoa and various larger organisms known collectively as mesofauna and macrofauna. We take up each in turn.

### Microbes

Much like the study of climate-driven changes in the distributions of plants and animals pre-dates the recent study of climate change biology by over a century (Chapters 5 and 6), there is a long history of microbial biogeography that pre-dates our current study of microbial responses to climatic change. In a well-known paper on microorganisms written in 1913, Martinus Beijerinck (1913) proposed that most microorganisms are cosmopolitan, which gave rise to the thinking that in the microbial world 'everything is everywhere, but the environment selects'. Although this oversimplification skewed the study of microbial biogeography for decades to come (i.e. everything is not everywhere; O'Malley, 2007), in many cases soil microbial communities may not be as dependent on migration as a means of adjusting to climatic change as the aboveground organisms we described in Chapter 6. Rather, soil microbial communities already exhibit high intra-annual variation in response to seasonal changes and pre-existing populations of soil microorganisms at a given location may rapidly increase in response to changes in their environment.

A range of methodological approaches can be taken to characterize microbial responses to climatic change. The soil can be treated as a black box, with a focus on changes in microbial-driven processes such as respiration or nutrient mineralization. The results from this type of approach were considered in detail in Chapter 7. These processes can be more finely divided to focus on the decomposition of specific organic substrates by microbial exoenzymes (extracellular enzymes excreted by microorganisms to decompose substrates in their vicinity). The latter is accomplished by linking the substrates of interest to compounds that can be identified using colorimetric or fluorimetric assays when cleaved from the substrate (Marx *et al.*, 2001).

Other techniques are available to characterize changes in microbial community composition in response to climatic change. While direct extraction and morphological characterization of fungal spores can provide useful community composition data, for bacteria, staining and direct counts can generally provide only total abundance data. Both bacteria and fungi can be divided into coarse groups using phospholipid and fatty acid profiling, and several powerful DNA-based techniques have been developed that provide microbial community compositional data at a much finer scale. However, a gulf still exists between microbial community composition data and the corresponding functional data needed to link community composition to the processes described above. The problem is that only a small percentage of soil microorganisms can be cultured in isolation in the laboratory (estimates range from 0.1 to 10%; Amann *et al.*, 1995). We therefore can describe changes in community composition in response to climatic change in great detail but, unlike for many aboveground communities, we often cannot describe the functional roles of the individual species.

Soil warming is typically associated with increased rates of organic matter decomposition by microorganisms (see Chapter 7). However, this trend can be complicated by shifts in microbial community composition and substrate usage at higher temperatures (Zogg *et al.*, 1997). In addition, after multiple years of experimental warming, the initial stimulation of microbial activity often diminishes and may even decrease relative to control plots (Frey *et al.*, 2008). There can be high seasonal variability in microbial sensitivity to warming, as observed for nitrogen-fixing microbes in the High Arctic (Deslippe *et al.*, 2005), and variability in the sensitivity of microbial communities to warming along climate gradients, as observed in the Antarctic (Rinnan *et al.*, 2009a). Warming can also decrease soil moisture, and several studies in forest systems have revealed that reductions in microbial biomass and changes in community composition in response to warming can be attributed directly to soil drying (Fig. 9.1; Arnold *et al.*, 1999; Allison and Treseder, 2008).

Studies of microbial responses to changes in soil moisture have focused on altered precipitation patterns and drought. Drought decreased fungal community diversity in a heathland (Toberman *et al.*, 2008) and decreased overall microbial diversity in bog and fen sites (Kim *et al.*, 2008), although microbial communities were generally resistant to water stress in drought-prone tallgrass prairie (Williams, 2007). Similarly, in a study of heathland soil communities along a climate gradient in Europe, the effect of drought on soil enzyme activity was more severe for the northern European sites than for the southern European sites, which are generally more moisture-limited (Sowerby *et al.*,

2005). It can, however, be difficult to generalize how water availability affects microbial communities as a whole. For example, in a study of responses to drought and precipitation variability in the Chihuahuan desert, the bacterial community was most sensitive to changes in precipitation frequency and timing at a mid-elevation grassland site, whereas the fungal community composition was most sensitive in a low desert scrub site (Clark *et al.*, 2009). When warming and precipitation treatments were combined in a single field experiment, changes in precipitation had the greatest effect on bacterial and fungal community composition in an old field ecosystem (Castro *et al.*, 2010) and water availability played a strong role in regulating microbial responses to warming in a semi-arid grassland (Liu *et al.*, 2009). Nevertheless, the relative effects of these two factors are likely to depend on the system in question.

Soil microbial community responses to elevated $CO_2$ have been mixed, with changes in biomass that are not accompanied by changes in species richness (Klamer *et al.*, 2002), changes in community composition not accompanied by changes in biomass (Janus *et al.*, 2005), the alteration of community succession (Zheng *et al.*, 2009) and no change at all (Bruce *et al.*, 2000; Austin *et al.*, 2009). Elevated $CO_2$ effects are also observed in interaction with other factors such as nitrogen addition (Lagomarsino *et al.*, 2007) or ground-level ozone (Kanerva *et al.*, 2008). With respect to functional changes in microbial communities, increased activities of nitrogen- and phosphorus-acquiring enzymes relative to carbon-acquiring enzymes have been indicative of increasing nutrient limitation (Finzi *et al.*, 2006) and shifts in the use of different carbon compounds can also occur (Billings and Ziegler, 2005). Although it has often been assumed that elevated $CO_2$ will lead to higher carbon storage in soils, declines in soil carbon associated with increased relative abundances of fungi and soil carbon-degrading enzymes have revealed that changes in microbial communities under elevated $CO_2$ can cause a potential carbon sink to become a source (Carney *et al.*, 2007). Shifts in the abundances and structures of key functional groups, such as methanotrophic bacteria (Kolb *et al.*, 2005) and ammonia-oxidizing bacteria (Bowatte *et al.*, 2008), have also been noted. However, as discussed in Chapter 3, exposure to sudden increases in $CO_2$ can have different effects on soil microbial communities than more

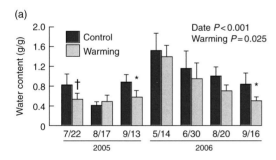

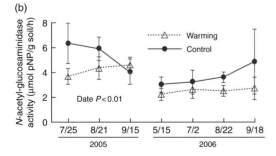

**Fig. 9.1.** (a) Soil water content and (b) activity of the chitin-degrading extracellular enzyme *N*-acetyl-glucosaminidase in warmed and control boreal forest soil plots. (Redrawn from Allison and Treseder, 2008.)

gradual shifts (Klironomos *et al.*, 2005; see Box 3.1), and complex, non-linear responses to different levels of elevated $CO_2$ have also been observed (Lipson *et al.*, 2006). Prior to the more recent use of free air enrichment systems, open-topped chambers used to test elevated $CO_2$ effects on soil microbial communities also produced significant chamber effects independent of the $CO_2$ effect (Mayr *et al.*, 1999).

In most assessments of microbial community response to climatic change, it is difficult to separate the direct effects of climate from indirect effects caused by changes in plant root exudates or, in the longer term, by changes in litter quality or the relative abundances of plant species. The exception is for elevated atmospheric $CO_2$, which almost exclusively affects soil microbial communities indirectly (since belowground $CO_2$ concentrations are far higher than ambient concentrations), as we discuss in further detail below in Section 9.4 on plant–soil feedbacks. There are some systems, however, such as biological soil crusts in arid systems, where soil microbial responses can be isolated, at least in part, from plant responses. In desert systems, these communities of cyanobacteria, lichens, mosses, algae, bacteria and fungi can contribute significantly to ecosystem and soil fertility, and their activities are particularly sensitive to precipitation pulse sizes (Cable and Huxman, 2004). Soil lichen cover and richness in these crusts have declined as the frequency of summer rainfall has increased (Belnap *et al.*, 2004). Declines in soil crusts in response to altered precipitation may be further exacerbated by warming and increased ultraviolet radiation (Belnap *et al.*, 2008), combined with trampling by livestock (Williams *et al.*, 2008).

### Protozoa

Soil protozoa, such as amoebae, flagellates and ciliates, are often grouped functionally with respect to their main prey items (e.g. as bacterivores or fungivores). Much like aboveground consumers, they respond both directly to the environment and indirectly in response to changes in the abundances of their prey or predators. Small, free-living amoebae are the main predators that control bacterial population in soils, helping maintain a high mineralization rate of organic matter through bacterial predation (Rodriguez-Zaragoza, 1994). In Arctic tundra, the abundance of testate (shelled) amoebae did not decline in response to an induced heat wave, although the heating shifted species abundances in favour of those belonging to bactivorous genera (Beyens *et al.*, 2009). Reciprocal transplants of soil among desert sites also revealed differential adaptation to specific temperature and precipitation regimes for protozoa associated with biological crusts, although protozoa that encysted were able to survive the upper end of daily temperatures in all systems (Darby *et al.*, 2006).

The abundance of gymnamoebae (naked amoebae) tends to be positively correlated with soil moisture (Anderson, 2000) and these species are very sensitive to drought (Bischoff, 2002). Species-specific drought tolerance has been observed for flagellate protozoa, and the ratio of flagellates to competing bacterivorous nematodes is also affected by drought (Bouwman and Zwart, 1994). In grass-dominated, open-topped elevated $CO_2$ chambers, flagellate density increased and amoebal density decreased, apparently in response to changes in the bacterial community (Fig. 9.2; Treonis and Lussenhop, 1997). However, under a wheat crop exposed to elevated $CO_2$, increases in bacterivorous protozoa did not correspond with increased bacterial abundance and the density of protozoa did not increase in the rhizosphere; rather, increases in protozoa appeared to be caused by overall increases in root production (Ronn *et al.*, 2003).

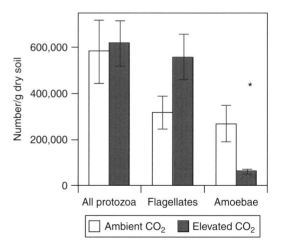

**Fig. 9.2.** Response of the total protozoan population, flagellates and amoebae in soils of the grass *Brassica nigra* grown under elevated $CO_2$. (Redrawn from Treonis and Lussenhop, 1997.)

## Mesofauna and macrofauna

Although the range of sizes of soil fauna forms a continuum, the microfauna described in the previous section are typically grouped separately from mesofauna (e.g. nematodes, collembolans and mites), macrofauna (e.g. gastropods, ants, millipedes, earthworms, potworms, isopods and some insect larvae) and megafauna (e.g. vertebrates). As described above, many of these organisms have important roles as predators and detritivores in soil food webs. Others, such as earthworms, have been described as 'ecosystem engineers' (Jones et al., 1994) because of their large direct contributions to the modification of their habitat.

Soil nematode abundance and species composition often change in response to soil warming, but these responses have largely been idiosyncratic. For example, in subarctic soils in Sweden, warming increased the dominance of fungal- and plant-feeding nematode species (Ruess et al., 1999a); yet in a Swedish tundra soil, the composition of the nematode fauna changed very little in response to warming, despite increased nematode numbers (Sohlenius and Bostrom, 1999). In dry valleys in Antarctica, warming treatments were associated with a decline in the density of a microbial feeding nematode that was the dominant soil animal at the site (Simmons et al., 2009). Meanwhile, in the same study, extreme soil wetting from the rapid melting of upslope subsurface ice increased the mortality of most species, but favoured populations of a nematode species that is commonly found in wet habitats and feeds on soil algae.

The effects of warming on soil arthropods generally have been minor (e.g. Bokhorst et al., 2008; Hagvar and Klanderud, 2009) and, as described for soil microbes above, notable warming responses are often linked to soil drying (Hodkinson et al., 1996; Briones et al., 1997). Soil moisture often dictates soil arthropod community composition in water-limited systems (Chikoski et al., 2006), but drought effects on soil arthropods can vary widely with the dominant vegetation type (Whitford and Sobhy, 1999). Similarly, in tallgrass prairie, forest and Antarctic systems, responses to changes in water availability have been strong but both complex and non-linear (O'Lear and Blair, 1999; Lindberg et al., 2002; Sinclair, 2002; Tsiafouli et al., 2005). Rapid recovery of soil arthropod communities from drought can occur, and this recovery may be accelerated by post-drought increases in the microbial populations that are their main food source (Holmstrup et al., 2007). Earthworms are sensitive to drought (Friis et al., 2004; Zaller et al., 2009) but, as commonly observed on pavements following large rain events, they also experience increased mortality in response to other hydrological extremes such as flooding (Plum and Filser, 2005). However, earthworm populations can recover because their cocoons survive floods (Thonon and Klok, 2007).

Over winter, snow cover decouples soil temperatures from air temperatures, and both warmer mean temperatures and increased climate variability are predicted to increase the frequency and intensity of soil freeze–thaw cycles as a result of decreased snow cover (Henry, 2007, 2008; see Box 9.1). Although soil fauna are often resistant to benign freeze–thaw cycles (Sjursen et al., 2005a,b), extreme freezing events can increase faunal mortality substantially (Sulkava and Huhta, 2003). Soil arthropods can also lose much of their cold and freezing tolerance rapidly on regaining activity in spring, leaving them vulnerable to episodic cold temperature events at this time (Coulson et al., 1995). Increased precipitation over winter can also affect soil temperatures. For example, microarthropod activity can be positively correlated with snowpack thickness because of increased soil surface temperatures (Addington and Seastedt, 1999).

In response to elevated atmospheric $CO_2$ altered nematode community structure has been predicted to occur as a result of bottom-up processes that alter their food sources, but changes in soil aggregate sizes caused by higher soil moisture under elevated $CO_2$ may also reduce the abundance of large nematode species that rely on the presence of large soil pores for locomotion (Niklaus et al., 2003). Numerous elevated $CO_2$ experiments in grasslands have failed to detect substantial changes in nematode communities (e.g. Ayres et al., 2008; Nagy et al., 2008). However, when changes in nematode communities have occurred, a wide range of responses have been reported, including increases in omnivorous and predacious species, with decreases in bacteria-feeding nematodes in a grassland (Yeates et al., 1997), increases in root feeders and microbial feeders in grazed pasture (Yeates et al., 2003) and reduced fungivore abundance in forest soils (Neher et al., 2004). The sensitivity of nematode feeding types to elevated $CO_2$ also can be crop-specific (Sticht et al., 2009) and it

**Box 9.1. Soil freeze–thaw dynamics.**

Predictions for changes in soil freezing dynamics with climatic change, and the effects of freezing on soils, were examined by Henry (2007, 2008), who addressed the following points. In winter, soil temperatures are dictated by interactions between air temperature and precipitation. Climate warming can lead to colder soil temperatures in winter when it reduces the thickness or cover of the insulative snowpack, exposing soils to cold temperatures overnight or with the subsequent arrival of cold weather systems that are not accompanied by more snow (Fig. 9.3).

Changes in soil freezing dynamics with climatic change are likely to vary by region. Although warming is predicted to be most severe at high latitudes, soils in systems at more southerly latitudes remain closer to the freezing point throughout the winter and experience less snow accumulation on average. These systems may therefore be most vulnerable to changes in soil freezing dynamics with climatic change.

Freezing effects on soil microorganisms and soil fauna have been studied extensively over the last decade, and the sensitivities of these organisms to freezing appear to vary widely. However, responses to freezing are typically non-linear, and variation in the severity of freezing treatments may explain much of the variation in results observed among soil freezing experiments.

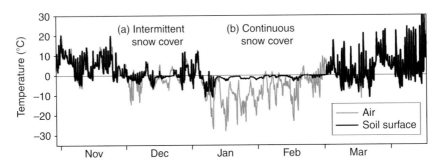

**Fig. 9.3.** Hourly air temperatures (grey line) and soil surface temperatures (black line) for bare soil in an agricultural field. The decoupling of air and soil temperatures during periods of (a) intermittent snow cover and (b) continuous snow cover are noted. (Adapted from Henry, 2007.)

can interact with nitrogen addition in agricultural fields (Li, Q. *et al.*, 2007, 2009). In addition to results obtained in multi-year field experiments (Yeates and Newton, 2009), the long-term responses of nematode populations have been studied in relation to their proximity to naturally occurring $CO_2$ vents (Yeates *et al.*, 1999).

Similar to nematodes, soil arthropod responses to elevated $CO_2$ have been predicted to occur based on changes in their food quality and quantity but no general patterns of response have occurred (Couteaux and Bolger, 2000). Changes in litter quality are not always accompanied by changes in detritivore communities (Kasurinen *et al.*, 2007) and in open-topped $CO_2$-chamber experiments, chamber effects have occurred that are independent of the elevated $CO_2$

effect. When responses do occur, they are often species-specific (Sticht *et al.*, 2006; see Chapter 14) or interact with the dominant vegetation type or crop.

## 9.2 Responses of the Biotrophic System

Species interactions can, and often do, modify the responses of individual species to climatic change. In this section we examine two classes of such interactions: (i) plant–micobe mutualisms; and (ii) root herbivory and disease.

### Plant mutualistic interactions

Plant–mycorrhizal associations are widespread and occur in approximately 95% of all plant families

(Trappe, 1987), with 85% of plant families associating with arbuscular mycorrhizae (Wang and Qiu, 2006). In addition to enhancing plant nutrient uptake in exchange for plant carbon, mycorrhizae can alter soil aggregation and stability by producing the glycoprotein glomalin (Rillig and Allen, 1999). Soil warming can stimulate root colonization by mycorrhizae, although the stimulation of decomposition can simultaneously decrease glomalin accumulation (Rillig et al., 2002). In other experiments, warming has had no effect on root colonization by mycorrhizae (Heinemeyer et al., 2004). The effects of drought on mycorrhizal colonization have also been mild, although in some systems drought has shifted mycorrhizal community composition (Shi et al., 2002; Bogeat-Triboulot et al., 2004).

Given the expectation that elevated $CO_2$ increases plant rhizodeposition, the effects of elevated $CO_2$ on root–mycorrhizal associations have received substantial attention over the last two decades. A meta-analysis of mycorrhizal responses to elevated $CO_2$ revealed almost a 50% increase in mycorrhizal abundance on average (Treseder, 2004), although there are numerous examples of studies where $CO_2$ effects were plant species-specific or altogether absent. In addition, it has been argued in many studies that $CO_2$ effects on mycorrhizal abundance are simply a function of faster plant growth (Staddon and Fitter, 1998). However, shifts in mycorrhizal community composition also occur in response to elevated $CO_2$ (Fransson et al., 2001; Treseder et al., 2003) and the effects of elevated $CO_2$ on mycorrhizal colonization can vary with fungal taxa (Fig. 9.4; Johnson et al., 2005).

The interactive effects of elevated $CO_2$ and other global change factors on mycorrhizae have also been studied extensively. Both elevated $CO_2$ and warming stimulated mycorrhizal hyphal production in constructed communities, but these effects were not additive (Staddon et al., 2004a). Not only do nitrogen additions tend to decrease mycorrhizal colonization (Treseder, 2004), but increases in mycorrhizal abundance under elevated $CO_2$ often occur only under low-nitrogen conditions (Klironomos et al., 1996; Staddon et al., 2004b). A similar trend is noted for phosphorus availability, where $CO_2$-induced increases in mycorrhizae are more common under phosphorus limitation (Syvertsen and Graham, 1999). Increased ground-level ozone can also offset the stimulatory effects of $CO_2$ on mycorrhizae (Kytoviita et al., 2001; Andrew and Lilleskov, 2009), and the effects of mycorrhizal fungi can interact with

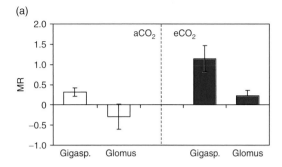

(a)

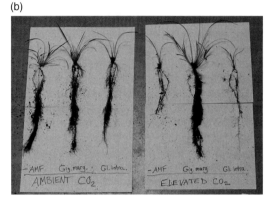

(b)

**Fig. 9.4.** (a) The effects of $CO_2$ enrichment (e$CO_2$) relative to ambient $CO_2$ (a$CO_2$) and arbuscular mycorrhizal (AM) taxa on mycorrhizal responsiveness (MR = the natural log of the ratio of dry masses for plants without AM fungi to plants with AM fungi) under the grass *Schizachyrium scoparium*. Positive MR values indicate AM mutualism and negative values indicate AM parasitism. (b) Photograph of *S. scoparium* that has been grown at ambient (left) or elevated (right) $CO_2$ in the absence of arbuscular mycorrhizal (–AMF), with *Gigaspora gigantea* (Gig marg.) or with *Glomus intraradices* (Gl. intra.). (Adapted from Johnson et al., 2005. Photo: J. Wolf, Biological Sciences, Northern Arizona University.)

those of root endophytes (Chen et al., 2007) or other soil fungi and bacteria (Drigo et al., 2007).

Nitrogen-fixing symbionts (rhizobia) are hypothesized to give leguminous plants a competitive advantage over other species under elevated $CO_2$; however, legumes have tended to be more responsive to elevated $CO_2$ only in managed systems, whereas nutrient availability (often phosphorus) can limit the responses of legumes in natural systems (Fig. 9.5; Rogers et al., 2009). Elevated $CO_2$ can increase the quantity of *Rhizobium* inoculum in the rhizospheres of host plants (Schortemeyer et al., 1996), increase nodule occupancy and shift the composition of *Rhizobium* strains present in roots

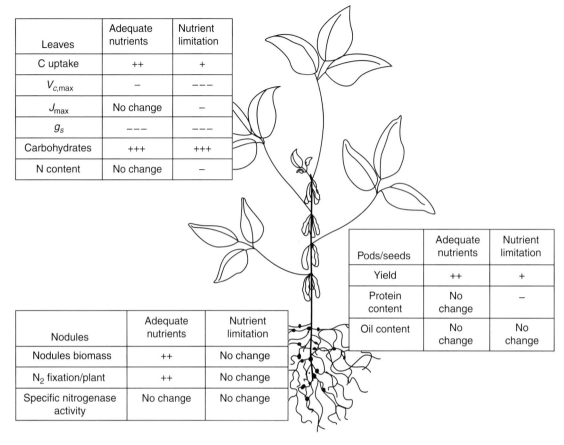

| Leaves | Adequate nutrients | Nutrient limitation |
|---|---|---|
| C uptake | ++ | + |
| $V_{c,max}$ | – | ––– |
| $J_{max}$ | No change | – |
| $g_s$ | ––– | ––– |
| Carbohydrates | +++ | +++ |
| N content | No change | – |

| Pods/seeds | Adequate nutrients | Nutrient limitation |
|---|---|---|
| Yield | ++ | + |
| Protein content | No change | – |
| Oil content | No change | No change |

| Nodules | Adequate nutrients | Nutrient limitation |
|---|---|---|
| Nodules biomass | ++ | No change |
| $N_2$ fixation/plant | ++ | No change |
| Specific nitrogenase activity | No change | No change |

**Fig. 9.5.** Effects of elevated $CO_2$ and nutrient supply on legume leaf, pod/seed and nodule characteristics, including stomatal conductance ($g_s$), maximum photosynthetic carboxylation capacity ($V_{c,max}$) and maximum capacity for electron transport leading to RuBP regeneration ($J_{max}$). '+' indicates increases in elevated $CO_2$, '–' indicates decreases in elevated $CO_2$, and the number of '+' or '–' symbols indicates the magnitude of change at elevated $CO_2$. (Redrawn from Rogers *et al.*, 2009.)

(Montealegre *et al.*, 2000). Plant responses to elevated $CO_2$ can also vary based on the rhizobial strains that are present (Bertrand *et al.*, 2007). In natural systems, the magnitude of nitrogen-fixation responses to elevated $CO_2$ can depend on soil nitrogen availability (West *et al.*, 2005), particularly for seedlings (Thomas *et al.*, 1991), and water limitation also limits $CO_2$ responses (Aranjuelo *et al.*, 2009).

### Root herbivores and disease

Much like our earlier discussion of soil fauna, climate change effects on root herbivores appear to be largely a function of the balance between warming effects and soil drying. For example, in the clover root weevil, warming rapidly shortens the time needed for egg development, but soil drying increases egg development time and decreases egg viability (Johnson *et al.*, 2010). Many studies of root herbivory in the context of climatic change have focused on plant responses to the combined effects of herbivory and climate, rather than directly on root herbivore populations. For example, in understorey tropical forest tree seedlings, damage by a coleopteran root borer was fatal only under drought conditions (Bebber *et al.*, 2002), and root-chewing insect larvae altered aboveground feeding by leaf miners, but only during drought (Staley *et al.*, 2007, 2008). Root herbivory can intensify under elevated $CO_2$ (Bezemer and Jones, 1998; Wilsey, 2001) and the benefits of increased nitrogen fixation by legumes under elevated $CO_2$

can even be undermined by increased root herbivory (Johnson and McNicol, 2010). However, these results are balanced against an equal number of studies where elevated $CO_2$ has had no effect on root tissue quality or belowground herbivory (e.g. Salt *et al.*, 1996; King *et al.*, 1997).

Climate change is also expected to influence the geographic range of root pathogens. For the oomycete *Phytophthora cinnamomi*, a severe tree root pathogen, climate warming is expected to enhance activity in its existing disease locations in the Mediterranean and coastal north-west Europe, although activity is not expected to extend into areas that will continue to experience cold winters (Brasier, 1996). The interactions of drought and forest tree root pathogens have been studied extensively and these two factors often operate synergistically as multiple stressors (see Chapter 13), such that pathogen resistance by the tree is reduced; the direct effects of drought on the pathogens are often negative, but the pathogens exhibit sufficient

plasticity to grow at water potentials below the minimum for growth of their host plants (Desprez-Loustau *et al.*, 2006). $CO_2$ enrichment can also alter the epidemiology of infected plants, as in the case of oat grass infected by the barley yellow dwarf virus, which exhibited increased root mass in response to elevated $CO_2$ when infected by the virus but did not in its absence (Malmstrom and Field, 1997).

## 9.3  Soil Food Web Responses

Many responses of soil organisms to climatic change may be indirect and mediated through interactions with other species. While we gave a brief description of the trophic structure of microbial communities earlier, such a description greatly oversimplifies the structure of microbial food webs. For example, as displayed in Fig. 9.6, nematodes not only function as microbivores, but some species also prey on protozoans or other nematodes. Likewise, as indicated by the dotted boxes in Fig. 9.6,

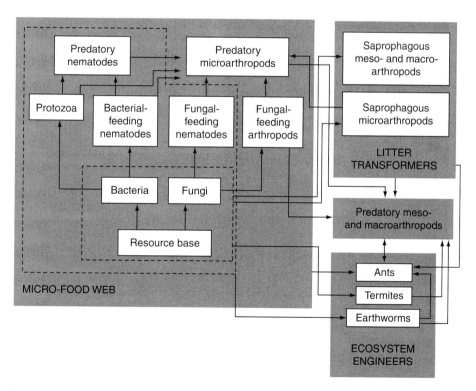

**Fig. 9.6.** Soil food web separated into the micro-food web, litter transformers and ecosystem engineers. Arrows indicate the direction of energy flow, and arrows starting at a dotted box indicate the simultaneous consumption of all organisms within the dotted box. (Redrawn from Coleman *et al.*, 2004.)

saprophagus microarthropods consume both plant litter (the 'resource base') and the fungal and bacteria colonists of this litter; and earthworms ingest whole communities of bacteria, fungi, protozoa and nematodes. Even Fig. 9.6 is oversimplified, as demonstrated by the functional roles of collembolans (described by Rusek, 1998), which are hosts of many parasitic protozoans, nematodes, trematodes and pathogenic bacteria, prey of many different predators, and themselves consume protozoans, nematodes, rotifers, worms, invertebrate carrion, bacteria, fungi, algae, plant litter, live plant biomass and some plant pathogens!

As with the study of aboveground community food webs, interest in climate change effects on belowground food webs has focused on separating the relative influences of top-down effects of climatic change (i.e. direct effects on consumers that cascade downwards to influence carbon and nutrient cycling) from bottom-up effects (i.e. environmentally driven changes in nutrient cycling that ultimately affect consumers). For example, in a study of soil microarthropods in three subarctic systems, changes in arthropod densities and community composition in response to warming were linked more to changes in food availability than to direct climate influences (Sjursen *et al.*, 2005a). In contrast, warming appeared to directly increase nematode grazing on microorganisms in the same systems, contributing to increased nitrogen and phosphorus mineralization rates and plant nutrient availability (Ruess *et al.*, 1999b). However, top-down and bottom-up effects in soils are not mutually exclusive, and it is not unlikely that both can contribute substantially to the abundance of a given species in a food web.

A question of particular interest to soil ecologists has been how changes in soil food webs in response to climatic change might alter soil carbon storage. In a temperate grassland, a decreased abundance of ecosystem engineers (large oligochaete and enchytraeid worms) coupled with an increase in fungivorous mites was predicted to have important implications for soil organic carbon turnover in response to sustained warming (Briones *et al.*, 2009). Conversely, although warming disrupted interactions between enchytraeid worms and the microbial community in a peatland system, it did not appear to be detrimental for ecosystem function (Cole *et al.*, 2002). An increased mechanistic understanding of food web responses to climatic change can be obtained by using litter bags (mesh pouches

containing small, pre-weighed samples of litter) with different mesh sizes to sequentially exclude macro- and mesofauna from the system (e.g. Taylor *et al.*, 2004). The small size of soil organisms also allows the dynamics of model systems, where initial community membership is controlled, to be monitored at the microcosm level (Kandeler *et al.*, 1998).

Under elevated $CO_2$, changes in resource availability often appear to dictate soil food web responses (e.g. Sonnemann and Wolters, 2005) and, in particular, the effects of changes in plant root exudation may directly affect the microbial community in the rhizosphere, rather than that of the bulk soil (Fig. 9.7; Drigo *et al.*, 2008). Nevertheless, organisms at higher trophic levels can respond directly to $CO_2$-induced changes in plant litter quality, potentially leading to top-down effects on microbial grazers (Mitchell *et al.*, 2003). In annual grasslands, elevated $CO_2$ effects on soil food webs were most pronounced early in the growing season, which was consistent with an increased flux of carbon from root to soil at this time because plants were undergoing rapid vegetative growth (Hungate *et al.*, 2000). Increased grazer biomass has also been observed with microbial biomass remaining stable, which has been interpreted as increased microbial turnover in response to elevated $CO_2$ (Lussenhop *et al.*, 1998; Frederiksen *et al.*, 2001, Allen *et al.*, 2005). The manipulation of food web complexity has interacted with elevated $CO_2$ in model soil systems (Couteaux *et al.*, 1991), and $CO_2$ effects on soil food webs are sometimes observed only in interaction with nitrogen addition (Klironomos *et al.*, 1997).

## 9.4   Plant–Soil Feedbacks

While we have already discussed how plant roots, microorganisms and soil fauna interact in response to climatic change, these interactions can be extended to changes in the quantity and quality of aboveground plant material. Soil organisms are affected by changes in litter quality and quantity within a given plant species and, over the longer term, changes in the relative abundances of plant species can further modify plant–soil interactions. Soil organisms play active roles in these feedbacks, with changes in decomposition and nutrient mineralization feeding back on plant responses.

An important role of plants in mediating soil responses to climate warming has been suggested

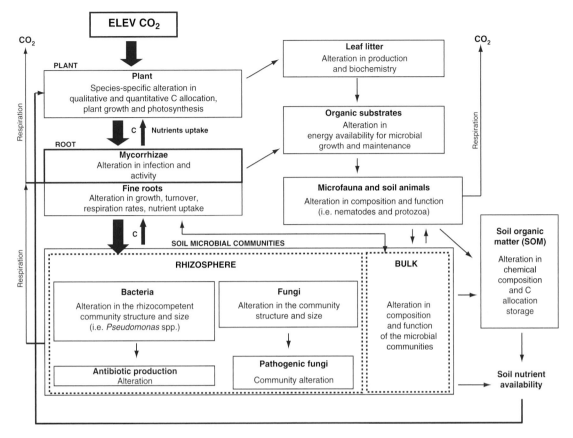

**Fig. 9.7.** Conceptual model describing plant species' responses to elevated atmospheric $CO_2$ and the resulting feedbacks with soil organisms in the rhizosphere and bulk soil. The increased plant carbon allocation is initially directed mainly to mycorrhizae, thereby changing the structure, size and activity of the rhizosphere microbial community (bacteria and fungi) to a larger extent than the community of the bulk soil. Soil microbial communities subsequently affect food web interactions. (Redrawn from Drigo *et al.*, 2008.)

by long lag times of response in some field experiments. For example, in a subarctic heath system, the soil microbial community exhibited a significant response to warming only after 15 years, and this response was attributed to gradual changes in plant biomass and community composition (Rinnan *et al.*, 2007). Warming effects on soil communities can also vary substantially depending on the presence or absence of aboveground herbivores (Rinnan *et al.*, 2009b) or leaf-clipping treatments (Zhang *et al.*, 2005). Nevertheless, large increases in plant biomass or shifts in rooting depth in response to warming do not always affect microbial activity or community composition (Bjork *et al.*, 2007; Biasi *et al.*, 2008).

Plant microclimate effects on soil in the context of warming have been demonstrated through the use of reciprocal soil transplants between areas with contrasting plant canopy cover. For example, the reciprocal transplantation of soils from under oak canopies and adjacent open grasslands revealed asymmetric responses of microbial communities, with those from under the oak canopy shifting substantially in composition in response to the extreme temperatures experienced in the open and those from the open environment remaining stable under the oak canopy (Waldrop and Firestone, 2006). Interactions between plant species effects and warming have been explored by examining soil community responses under specific plant functional groups (Malchair *et al.*, 2010). In addition, the assessment of microbial responses to warming in soil mesocosms in the presence and absence of plants and litter has further revealed an important role of plants in soil warming responses (Jonasson *et al.*, 2004). While the effects of changes in litter quality are often discussed in the context of climatic change, changes in soil microclimate caused by variation in litter mass can also strongly affect soil microbial biomass (Rinnan *et al.*, 2008).

As we described above, plant–soil feedbacks in response to elevated atmospheric $CO_2$ have received an overwhelming amount of attention from soil biologists. Given that the concentration of $CO_2$ in the soil is approximately 10–50 times higher than in the atmosphere, it is unlikely that soil organisms respond directly to changes in atmospheric $CO_2$ concentrations; instead, changes in plant root growth and rhizodeposition in response to elevated $CO_2$ can affect soil organisms, and elevated $CO_2$ also can decrease the soil water deficit by allowing plants to keep their stomata closed more frequently (Sadowsky and Schortemeyer, 1997). Important feedbacks between aboveground plant litter quality and decomposers have been hypothesized whereby decreased plant litter quality under elevated $CO_2$ prompts soil microorganisms to outcompete plants for soil nutrients, leading to further declines in plant and litter nutrient content. However, a meta-analysis revealed that declines in litter nitrogen content and increases in lignin content under elevated $CO_2$ had no consistent effects on litter decomposition rates, and it was concluded that any

changes in decomposition rates resulting from exposure of plants to elevated $CO_2$ are small when compared with other potential impacts of elevated $CO_2$ on carbon and nitrogen cycling (Norby *et al.*, 2001).

The effect of elevated $CO_2$ on microbial communities is highly dependent on plant cover type (Fig. 9.8; Montealegre *et al.*, 2002; Jin and Evans, 2007; Drigo *et al.*, 2009) and these two factors can interact together with nitrogen availability (Chung *et al.*, 2007). Null responses of soil microorganisms to elevated $CO_2$-induced changes in plant growth are not uncommon, although these results are often only published when accompanied by other factors (e.g. plant diversity or growth stage) that have significant effects on microbial communities within the same study (Gruter *et al.*, 2006; Bowatte *et al.*, 2007). Null responses may be particularly common in young developing ecosystems, where soil microorganisms are regulated more by existing pools of soil organic matter than by underdeveloped plant root systems (Zak *et al.*, 2000a). In one unusual case, root biomass responded to elevated $CO_2$ under one plant cover type but soil microorganisms did not, whereas the opposite pattern occurred under a different plant cover type (Kao-Kniffin and Balser, 2007).

Plant–soil feedbacks in response to elevated $CO_2$ can be extended beyond soil microorganisms to the

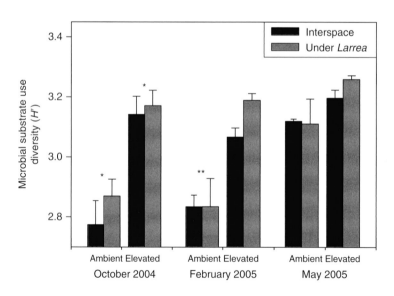

**Fig. 9.8.** Diversity of substrate use by soil microbial communities in interspace soils and soils under the shrub *Larrea tridentata* during the growing season in the Mojave Desert. (Redrawn from Jin and Evans, 2007.)

whole soil community. Soil food web complexity can enhance the decomposition of low-quality litter produced under elevated $CO_2$ (Couteaux *et al.*, 1996). In turn, declines in the quality of root and leaf litter under elevated $CO_2$ have altered nematode and enchytraeid worm abundance (Markkola *et al.*, 1996; Hoeksema *et al.*, 2000) and collembola have benefited from the improved soil water status under elevated $CO_2$ (Niklaus *et al.*, 2007). Soil arthropods modify microbial activity by mixing and fragmenting plant litter, and elevated $CO_2$ can modify fly oviposition preferences and fungal interactions on leaf litter (Frouz *et al.*, 2002). Pulse-labelling experiments with $^{13}CO_2$ are particularly useful for tracking carbon fluxes from plant roots through different compartments of soil food webs (Leake *et al.*, 2006).

In temperate grasslands, elevated $CO_2$ can indirectly increase the surface casting activity of earthworms, with large effects on soil carbon and nitrogen cycling (Zaller and Arnone, 1997). Similarly, proximity to earthworm casts can influence plant $CO_2$ growth responses (Fig. 9.9; Zaller and Arnone, 1999). Nevertheless, in grazed pasture, elevated $CO_2$ decreased nitrogen concentrations in earthworm casts, probably as a result of decreased nitrogen concentrations in animal dung (Chevallier *et al.*, 2006).

## 9.5 Conclusions

Overall, in this chapter we have demonstrated that although the responses of soil organisms to climatic change can be rather idiosyncratic, clear mechanisms have emerged whereby the direct and indirect responses of soil communities can often have important consequences at the plant and ecosystem levels. Soil communities are very complex and questions remain as to whether the responses of these complex communities can be predicted based

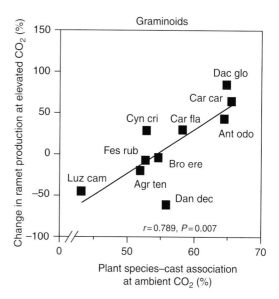

**Fig. 9.9.** Response of grasses and sedges to elevated $CO_2$ as a function of their mean degree of association with surface casts at ambient $CO_2$. Plant species key: Agr ten = *Agrostis tennis*, Ant odo = *Anthoxanthum odoratum*, Bro ere = *Bromus erectus*, Car car = *Carex caryophyllea*, Car fla = *Carex flacca*, Cyn cri = *Cynosurus cristatus*, Dac glo = *Dactylis glomerata*, Dan dec = *Danthonia decumbens*, Fes rub = *Festuca rubra*, Luz cam = *Luzula campestris*. (Redrawn from Zaller and Arnone, 1999.)

on highly simplified experimental set-ups (Kampichler *et al.*, 1998). Transient responses of soil communities also need to be considered in the context of longer-term responses to cumulative changes in soil organic matter and plant species composition. Finally, as we move from phenomenological studies of soil responses to climatic change to mechanistic studies, strengthening the links between community structure and function remains an important challenge.

# 10 The Future of Forest Productivity

Forests cover about 30% of the planet's land area and are responsible for an even greater proportion of global terrestrial NPP (about 50%) and carbon stocks (about 45%). The world's forests, collectively, are currently carbon sinks, meaning that they help draw down atmospheric $CO_2$, but, as we saw in Chapter 7 with regard to terrestrial vegetation as a whole, may become sources of carbon as climatic change proceeds. Forest productivity is, of course, also important economically, particularly for lumber, paper and fuel (the forestry sector contributes nearly US$500 billion per annum to the global economy; FAO, 2009); in the near future, the economic benefits of forests as sources of carbon credit may also materialize.

## 10.1.  Impacts on Net Primary Production

How has forest growth been affected by the recent directional changes in climate? Since temperature increases have been largest at high latitudes, much attention has been paid to boreal forests. Goetz et al. (2007) used satellite observations to detect changes in boreal forest productivity over the past 20 years, using indices of photosynthetic carbon uptake. They found that less than 4% of undisturbed boreal forest in North America showed productivity increases, while more than 25% showed productivity declines. These findings were attributed to the negative effects of warming-induced late-season drought conditions on dense conifer forests. This result contrasts with a survey of 49 studies of forest growth over the past 50 years; most of these studies indicated increases in forest productivity, including in the boreal forest.

In both temperate and boreal forests, increases in growth (most often reported as stem or basal area growth) observed over the past several decades have been attributed mainly to lengthening of the growing season with rising temperatures (Boisvenue and Running, 2006).

It is more difficult to discern recent responses of tropical forests to climatic change and much debate exists on this issue. While some researchers have reported climate change-driven increases in aboveground biomass in undisturbed, mature tropical forests over the past few decades (e.g. Lewis et al., 2004), others claim that the productivity losses brought on by recent drought events illustrate the negative effects that increased temperatures are having on tropical forests (Clark, 2004). Most of the evidence suggests that, in the future, increasing temperatures will have a negative effect on forest NPP due to water stress imposed by increased evapotranspiration rates. The combination of rising temperatures and elevated $CO_2$, however, has the potential to increase forest productivity.

The IPCC's Fourth Assessment Report (Working Group II; Easterling et al., 2007) concludes that despite likely declines in some regions, global timber production is predicted to increase due to higher growth rates and shifts in forest ranges (e.g. poleward shifts of productive timber species). Various forest yield models (those used to estimate timber resources) predict increases in global stemwood production of about 30% by the end of the century, particularly when elevated $CO_2$ is taken into account (Easterling et al., 2007). We must use caution, however, when interpreting the results of timber production models as most do not incorporate ecological processes such as nutrient cycling and disturbances (Kirilenko and Sedjo, 2007) and they assume that increases in stemwood growth will be proportional to increases in NPP (Sohngen *et al.*, 2001). The rare timber models that are based on ecological processes show more modest effects; for example, 18% increases in stemwood growth are predicted for European forests by 2030 by Nabuurs *et al.* (2002) using a physiologically based mechanistic modelling approach (Chapter 3).

Modelling studies that consider only climatic changes tend to find negative impacts on forest productivity in the future. For example, Battles *et al.* (2008) modelled growth of commercial conifer species in the Sierra Nevada region of the USA under projected changes in climate and found reduced stemwood production over the coming century in all species. This region is characterized by summer drought conditions; over the next 100 years, the authors estimate that increases in summer temperatures will essentially shorten the growing season by inflicting moisture stress earlier in the summer than usual, thus impacting tree growth. Similarly, modelling of future NPP in forests of north-eastern China suggests declines in NPP over the coming century in both conifer and deciduous forests, linked to rising temperatures (Xiong *et al.*, 2007). Again, moisture shortage was the culprit; elevated $CO_2$ was not included in either of these studies and may alleviate moisture deficits somewhat.

Forest ecosystem models show the potential for increases in productivity if adequate nutrients are available and moisture stress is not an issue. Thornley and Cannell (1996) modelled a temperate conifer plantation forest under climate and atmospheric change, and found that while increasing temperatures alone led to declines in productivity due to water stress, the inclusion of elevated $CO_2$ in the model could lead to greater productivity by ameliorating water stress (Fig. 10.1).

As seen in the forest FACE studies, NPP has been found to increase under elevated $CO_2$ alone, although this is not always reflected in increasing tree biomass, depending how the extra carbon is allocated (see Section 10.2). Across a range of temperate forest FACE sites, young conifers showed increases in stemwood production while deciduous hardwoods showed increases in fine root production at the expense of stemwood (Norby *et al.*, 2005). NPP was found to increase by 19 to 27%, with the lowest increases seen in low-nitrogen sites and the higher increases seen in more fertile sites. Increases in NPP were attributed to both increases in light capture (as leaf area increased) and to improved efficiency of light use. Elevated $CO_2$ allows for more carbon capture by trees and can alleviate water stress by reducing stomatal opening (see Chapter 4). Most $CO_2$ enrichment studies have taken place in young forests; the few results from mature forests indicate that these will likely show little tree biomass response to elevated $CO_2$ (e.g. Rasmussen *et al.*, 2002). Some suggest that there is little evidence for long-term increases in forest biomass production under elevated $CO_2$, due to widespread nitrogen

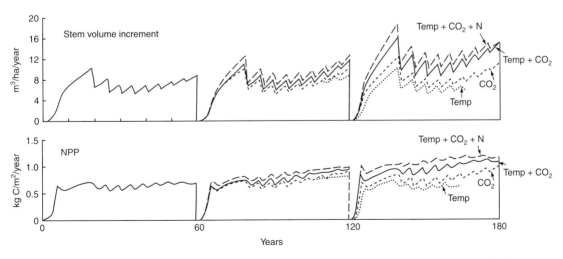

**Fig. 10.1.** Stemwood volume increment and NPP of a modelled temperate conifer plantation forest over the last of three 60-year rotations. The climate and atmospheric change treatment begins in the second rotation. (Redrawn from Thornley and Cannell, 1996.)

limitation in forest ecosystems and other external stresses such as ozone pollution.

Compared with temperate and boreal forests, much less research has been done on the carbon dynamics in tropical forests and their responses to perturbation, rendering the future response to climatic change of tropical forest productivity even more uncertain. Drought events in the Amazon basin are expected to become more common and more severe in a warming world, with likely negative effects on tropical forest productivity. An experimental drought (40% reduction in the rainfall allowed to reach the forest floor) in an Amazon basin forest (Brando et al., 2008) revealed significant declines in stemwood production (up to 60% per year) and overall carbon stocks. The declines in wood production were proportionally larger than the declines in aboveground NPP due to changes in the way carbon was being allocated by the trees (see Section 10.2).

Local-level changes in tree growth will combine with shifts in tree species ranges to determine larger-scale changes in forest productivity. At present, there is little evidence of northward boreal forest expansion into tundra (Masek, 2001) in response to recent climatic change. However, models suggest that the northward shift of boreal forests over the coming century will be one of the most significant alterations to current vegetation distribution patterns (e.g. Cramer et al., 2001).

A study of vegetation shifts in the USA projected by several different climate models under doubled $CO_2$ shows overall declines in total forest area by the end of the century under all but the mildest temperature increases. This decline is due mainly to losses in north-eastern mixed forest that will not be made up for by increases in southern forest types as they advance north (Fig. 10.2). Instead, the area of savannah-like vegetation will increase in place of

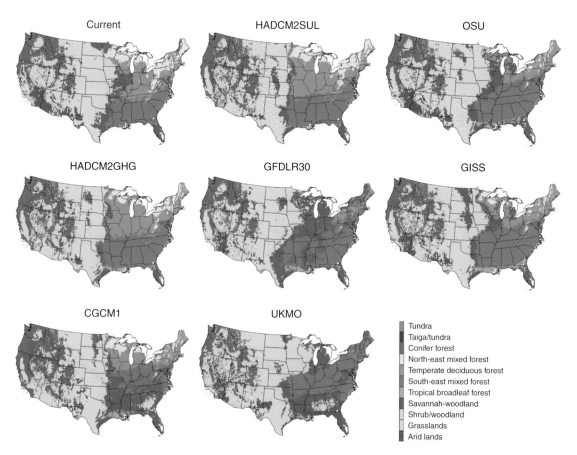

**Fig. 10.2.** Shifts in vegetation distribution in the USA under various climate change projections by different GCMs. (From Bachelet et al., 2001.)

Chapter 10

forest (Bachelet *et al.*, 2001). The effects on forest vegetation carbon stocks depend on the magnitude of warming, with moderate warming (3°C) leading to carbon gains in some forest types, due to growing season extensions and $CO_2$-mediated amelioration of water stress, and more extreme warming (5°C) resulting in carbon losses across the board, due to increased drought stress.

There will likely be much regional variation in changes to forest distribution in response to climatic change. One recent study found both positive and negative effects of forest distribution shifts on NPP in different regions of the USA. In some areas NPP increases by more southerly species outweighed the declines of northerly species, while elsewhere the opposite was true. Overall, though, this study found that the local-level direct effects of climatic change were more important than species shifts in influencing future NPP (Chiang *et al.*, 2008). In another recent modelling study, the response of northern Wisconsin forests to climatic change found overall increases in aboveground tree biomass over the coming century (by 5 to 30% depending on the climate projection used; elevated $CO_2$ not included) but differential responses among species, with some showing strong increases and others becoming extirpated from the region. In general, tree community composition was shifted towards greater dominance by southerly species (Scheller and Mladenoff, 2005).

Ultimately, changes to NPP due to shifts in forest distribution and direct effects on growth are not the whole story when it comes to climate change impacts on wood production and overall forest carbon storage: how the NPP is allocated is important, as is the fate of the carbon once the tissues are shed as detritus. Forest NPP also depends greatly on the stage of development a forest finds itself in. In forest regions where stand-replacing disturbance (e.g. fire) is common, this is particularly crucial to landscape- and regional-level rates of NPP. The following sections address these topics.

## 10.2 Carbon Allocation Patterns

Both the economic and ecological impacts of NPP responses to climatic change depend on how plants 'choose' to allocate the extra available carbon from rising atmospheric $CO_2$ levels and/or how changing environmental conditions (warming, droughts) might also alter how carbon is allocated in plants. It is not a given fact that, for example, enhance-

ments in NPP will result in increased bole (trunk) volume in trees. Such an increase would be a boon from the commercial perspective, and also from the carbon sequestration perspective as woody stems represent relatively long-term pools of carbon.

There are various reasons why trees, and other plants, might change their carbon allocation patterns in response to climatic change. The most important kind of change from a wood production or carbon sequestration perspective would be a shift from long-lived to short-lived tissues or vice versa. For example, if trees respond to elevated $CO_2$ by using the additional carbon to beef up their contribution of carbohydrates to the rhizosphere (i.e. to mycorrhizae and soil bacteria, via root secretions), this would represent a fast turnover of the extra carbon since these materials are quickly decomposed and respired. We would see a similar outcome if the tree invests in more fine roots and/or leaves – short-lived tissues that are shed regularly and decompose relatively quickly (although there may be a time lag of years to decades in cold and/or very wet environments). The indirect effects of climatic change, such as changing water and nutrient availabilities, may trigger changes in allocation patterns regardless of $CO_2$ levels – faced with changing soil resources, trees will alter their investment in belowground resource acquisition accordingly (i.e. shunt more or less carbon to fine roots and/or rhizosphere carbohydrates).

There is evidence that trees will allocate more of the excess carbon to short-lived tissues such as leaves, fine roots and reproductive structures, rather than to stemwood production (Beedlow *et al.*, 2004). Some results of the FACE experiments seem to support this contention. A forest FACE experiment in the south-eastern USA (Oak Ridge, Tennessee) found that while NPP increased by an average of 22% over 6 years under doubled $CO_2$, most of the increase was allocated to fine root production (Fig. 10.3; Norby *et al.*, 2004). But since fine root turnover rates did not change (mortality rates kept pace with the increased production rates), most of the extra assimilated carbon was turned over quickly rather than being directed towards long-term pools like stemwood. Other researchers have found that the flow of carbohydrates from tree roots to fungal associates (e.g. ectomycorrhizae) may act as an 'overflow tap' for excess carbon taken up by trees growing under elevated $CO_2$, providing another example of a response that results in faster throughput and

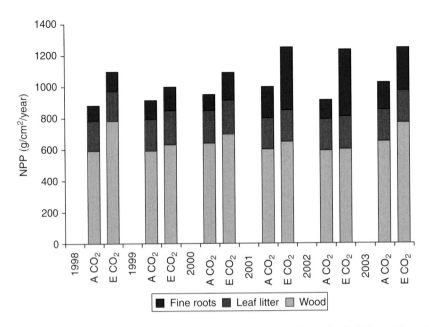

**Fig. 10.3.** Fine root production dominates response of a deciduous forest to atmospheric $CO_2$ enrichment. (Adapted from Norby *et al.*, 2004.)

turnover of assimilated carbon rather than increased long-term storage in tree tissues (Heinemeyer *et al.*, 2007).

So the response of a forest, in terms of stemwood production and carbon sequestration, is likely to depend on whether and how carbon allocation patterns change. There are still unanswered questions, however, about how these allocation patterns will change in response to elevated $CO_2$. Across a series of forest FACE sites, the proportion of increased NPP allocated to stemwood ranged from as little as 11% to as much as 93%, with no obvious patterns among sites despite differences in tree species and site fertility (Norby *et al.*, 2005). Similarly, the fraction of NPP allocated to stemwood increased under elevated $CO_2$ in young conifers, but not in deciduous trees (McCarthy *et al.*, 2006).

Changing environmental conditions will also likely influence carbon allocation patterns as trees adjust to limiting resources. In an Amazon basin forest, trees responded to experimental drought by reducing carbon allocation to wood production but maintaining belowground allocation, likely to maximize water uptake (Brando *et al.*, 2008). Similar findings on a more general scale were reported in a review of studies from forests around the world. In temperate and tropical forests,

increasing mean annual temperature led to more carbon being allocated belowground, likely to seek out increasingly limited resources like water. In boreal forests (for which limited data were available) belowground allocation of captured carbon seemed to decline with increasing temperatures; this makes sense, as increased decomposition and nutrient release rates with warming at high latitudes can relax soil resource limitations (Litton and Giardina, 2008).

## 10.3  Changes in Disturbance Regimes

Disturbances are integral processes in forest ecosystems, shaping forest structure and function and acting as important renewal agents. Fire, for example, is a natural and relatively frequent occurrence in many forest types. In the boreal forest, fire often results in 'stand replacement', i.e. high levels of tree mortality and the re-initiation of succession. Boreal trees have thus developed adaptations to deal with fire, with some species requiring fire for regeneration. For example, jack pine (*Pinus banksiana*) and lodgepole pine (*Pinus contorta*) have serotinous cones; these are resin-sealed and open only with extreme heat (Fig. 10.4), to ensure that seeds are not released until the right conditions for regeneration are in place. In

**Fig. 10.4.** A resin-sealed, serotinous cone from a lodgepole pine (*Pinus contorta*) opens in response to fire. (Photos: S. Hunt, School of Environmental Sciences, University of Guelph.)

temperate and tropical forests fire is usually much less frequent, and the prevalent disturbances usually result in the scattered mortality of single trees or groups of trees (e.g. windthrow or attack by a host-specific insect or pathogen), creating small gaps in the forest canopy. Whatever the spatial scale, disturbance events act to alter microclimate and resource availability at the forest floor, altering process rates (e.g. nutrient cycling) and providing new opportunities for young trees and other vegetation to access resources. While the forest biota is finely tuned to a particular disturbance regime (i.e. the frequency, intensity, severity and spatial extent of a disturbance), disturbances that fall outside the natural range can change forest structure and function. Climatic change has the potential to affect both biotic disturbances (insect and pathogen outbreaks) and abiotic disturbances (e.g. fire and wind). In Chapter 13 we examine further how climate change and disturbance can interact to produce often surprising outcomes in ecosystems more generally. Increased rates of disturbance on the forest landscape have the potential to reduce overall NPP and carbon storage despite any positive direct effects that temperature and/or elevated $CO_2$ may have.

### Fire

Several studies have shown that in some forest regions (e.g. in the Canadian boreal and the western USA) the amount of area burned has been increasing over the past few decades in concert with rising temperatures (Gillett *et al.*, 2004; Westerling *et al.*, 2006; Fig. 10.5). Over the next century, the occurrence of fire and the total area burned in many forest regions will likely continue to rise as a result of climatic change (Lenihan *et al.*, 2008).

When it comes to climatic factors, temperature seems to have the strongest influence on fire activity. Increasing temperatures can increase fire activity by increasing evapotranspiration rates (and therefore creating drier conditions and drier fuels), increasing lightning activity (a major ignition source) and lengthening the fire season (Flannigan *et al.*, 2009). Indirectly, the shifts in tree species distributions predicted to occur with climatic change will also influence fire activity by changing fuel characteristics (for example, deciduous, broad-leaved species tend to be less flammable than evergreen conifers). And fuel quantity may increase if

temperature combines with elevated $CO_2$ to increase forest NPP, further affecting fire behaviour.

Modelling studies predict that projected changes in climate will increase the occurrence of weather conditions conducive to fire in many parts of the world, including Australia, the Mediterranean, and the boreal forests of North America and Eurasia. Most of the efforts to predict changes in total area burned relate to North American forests, with most estimates ranging from increases of 20 to 100% above historical values by the end of this century (Flannigan *et al.*, 2009). Less is known about how climatic change may affect fire regimes in tropical forests. While these forests have experienced increased fire activity over the past few decades, this is tied to human activity and the use of fire as a tool for forest clearing and land-use change (see Chapter 13).

An increase in forest fire activity has strong potential for feedbacks to climatic change, since fire frequency, severity and spatial extent influence the forest carbon balance. Fire results in immediate carbon losses from forests as organic matter is oxidized and $CO_2$ is emitted to the atmosphere. Fire frequency also determines the amount of carbon being stored on the forest landscape, through its influence on forest age-class structure. More frequent stand-replacing fire leads to a higher proportion of young

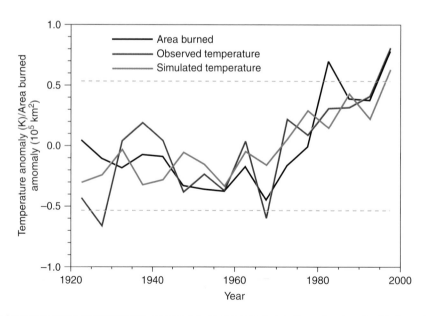

**Fig. 10.5.** Relationship between temperature and area burned anomalies in Canadian forests. (Redrawn from Gillett *et al.*, 2004.)

forest on the landscape, and hence lower carbon stocks overall due to the small stature of young trees and the smaller pools of dead organic matter (immediately after fire dead organic matter pools are large, but decomposition rates are also accelerated, with carbon losses outweighing photosynthetic gains for several years to decades post-fire). So more frequent fires could initiate a positive feedback loop with climate wherein more burning leads to less carbon sequestration, more carbon in the atmosphere and hence more warming. On the other hand, this runaway response could be dampened if the younger, regenerating landscape were less susceptible to further burning, e.g. if a higher proportion of young deciduous stands were to occupy the landscape (Amiro et al., 2001). Young stands have also been shown to have higher albedo (reflectance of solar energy) than mature conifer stands, and so could counterbalance at least to some degree the positive climate forcing of greater $CO_2$ emissions with a cooling effect (Goetz et al., 2007).

As we can see, the relationships between forests and climate are very complex, which explains the degree of uncertainty in predictions of future forest conditions. Overall, at least for the relatively well-studied Canadian boreal forest region, increases in the total area burned will likely reduce the sink strength of this region and potentially render it a carbon source (Kurz et al., 2008c).

## Insect outbreaks

It has long been recognized that insects are important organisms in forest ecosystems, contributing to and affecting forest processes such as nutrient cycling and net primary productivity, and influencing forest structure, successional patterns and species composition. The impact of climatic change on the relationship between insects and forests represents an important unanswered question with regard to understanding the effects of climatic change on forest ecosystems (Gower, 2003). A changing climate will likely alter the disturbance dynamics of native forest insects and diseases, as well as facilitating the establishment and spread of non-indigenous species.

The mechanisms by which climatic change could alter disturbance patterns from insects and pathogens include: (i) direct effects on the development and survival of herbivores and pathogens; (ii) physiological changes in tree defences; and (iii) indirect effects from changes in abundances of natural enemies, mutualists and competitors (Ayres and Lombardero, 2000). The first mechanism will presumably be most important in the case of invading, non-native species. Their short life cycles, physiological sensitivity to temperature, mobility and reproductive potential will allow forest insects and diseases (native and non-native) to respond rapidly to climatic change (Ayres and Lombardero, 2000).

Direct effects of climatic change on forest pests in temperate and boreal forests will likely be manifested as increased survival rates due to predicted warmer winter temperatures and increased developmental rates due to warmer summer temperatures (see Chapters 4 and 5). Ungerer et al. (1999) predicted that warmer winter temperatures as well as less variance in temperature would be more important than warmer summers for the northward expansion of the southern bark beetle (Dendroctonus frontalis) in the USA. Increased climatic variability in the form of more extreme weather events could also possibly benefit forest pests by increasing the vulnerability of trees to infestation (e.g. more extreme weather events such as windstorms can produce damage on trees and allow entry to pathogens, while more lightning strikes can encourage bark beetle infestation; Ayres and Lombardero, 2000).

In a modelling study of the response of mountain pine beetle (Dendroctonus ponderosae, a native species in north-western North America) to climate warming, researchers found that the temperature is currently strongly limiting to the beetle, meaning that small changes in climate could remove significant limits on this species (see Box 10.1; Williams and Liebhold, 2002). Some speculate that a warming climate may already be partly responsible for the apparent northward expansion of mountain pine beetle in British Columbia, Canada (Logan et al., 2003). However, there is a great deal of complexity involved in species' response to climatic change. For example, while an increase in temperature of just over 2°C would shift mountain pine beetle populations from semivoltine (requiring 2 years to complete a generation) into synchronous, univoltine (requiring 1 year to complete a generation), further warming could shift populations into an unfavourable (for the insect) asynchronous life cycle due to disrupted seasonality requirements (Logan and Powell, 2001). Other insect species may be negatively affected by climatic change if diapause (cold-weather dormancy) requirements have a lower chance of being met,

**Box 10.1. The mountain pine beetle epidemic in British Columbia, Canada.**

The mountain pine beetle (*Dendroctonus ponderosae*) is an important forest insect in lodgepole pine (*Pinus contorta*) forests of western North America. It is a naturally occurring insect and has always been present in these forests in at least small population sizes. The beetle is a bark-boring insect, and successful attacks on a host tree are usually fatal. Attacks begin when female beetles swarm a tree and bore through its bark. The females then excavate galleries, mate and lay eggs. The beetles also introduce a fungus into the tree which quickly spreads into the sapwood, inhibiting the flow of nutrients and water and increasing the likelihood that the eggs will survive. When the eggs hatch, the newly born larvae feed on the phloem of the tree, effectively girdling it and, with help from the fungus, usually killing it.

The beetle larvae overwinter beneath the bark of the tree and then pupate, emerge and start the cycle over in the following spring. Most of the time, mountain pine beetles are present on the landscape in low numbers, and they usually survive by attacking only weakened and stressed trees.

Occasionally, though, the population explodes and huge tracts of trees are killed during the resulting outbreak (Fig. 10.6). While predators such as woodpeckers play a small role in regulating beetle population sizes, it turns out that winter climate is the critical factor in determining when the beetles persist at small population sizes and when outbreaks occur. When winter temperatures drop below −40°C, or when spring or autumn temperatures drop below about −25°C, significant mortality of larvae occurs and outbreaks are normally arrested (Safranyik and Linton, 1991, 1998). Winter precipitation also plays an important role, because deep snowpacks can insulate the lower boles of trees against these cold temperatures.

Since 2003 winters in British Columbia have been consistently warm, and the mountain pine beetle population has exploded. As of 2009, 16.3 million ha of lodgepole pine forest had been affected by mountain pine beetle and 80% of the mature pine forest is predicted to be killed by 2013. This recent period of exceptionally warm winters has allowed the mountain pine beetle to attack

Fig. 10.6. This forest, near Nelson, British Columbia, has been attacked by the mountain pine beetle. The trees entirely without foliage (called 'grey-attack trees') were probably attacked in the last 4–7 years, the trees with red needles ('red-attack') were probably attacked in the last 2–5 years, and the trees with the yellow foliage were probably attacked in the last 1–2 years. In all likelihood all the green pine trees in the photo are currently under attack. (Photo: Z. Gedalof, Department of Geography, University of Guelph.)

hosts that are either normally resistant or that live outside the beetle's climatic optimum (Logan *et al.*, 2010). For example, severe droughts in 2003 and 2004 coupled with a very high density of beetles led to successful attacks on ponderosa pine (*Pinus ponderosa*) in the Kamloops area of British Columbia. Ponderosa pine is normally resistant due to its thick bark and abundant resin, but the drought reduced its capacity to 'pitch out' the beetle and widespread mortality of trees resulted. At very high elevations throughout western North America whitebark pine (*Pinus albicaulis*) is an ecologically important treeline species. Historically it was not attacked by the mountain pine beetle due to the particularly cold climate in the subalpine zone. In recent years, however, the mild climate has allowed the mountain pine beetle to expand its range upslope and large numbers of whitebark pine have been killed or are currently under attack.

Of perhaps greater ecological concern, continued warming will give the mountain pine beetle access to jack pine (*Pinus banksiana*) forests that extend from the northern Rocky Mountains to the Atlantic Ocean. Much of this forest is already south of the winter −40°C isotherm and is currently buffered against the mountain pine beetle only by the Canadian prairies, which are largely treeless. Unlike lodgepole pine, these other pines have not co-evolved with the mountain pine beetle and the outcome of the beetle's immigration into their range is likely to be severe and unpredictable.

While it is tempting to blame the current outbreak on global warming, it is no more plausible to say that this warm interval is caused by anthropogenic emissions of greenhouse gases than it is to say that a freak snowstorm is proof that climate change is a hoax. However, we can reasonably argue that this sort of anomalous warm weather is predicted to occur more frequently in a greenhouse world, and regardless of whether the current outbreak is caused by global climatic change, future outbreaks are likely to be larger and more frequent than those of the past and are likely to expose species previously protected from attack to the mountain pine beetle.

thereby disrupting developmental cycles (Ayres and Lombardero, 2000; see Box 10.1 for more on the mountain pine beetle). Hunt *et al.* (2006) found that climatic change would result in the potential for several non-native, invasive forest insect species to increase their range sizes in Canada.

As we have seen in Chapter 4, physiological changes in host trees may include increases in the amounts of defensive chemicals produced as a response to increased atmospheric $CO_2$ concentrations and/or moderate drought, increasing host resistance to pest attack. Raised $CO_2$ levels may also make conditions less favourable to insects by increasing the C:N ratio of tissues, which could result in nutritional deficiencies, but could also result in more feeding as the organisms attempt to gain enough nitrogen (see Chapter 5). Increased drought could be potentially advantageous to insect herbivores if sugar concentrations in foliage are increased as a result, making it more palatable, or if stomatal conductance declines, raising microclimate temperatures on tree foliage and thereby benefiting insects (Fleming and Candau, 1998). Drought conditions can also indirectly influence insect populations through changes in resource allocation by host trees; for example, drought causes jack pine to produce more reproductive tissue, which results in greater survival of the jack pine budworm (*Choristoneura pinus*), a native forest pest (Fleming, 2000). Canada's boreal forest region in particular is expected to become warmer and drier, and while moderate drought may increase production of defence compounds in trees, drought stress can also make trees more vulnerable to mortality from insect attack (Fleming, 2000; Hogg *et al.*, 2002).

There are three relevant time frames in the consideration of insect outbreak patterns in a changing climate: (i) the short term, in which insect populations will likely respond to climatic change more rapidly than their hosts; (ii) the medium term, which may see widespread disruption to forest ecosystems, as trees experience difficulties regenerating and shifts in tree species' ranges occur; and (iii) the long term, where a new equilibrium will develop between insects and forest ecosystems (MacLean *et al.*, 2002).

On balance, Kurz *et al.* (2008c) concluded that for the Canadian boreal forest (which holds about 15% of global forest carbon stocks), any NPP increases due to climate and atmospheric change would not be likely to outweigh the increases in carbon loss caused by increased disturbance rates.

### Resilience to disturbances

In addition to changes in the characteristics of the disturbances themselves, climatic change has the potential to alter a forest's ability to 'bounce back' after disturbance. Changes to disturbance severity (e.g. the amount of organic matter consumed by fire) can alter successional pathways through effects on tree seedling recruitment. For example, in boreal forests, greater severity can lead to more deciduous tree recruitment because boreal deciduous species have more difficulty germinating on thick organic layers. Less severe fires favour conifers like black spruce (*Picea mariana*), which can successfully establish themselves on these thick forest floors.

A study in an Alaskan black spruce forest, comparing ecosystem recovery after fire in dry versus mesic sites, found that while recovery of tree biomass was initially faster on the dry sites, later on in stand development biomass production was lower than in mature stands on mesic sites (Mack *et al.*, 2008). Climatic change may convert many mesic, permafrost-underlain sites to dry sites as permafrost disappears in a warming climate, and this may result in more fast-growing deciduous vegetation in early succession but lower biomass due to lower spruce density as stands mature. Ultimately, carbon storage on the forest landscape in Alaska may be reduced due to climatic change not only because of increased fire frequency and intensity, but also because of altered post-fire vegetation dynamics (Mack *et al.*, 2008).

Climatic change may have the potential to change the physiognomy of the ecosystem, by preventing or slowing the re-establishment of forest vegetation after disturbance. Prieto *et al.* (2009) found evidence of stalled succession after a Mediterranean pine forest had been subjected to fire. Shrubby vegetation persisted longer and establishment of pine seedlings was slower in the plots subjected to experimental warming and drying.

So, to understand climate change effects on disturbance rates and the resulting impacts on forest productivity and carbon balance, we need to consider not just the immediate effects, but also the potential long-term effects of altered recovery rates and pathways.

## 10.4 Conclusions

In this chapter we have explored the various ways that the world's forests may be affected by climatic and atmospheric change. While there is some evidence to suggest that, globally, forest productivity may increase, at least for the next few decades, due to range expansions, longer growing seasons and the combined effects of temperature and elevated $CO_2$ on stand-level NPP, there are complicating factors that make it difficult to embrace the optimistic view that forest carbon stocks and the forest products industry will benefit from climatic change. First, carbon allocation patterns matter. NPP may increase, but unless stemwood is the repository for the increased carbon uptake, the industry will not necessarily benefit. Increased NPP also does not necessarily translate into increased carbon storage in forests, if the additional carbon is simply turned over quickly (i.e. via short-lived tissues like fine roots and decomposition rates that keep pace with production increases) and returned to the atmosphere. Second, we do not get the full picture of climate change impacts on forests until we 'zoom out' to consider larger spatial scales. Changes in forest productivity at the local level can be eclipsed by landscape-scale changes, such as changes to disturbance regimes. Lastly, while the research community has managed to amass a fair amount of information on how boreal and temperate forests may respond to climatic change and elevated $CO_2$, tropical forests have received much less attention despite the fact that they dominate other forest types when it comes to land area, NPP and carbon stocks. The few studies that do exist on tropical forest responses to climatic change have provoked much debate, which it is hoped will lead to intensified efforts to understand how these ecosystems will be affected.

# 11 The Future of Agricultural Production

In this chapter we consider how agricultural production might be impacted by climatic change, what options are available for mitigating those impacts and what the prospects are for adapting agricultural practice to the new climate reality. Agricultural scientists have been interested in the impacts of climatic change even longer than ecologists and, owing to the simplification of the environment and ecosystem, impacts on agriculture are probably better known than most biological impacts of climatic change.

## 11.1 Impacts on Crop Systems

In some ways, answering questions about the potential or likely impacts of climatic change on agricultural systems is actually harder than for 'natural' ecosystems. This is because any answer will likely have lots of caveats about human behaviour. One could give a straightforward answer, assuming that the same crops are grown in the same places using the same techniques in the future as they are now. Then it would be relatively simple to say how the projected changes in climate will impact production. However, that answer would be pretty meaningless because we will not continue to grow the same crops in the same places using the same techniques; we will adapt our agricultural systems, to the extent technically possible. And there are many possible adaptation and mitigation strategies, from simple changes in the timing of planting and harvesting, to choices of cultivars and crops, to the development of new crops and/or new crop traits. We consider some of these below; but

before considering the potential for adaptation, we first consider four aspects of climate impacts: (i) location of agroecological zones; (ii) $CO_2$ fertilization; (iii) water availability; and (iv) extreme weather events (Kurukulasuriya and Rosenthal, 2003).

### Agroecological zones

Agroecological zones are a classification system, of the Food and Agriculture Organization of the United Nations (FAO; http://tinyurl.com/ch11 AEzones), that has been widely used to consider the impacts of climatic change on agricultural production, particularly at the regional, continental and global spatial scales. Agroecological zones are identified using a standard methodology that considers an area's climate, soil and terrain conditions as they pertain to agricultural production (example maps can be found on the FAO web site). Since climate is an integral part of the assessment and climate is expected to change, so too are the locations of current agroecological zones. Rosenzweig and Hillel (1998) point out that rising temperatures in middle and higher latitudes will tend to be beneficial, lengthening the growing season and generally shifting areas suitable for agriculture towards the poles. On the other hand, increasing temperature in the lower latitudes will tend to be detrimental due to increasing heat stress and reduced soil moisture. Just because the *climate* may become suitable for agriculture in regions where it does not currently exist does not necessarily mean that agriculture will be possible, as soil type may still be limiting or at least challenging.

## CO$_2$ fertilization

We considered the impacts of CO$_2$, temperature and precipitation on photosynthetic efficiency in some detail in Chapter 4. In general, at least for C$_3$ plants, increasing concentrations of CO$_2$ lead to greater plant growth, the so-called 'CO$_2$ fertilization effect'. Agricultural production operations based in greenhouses have exploited this result to their advantage for decades, by flooding greenhouse bays with additional CO$_2$. It is not uncommon to find greenhouse operations enriching CO$_2$ concentrations to 1000 ppm or more. So in general, all else being equal, we expect increased atmospheric CO$_2$ concentrations to enhance plant growth and agricultural yield, at least for C$_3$ crops. The real question is: 'By how much will elevated CO$_2$ increase crop yield?' Long *et al.* (2005, 2006) have argued that crop models use artificially high estimates of the CO$_2$ fertilization effect, obtained from growth room-type studies. These values, while perhaps representing the impact of elevated CO$_2$ per se, nevertheless overestimate what can be achieved under field conditions. They argue that when one looks solely at the more externally valid FACE studies (see Chapter 3), much lower and more realistic fertilization effects are seen. However, Ziska and Bunce (2007) have carefully reviewed all the evidence for rice (*Oryza sativa*), wheat (*Triticum aestivum*) and soybean (*Glycine max*) and conclude that FACE studies give remarkably similar results to those obtained by other methodologies. Their results are summarized in Table 11.1.

Overall, experiments validate the theory that the impact of elevated CO$_2$ should be positive. Table 11.1 shows the relative yields for these three important crops (but note that all of these are C$_3$ crops). These results suggest about a 20–30% increase in rice yield, a 30–40% increase in soybean yield and a 25–35% increase in wheat yield.

Leakey (2009) reviewed the evidence for C$_4$ crops and concluded that there is no stimulation of photosynthesis or yield in these crops due to elevated CO$_2$ *when adequate soil moisture is available*. As we saw in Chapter 4, C$_4$ photosynthesis is approximately CO$_2$-saturated at ambient CO$_2$ concentrations, so all else being equal we do not expect a fertilization effect as we saw in C$_3$ crops. However, Leakey's review (2009) finds experimental support for the hypothesis that increased CO$_2$ concentrations improve WUE in C$_4$ crops as well as C$_3$ crops (see next section) and that this results in increased yields even in C$_4$ crops, especially under drought conditions.

## Water availability

The effect of CO$_2$ on plant growth is not only through the mechanism of fertilization (i.e. reduced competition for binding sites on Rubisco and PEP), but also through increased WUE. As we discussed in Chapter 4, WUE is the plant yield relative to the plant water use (Brouder and Volenec, 2008). Under higher concentrations of CO$_2$ the plant can reduce its stomatal openings, greatly reducing its rate of water transpiration. This tends to increase WUE, but also increases leaf surface temperature because the transpiration also causes evaporative cooling. Figure 11.1 illustrates this point for soybeans grown in the FACE system in Illinois.

WUE has been widely studied and the results are fairly consistent with the expectations discussed above. Figure 11.2 shows the results from a study by Manderscheid and Weigel (2007) on wheat growth in open-topped chambers fitted with rain shelters. We see that WUE is approximately 13% higher in elevated CO$_2$ compared with ambient CO$_2$ when sufficient water is available, but is approximately 42% greater under drought conditions. It is important not to confuse WUE with actual growth (e.g. biomass production). Drought

**Table 11.1.** Relative yields. The entries show the mean change in relative yields (all comparisons were normalized to 700 ppm/370 ppm) with the range given in parentheses and the number of studies given in square brackets. (Adapted from Ziska and Bunce, 2007; see that study for details.)

| Methodology | Rice | Soybean | Wheat |
|---|---|---|---|
| Glasshouse | 1.44 (0.91–3.41) [20] | 1.34 (1.09–1.91) [17] | 1.47 (1.12–1.85) [10] |
| Growth chamber | 1.19 (1.04–1.27) [10] | 1.36 (1.08–1.59) [11] | 1.33 (0.96–1.58) [12] |
| Open-topped chamber | 1.26 (0.91–1.57) [6] | 1.37 (1.19–1.50) [30] | 1.31 (0.99–1.99) [49] |
| FACE | 1.20 (1.11–1.22) [6] | 1.40 (1.24–1.87) [4] | 1.23 (1.22–1.23) [2] |
| | | | 1.19 (1.13–1.23) [4] |

NB: The double entry for wheat denotes analyses based upon different assumptions, the details of which are not for our purposes.

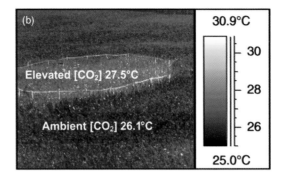

**Fig. 11.1.** (a) Illinois soybean FACE and (b) false-colour infrared photographs showing the increase in leaf surface temperature due to reduced evapotranspiration. (From Long *et al.*, 2006.)

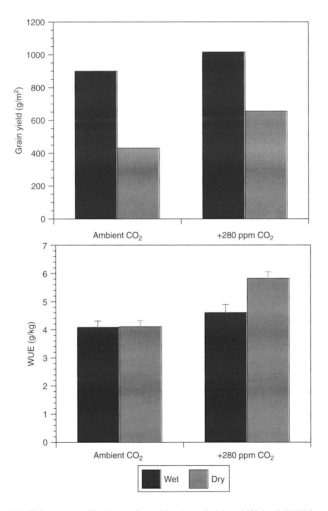

**Fig. 11.2.** Grain yield and WUE for wheat. (Redrawn from Manderscheid and Weigel, 2007.)

is not good for plant growth, regardless of the $CO_2$ concentration. In the Manderscheid and Weigel (2007) study drought significantly reduced biomass production (Fig. 11.2), but under drought conditions WUE was much greater under elevated rather than ambient $CO_2$ concentrations.

As we have emphasized elsewhere in this book, the effects of $CO_2$ often interact with the effects of nitrogen. For instance, Li, J. *et al.* (2007) used growth chambers to investigate the interaction between $CO_2$ concentration and quality of nitrogen on the growth of tomato plants. Quality of the nitrogen in this case was manipulated by altering the ratio of $NH_4^+$ (ammonium) to $NO_3^-$ (nitrate) ions. Both types of nitrogen are available to the tomato plant. Nitrate tends to be absorbed more quickly by tomato roots, but it also leaches more quickly from the soil than ammonium. In this study the total nitrogen was kept constant but the ratio of ammonium to nitrate was varied. In Fig. 11.3 we once

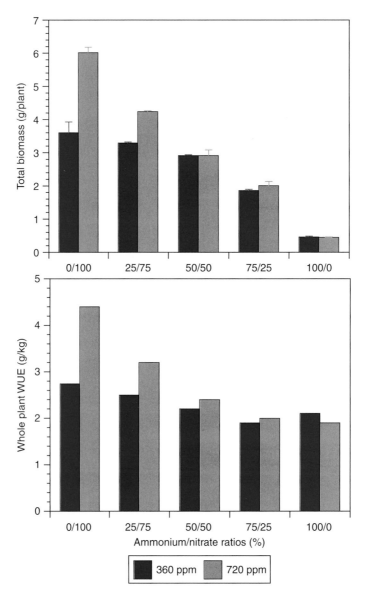

**Fig. 11.3.** Interaction between $CO_2$ concentration and the quality of nitrogen available. (Redrawn from Li, J. *et al.*, 2007.)

Chapter 11

again see that WUE increases with elevated $CO_2$, but this time the interaction is with nitrogen rather than water availability.

A real difficulty with answering definitive questions about the impacts of climatic change on water availability is that there is considerably less agreement among GCMs as to the changes in precipitation than there is about changes in temperature. Indeed, it is common for GCMs to disagree even as to the direction of precipitation change for any given location. Much of this disagreement disappears when climate models are statistically downscaled, but such results do not exist for all regions of interest worldwide, let alone for multiple scenarios or models. Bear this in mind when we consider the results of the agricultural production models below.

### Extreme events

As we discussed in Chapter 2 (Fig. 2.13), the mean values of climate variables are expected to change, as is the variability around those mean values. The impacts of climatic variability, particularly on agricultural production, are much less well studied than those on mean changes in climate. Yet we know that variability and timing are critically important to agricultural production, sometimes more important than the mean climate.

For example, Fig. 11.4 shows how the impact of a drought event on wheat yield depends crucially on when that event occurs in the plant's phenological development.

## 11.2 Potential for Adaptation

'Adaptation' refers to methods and policies to reduce the impacts of climatic change on agriculture, and these impacts are not insubstantial. Adaptation incorporates not just the notion of changing agricultural practices and technology, but also social and political changes that affect the adoption of adaptation options (Howden *et al.*, 2007). Here we briefly point to some of the work going on in these fields of study. Our intent is not to give a complete account of adaptation strategies, but to give some idea of the range of possibilities.

### Changing agronomy practices for adaptation

Changing practices for adaptation include changing the timing of various operations like sowing and harvesting, choice of crops, and conversion of rainfed agriculture to irrigated agriculture. We elaborate on some of these options below.

Changes in the start and length of the growing season can alter crop production. Agronomic responses

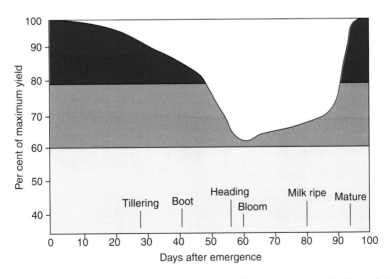

**Fig. 11.4.** The effects of water stress on wheat production depend critically on when in the life cycle that stress occurs. (Redrawn from Bauer, 1972.)

include early planting and choices of cultivars that either take longer to mature (increasing the time available for grain-filling) or require less vernalization (Rosenzweig and Tubiello, 2007). These approaches are potentially very effective, converting a serious negative impact into an overall gain, or at least reducing the negative impacts of climate change. Some examples are shown in Fig. 11.5.

Breeding for desirable traits is another adaptation strategy available to the agricultural sector but, unlike changing agronomy practices, breeding programmes take time. Some traits that might be useful in the future are also useful today, like drought tolerance, and are already the goal of breeding programmes for various crops. However, other crop traits have great potential for use in adaptation programmes. Responsiveness to atmospheric $CO_2$, for example, holds great promise, but is largely unexamined. There exist tens of thousands of cultivars of the main crops and yet literally only a handful of these have been used in elevated $CO_2$ experiments. Ziska *et al.* (2004) compared the performance of four spring wheat cultivars that were developed over the course of the 20th century (i.e. during a period of rapidly increasing $CO_2$ levels) at three concentrations of atmospheric $CO_2$ (Fig. 11.6). Whatever the differences among these cultivars, we would expect those differences to be the same regardless of the $CO_2$ concentration. However, the authors found that the differences were not consistent when compared at different $CO_2$ concentrations. This result suggests that there may be considerable scope to breed for increased responsiveness to $CO_2$. There are thousands of cultivars of spring wheat, so the genetic variation is available.

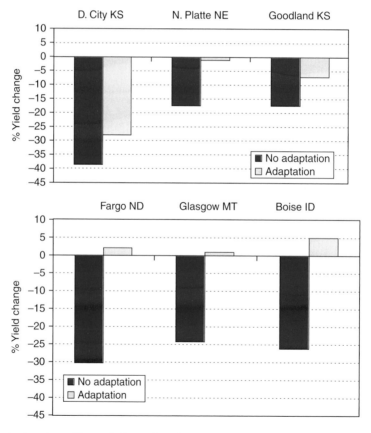

Fig. 11.5. How simply changing cultivars of winter wheat (top) can significantly reduce the impacts of climate change on crop yield. Also depicted (bottom) are the impacts of planting spring wheat earlier in the year. Here we see that adaptation not only reduces the negative impacts of climate change, it turns those impacts into net gains. (Redrawn from Tubiello *et al.*, 2002.)

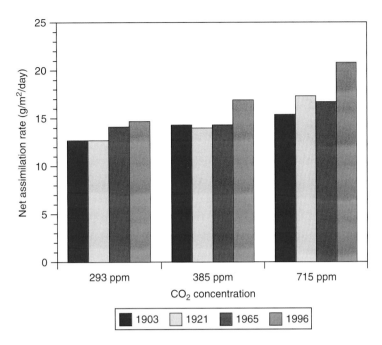

**Fig. 11.6.** The difference in the performance of four spring wheat cultivars depends on the atmospheric concentration of $CO_2$ in which they are compared. This interaction indicates the potential to exploit genetic differences among existing cultivars for breeding responses to $CO_2$. (Redrawn from Ziska *et al.*, 2004.)

Beyond breeding for desirable traits, the next option is perhaps to substitute crops. Wu *et al.* (2007) modelled crop changes due to climatic change, assuming that farmers are relatively autonomous, seeking to maximize their long-term profits subject to their constraints, and accounting for uncertainty in price and yield. They used the output from an older climate model (the Canadian CGCM1) in conjunction with an older emissions scenario (the IS92a). If this model were re-run with climate projections from AR4, undoubtedly there would be differences, but the general pattern would likely be similar. In any case, this work makes the general point that one way to adapt to climatic change is to change what one grows and/or where one grows it. Wu *et al.* (2007) investigated only as far into the future as 2035, but found some major shifts in crop choices. Worldwide, they found that the land devoted to wheat and rice would increase significantly and more or less linearly to 2035, particularly in Europe and North America in the case of wheat and Asia in the case of rice. Viewed at the global scale there is little insight except that agricultural land cover for most of the major crops will increase substantially. However, viewed at a

more local or regional scale, significant shifts become apparent. For example, the central mid-western part of the USA, which is currently heavily cultivated with soybean, will generally shift to maize and wheat. A similar trend will occur, according to the model, across much of the northern half of India, and so on. See Wu *et al.* (2007) for more details.

Genetically modified organisms may also play a role in our adaptation strategy. All of the breeding traits listed above may also be achieved via genetic modification technology. In Box 11.1 we consider some more specific traits and how biotechnology might be used to adapt our agriculture to climatic change, with a particular emphasis on elevated $CO_2$.

## 11.3 Future Agricultural Production

Attempts at comprehensive assessments of the impacts of climatic change on agricultural production are largely performed with physiologically based crop models (see Chapter 3) that have been parameterized from experiments on individual crops. There is a second approach, based on the

## Box 11.1. Traits for adaptation to elevated $CO_2$.

Ainsworth *et al.* (2008) suggest that the biggest gains from genetic modification and/or conventional breeding research on $C_3$ crop responses to elevated $CO_2$ would be from improvements to $CO_2$ uptake, carbohydrate sink strength and nitrogen-use efficiency (NUE). We consider each of these in turn.

We have already seen, at least for $C_3$ plants, that elevated $CO_2$ stimulates additional plant growth. One target for crop improvement would be to target the $CO_2$ uptake mechanism (see Chapter 4), Rubisco. As Ainsworth *et al.* (2008) explain, perhaps counter-intuitively, that gain would be achieved by reducing Rubisco's specificity for $CO_2$, rather than by increasing its specificity. The rate of photosynthesis depends on Rubisco's specificity – the higher the specificity, the less carbon loss occurs through photorespiration. However, because there exists an inverse relationship between the specificity of Rubisco for $CO_2$ and its catalytic rate, there is a trade-off necessary to achieve an optimal rate of photosynthesis under any set of environmental conditions. Zhu *et al.* (2004) demonstrate, via the use of a model, that the optimal solution to this problem is to actually aim for lower specificity under elevated $CO_2$ (Fig. 11.7). Thus one solution would be to get crops to express Rubisco with lower specificity. Zhu *et al.* (2004) show at the whole canopy level that such a modification could very substantially increase net carbon gain (>25%).

As Ainsworth *et al.* (2008) also point out, it is well known that the rate of photosynthesis is often down-regulated in the presence of leaf carbohydrate accumulation (see Chapter 4). Plants accumulate carbohydrate as a carbon storage compound when they cannot use it for maintenance, growth or reproduction. Experimental observations usually confirm this, but even accounting for this down-regulation, photosynthetic rates in elevated $CO_2$ concentrations often remain higher than they are in ambient $CO_2$ (see Ainsworth *et al.*, 2008 for more discussion and references). Nevertheless, targeting increased sink strength in the plant could reduce this down-regulation effect and keep photosynthetic rates even higher.

An indirect way of achieving increased sink strength would be to increase NUE, as this would also reduce the accumulation of foliar carbohydrates (because they would be used rather than stored). Ainsworth *et al.* (2008) note that all the major bio-tech companies are already working on improving this trait in the major crops, and that one way to do this for future $CO_2$ concentrations is to reduce the plant's investment in Rubisco. Because Rubisco can account for a substantial fraction of leaf nitrogen (about 25%), and because photosynthetic rates can be maintained at higher $CO_2$ concentrations with substantially less Rubisco (also about 25%), this means that genetic engineering or conventional breeding might save as much as 6% ($0.25 \times 0.25 = 0.0625$) of leaf nitrogen. Still, with up to 40% increases in carbon acquisition, there is room for improvement in NUE (Ainsworth *et al.*, 2008). A goal of reducing Rubisco concentrations shows how complex is the optimization problem discussed at the start of this box. Not only is there a trade-off between specificity and catalytic rate, but also between the amount of Rubisco and NUE, and both must be accounted for in order to optimize yield.

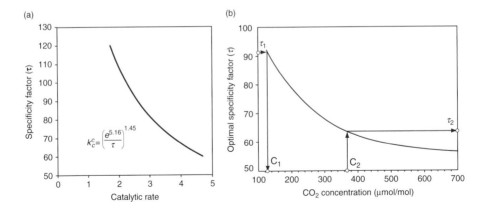

**Fig. 11.7.** (a) The fitted relationship between Rubisco's specificity and its catalytic rate. (b) The optimal solutions for different ambient $CO_2$ concentrations. Notice that the higher the $CO_2$ concentration, the lower the optimal specificity. (Redrawn and modified from Zhu *et al.*, 2004.)

economics of international trade, sometimes referred to as 'Ricardian models' (Cline, 2007: 7), but these are beyond the scope of this text. Interested readers should see Mendelsohn *et al.* (1994) as a starting point. For our purposes it will be sufficient to identify this family of models as 'statistical' models rather than physiologically based mechanistic models.

There has been a variety of assessments, which perhaps not surprisingly come to a variety of conclusions. These assessments differ in several ways, including which physiological effects they include, which GCM(s) they use to project future climate, which adaptations they incorporate and which part(s) of the world they focus on. Table 11.2 is adapted from Cline's (2007) attempted synthesis of the available literature and the results of Rosenzweig and Iglesias' (2006) update of an older study (Rosenzweig *et al.*, 1993). Cline's synthesis included five Ricardian statistical model approaches and several crop modelling studies. However, these latter studies are dominated by Rosenzweig and Iglesias' (2006), as this is the only crop model study that produced results for the whole world. As we can see, the results are indeed highly variable. It is possible, by cherry-picking from these results, to paint either a grim or a more positive picture of future agricultural production. Knowing this, one should be suspicious of comments about agricultural production under climate change projections unless the authors are forthcoming about the assumptions and limitations of their analysis.

Recognizing that the countries/regions in Table 11.2 are not equal contributors to global agricultural output, glancing at the numbers does not immediately yield any great insight into the overall impact of climatic change. Cline's (2007) preferred estimate for the whole world yields −15.9% without $CO_2$ fertilization and −3.2% when $CO_2$ fertilization effects are included (calculated as a 15% bump in yield). As we said, individual studies yield higher or lower estimates and we could quibble with Cline's method of combining these studies, but these numbers reflect the current view that overall climate change will have a modest, negative impact on agricultural production (but see Box 11.1). It is, however, important to note that none of these global or even regional estimates incorporates the impacts of plant pests, competitors or pathogens and how these organisms will respond to climatic change. Rosenzweig and Iglesias' (2006) study, for example, assumed no change in these pressures

(i.e. a loss of 30% of yield). We consider the impacts of climatic change on crop protection in Section 11.4.

## 11.4   The Future of Crop Protection

As we said above, many of the predictions about the impacts of climatic change on agricultural production rely on crop models or experiments where $CO_2$ concentrations or temperatures are manipulated (although rarely are the two manipulated simultaneously in the same experiment). However, every farmer knows that crop production depends upon crop protection. How will insect pests, crop diseases and weeds respond to climatic change? Will they become more problematic or less problematic? Surprisingly, there has been very little integration of crop protection into considerations of the impacts of climatic change on agricultural production, despite the relative abundance of research on the topic (Gregory *et al.*, 2009). In this section we take a look at the literature on climatic change and crop protection, but we note that the literature is unbalanced, with little known about diseases, particularly viral and bacterial pathogens, and relatively speaking a lot is known about insect pests. The literature is also geographically biased with far more work coming from North America than, say, Africa.

### Insect pests

Native and invasive agricultural pests will have two reactions to climatic change. First, they will respond directly to changes in temperature and moisture availability. Second, they will respond indirectly via changes induced in both the quantity and the quality of their host plants.

Insect performance depends at least in part upon how climatic change alters overwinter survival, development rates, fecundity and longevity. It also depends on how higher trophic levels respond to the same climatic changes. There is a rich literature on the responses of all of these life-history parameters to changing climates (see Chapter 5). One measure that population ecologists often use to integrate these life-history parameters into a single overarching variable is the intrinsic rate of increase ($r_m$). The higher the intrinsic rate of increase, the faster the population grows. Figure 11.9 shows an example for the coffee borer, *Hypothenemus hampei*, as it responds to changes in temperature (Jaramillo *et al.*, 2009). It shows that the effect of rising temperature

## Box 11.2. Global food security in a changing climate.

'Food security' and 'agricultural production' are not the same thing. As of 2010 the world actually produced enough food to feed the entire human population and yet around the world a significant proportion of humanity does not get enough to eat. And this problem is not restricted to the developing world; many people in G8 countries cannot afford enough food to maintain good health. The FAO defines food security as follows:

> 'Food security' means that food is available at all times; that all persons have means of access to it; that it is nutritionally adequate in terms of quantity, quality and variety; and that it is acceptable within the given culture. Only when all these conditions are in place can a population be considered 'food secure'.

Put simply, there is a tension between the causes of climate change and the consequences of those causes for other factors of life besides climate. One way to understand the relationship between climate change and humans is that greenhouse gas emissions arise from economic activity (and our dependence on fossil fuels). Those greenhouse gas emissions may be bad for the climate, but one can argue that the economic activity that generates them might be good for the quality of life in the developing world, including enhancing food security.

Questions about food security in a changing climate are addressed by taking the outputs from agroecological zone models and/or crop physiology models and running them through complex socio-economics/trade models (Easterling et al., 2007). The IPCC AR4 concluded that despite the large degree of uncertainty in these food security assessments, several robust conclusions emerge. First, considered in isolation, climatic change will increase the numbers of people at risk of hunger compared with the case of no climatic change. Depending on the scenario, the numbers would increase from 5 to 25%.

Second, those impacts of climatic change per se are likely to be small relative to the large and generally positive benefits of economic development and declining rates of population growth on food security. Schmidhuber and Tubiello (2007) among others have suggested that the number of people at risk of hunger may decline by >75% for all scenarios except A2; with its high population growth rate, the per capita income growth is much lower under A2 and so hunger persists. Third, changes will not, of course, be equally distributed about the world, with sub-Saharan Africa surpassing Asia as the most insecure region (Easterling et al., 2007; Battisti and Naylor, 2009). And fourth, there is considerable uncertainty with regard to the strength of the $CO_2$ fertilization effect (see above), but this uncertainty does not seem to negate the robust conclusions just listed.

Much of the above can be seen in the example depicted in Fig. 11.8. First concentrate on the 'no climate change' results and compare these with the observed level of food insecurity in the year 2000. These projections depict the generally positive effects of economic development on food security, particularly in the developing world. The small decrease for the A2 scenario is due to the large population growth assumed in that scenario. Now compare the impacts of 'climate change' with the 'no climate change' results. The negative effects of climatic change are apparent, but the overall effect is still a reduction in the number of people at risk. Even if the $CO_2$ fertilization effect is not as great as assumed in this study, the overall prediction is still greater food security than we currently enjoy. Although this is the general conclusion reached by the IPCC AR4, there are conflicting assessments. For example, Parry's (2007) analysis reaches the conclusion that: 'climate change will generally reduce production potential and increase risk of hunger'.

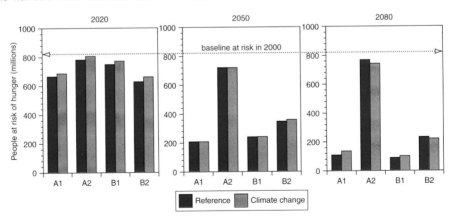

**Fig. 11.8.** Climate change impacts on the number of people at risk of hunger (millions) in the year 2080 compared with that in 2000 (824 million). The SRES scenarios are discussed in detail in Chapter 2. (Based on results from Schmidhuber and Tubiello, 2007.)

**Table 11.2.** Combined results from Rosenzweig and Iglesias (2006) and Cline (2007). The results show the percentage change in agricultural capacity for various countries and regions.

| | No CO$_2$ effects, no adaptation | | | 550 ppm CO$_2$, no adaptation | | | 550 ppm CO$_2$, level 1 | | | 550 ppm CO$_2$, level 2 | | | Preferred | |
|---|---|---|---|---|---|---|---|---|---|---|---|---|---|---|
| | GISS | GFDL | UKMO | GISS | GFDL | UKMO | GISS | GFDL | UKMO | GISS | GFDL | UKMO | No CO$_2$ | CO$_2$ |
| Africa LICX | -34 | -45 | -40 | -24 | -34 | -29 | -16 | -26 | -21 | -8 | -13 | -10 | -41.8 | -33.1 |
| Africa LICI | -40 | -40 | -45 | -29 | -29 | -34 | -21 | -21 | -26 | -10 | -10 | -13 | -29.1 | -18.4 |
| Africa MICX | -21 | 3 | -32 | -10 | 14 | -21 | -1 | 4 | -12 | 1 | 5 | -6 | -24.2 | -12.8 |
| Africa MICI | -33 | -33 | -38 | -22 | -22 | -27 | -13 | -13 | -18 | -6 | -6 | -9 | -39.6 | -30.5 |
| Africa OilX | -26 | -31 | -36 | -15 | -20 | -25 | -7 | -12 | -17 | -2 | -5 | -8 | -32.2 | -22.0 |
| Argentina | -32 | -36 | -39 | -21 | -25 | -28 | -3 | -7 | -10 | -4 | -3 | -4 | -11.1 | 2.2 |
| Australia | -16 | -17 | -16 | -5 | -6 | -5 | 6 | 6 | 6 | 6 | 6 | 6 | -26.6 | -15.6 |
| Brazil + | -35 | -26 | -44 | -24 | -15 | -33 | -12 | -7 | -21 | -4 | -2 | -10 | -16.8 | -4.4 |
| Canada | -7 | -5 | -35 | 4 | 6 | -24 | 24 | 25 | 2 | 24 | 25 | 2 | -2.1 | 12.6 |
| China + | -16 | 19 | -21 | -5 | -8 | -10 | 8 | 2 | 3 | 8 | 6 | 3 | -7.2 | 6.7 |
| Egypt + | -31 | -26 | -46 | -20 | -15 | -35 | -17 | -13 | -32 | -8 | -6 | -16 | 11.3 | 28.0 |
| Europe | -11 | -21 | -17 | 1 | -10 | -6 | 8 | 2 | 3 | 8 | 2 | 3 | -5.7 | 8.4 |
| Far East HMICX | -43 | -35 | -38 | -32 | -24 | -27 | -24 | -16 | -20 | -12 | -8 | -10 | -23.0 | -11.4 |
| Far East HMICI | -16 | -18 | -24 | -5 | -7 | -13 | 4 | 2 | -2 | 6 | 4 | 1 | -11.3 | 2.0 |
| Far East LI | -30 | -29 | -41 | -19 | -18 | -30 | -12 | -11 | -23 | -5 | -4 | -11 | -24.3 | -12.9 |
| Former USSR | -8 | -21 | -35 | 3 | -10 | -24 | 16 | 5 | -7 | 16 | 5 | -3 | -6.7 | 7.3 |
| India | -27 | -33 | -51 | -16 | -22 | -40 | -4 | -13 | -27 | -1 | -5 | -14 | -38.1 | -28.9 |
| Indonesia | -34 | -26 | -15 | -23 | -15 | -4 | -14 | -6 | 1 | -7 | -2 | 2 | -17.9 | -5.6 |
| Japan | -10 | -16 | -22 | 1 | -5 | -11 | 14 | 10 | 7 | 14 | 10 | 7 | -5.7 | 8.4 |
| Kenya | -33 | -33 | -33 | -22 | -22 | -22 | -14 | -14 | -14 | -7 | -7 | -7 | -5.4 | 8.8 |
| Latin America HICX | -33 | -28 | -43 | -22 | -17 | -32 | -14 | -9 | -25 | -7 | -4 | -12 | -32.3 | -22.1 |
| Latin America HICI | -34 | -29 | -45 | -23 | -18 | -34 | -15 | -11 | -12 | -8 | -8 | -13 | -28.3 | -17.6 |
| Latin America MLI | -36 | -27 | -45 | -25 | -16 | -34 | -17 | -8 | -27 | -8 | -3 | -13 | -25.8 | -14.7 |
| Mexico | -46 | -39 | -48 | -35 | -28 | -37 | -27 | -20 | -31 | -13 | -10 | -15 | -35.4 | -25.7 |
| New Zealand | 4 | -1 | -9 | 15 | 10 | 2 | 29 | 24 | 14 | 29 | 24 | 14 | 2.2 | 17.5 |
| Nigeria | -24 | -34 | -24 | -13 | -23 | -13 | -6 | -16 | -6 | -2 | -8 | -2 | -18.5 | -6.3 |
| North-east Asia MLI | -41 | -41 | -46 | -30 | -30 | -35 | -20 | -20 | -25 | -10 | -11 | -12 | -26.1 | -15.0 |
| North-east Asia OilX | -31 | -36 | -41 | -20 | -25 | -30 | -10 | -15 | -20 | -5 | -7 | -10 | -26.9 | -15.9 |
| Pakistan | -52 | -24 | -68 | -41 | -13 | -57 | -29 | -5 | -50 | -14 | -1 | -25 | -30.4 | -20.0 |
| Thailand | -40 | -29 | -42 | -29 | -18 | -31 | -19 | -8 | -24 | -9 | -4 | -12 | -26.2 | -15.1 |
| Turkey | -28 | -38 | -38 | -17 | -27 | -27 | -5 | -15 | -15 | -3 | -8 | -8 | -16.2 | -3.6 |
| USA | -18 | -26 | -42 | -7 | -15 | -31 | 5 | -2 | -13 | 2 | 1 | -6 | -5.9 | 8.2 |

Level 1, adaptation at small additional cost to the farmers (such as shifts in planting dates, variety and crop, and increases in water application to irrigated crops); level 2, adaptations that imply significant additional costs to the farmers (such as large shifts in crop production timing, increased fertilizer application, installation of irrigation systems and development of new varieties) (Rosenzweig and Iglesias, 2006). Preferred, Cline's (2007) attempt to combine all of the various crop production estimates into a single weighted value, using level 1 adaptation. GISS, Goddard Institute for Space Studies; GFDL, Geo Fluid Dynamics Laboratory; UKMO, UK Meteorological Office.

LICX, low-income calorie exporters; LICI, low-income calorie importers; MICX, medium-income calorie exporters; MICI, medium-income calorie importers; OilX, oil exporters; LI, low income; MLI, medium–low income; +, similar countries. The former USSR includes Eastern Europe.

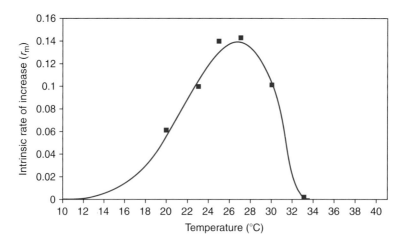

**Fig. 11.9.** Relationship between the intrinsic rate of increase and air temperature for the coffee borer, *Hypothenemus hampei*. (Redrawn from Jaramillo *et al.*, 2009.)

on the population dynamics of an insect depends strongly on the baseline temperature from which that change occurs. In many parts of a species' range increasing air temperatures can be beneficial, while in other parts of the range such increases are detrimental.

As either populations grow more quickly or the length of the growing season becomes long enough, insect pest species may be able to increase the number of generations they produce in a single season (see Box 4.1 for an example). This type of response can be particularly damaging to agricultural production. However, Tobin *et al.* (2008) point out that this effect is complicated by the fact that both temperature and photoperiod influence voltinism (i.e. the number of generations per year) in many species and, while temperature may be increasing, photoperiod is independent of climate change. Consequently, the response is likely to be complex. Tobin *et al.* develop a phenology model (see Chapter 4) for the grape berry moth, *Paralobesia viteana*, a native to North America and an important pest of grape crops. Their model incorporates both the temperature-dependent effects and the photoperiod-dependent effects. Figure 11.10 shows the forecasted number of moth generations as a function of the climate change scenario for three locations. The model shows just how sensitive generation number can be for this species. Although year-to-year variation in temperature is likely to result in large fluctuations in generation

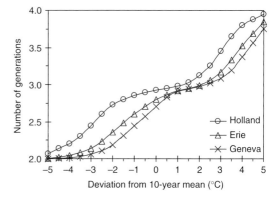

**Fig. 11.10.** Forecasted number of generations for grape berry moths under climate change scenarios based upon temperature deviations from a 10-year mean (1996–2005). (Redrawn from Tobin *et al.*, 2008.)

number, the long-term average is likely to be more generations per year.

Insect responses to changes in plant nutritional and defensive chemistry are also likely to be complex. One fairly general response of plants to elevated $CO_2$ has been a decrease in the nitrogen concentration in the plant (see Chapter 4). In general, insects respond to this reduction in quality by increasing their consumption rate in an attempt to compensate (Stiling and Cornelissen, 2007), sometimes successfully and sometimes not. Increased per capita consumption would cause more harm to the plant assuming that plant sizes and insect population sizes stayed constant. However, plants usually produce more biomass under elevated $CO_2$ (see Chapter 4) and so the increased per capita consumption may be inconsequential. Insect population size responses are more complex and will be discussed shortly.

Many researchers have attempted to link plant chemical defence responses to elevated $CO_2$ to the Optimal Defence Hypothesis, the Carbon–Nutrient Balance Hypothesis or the Growth–Differentiation Balance Hypothesis (see Stamp, 2003, for review of these hypotheses). In a recent critical review of the literature, Ryan *et al.* (2010) show that there is little support for any of the predicted responses to elevated $CO_2$. See Section 4.6 for more details.

Insect responses to $CO_2$ and changes in temperature have so far defied simplistic classifications: e.g. always increase, always decrease, etc. Indeed, some researchers have gone so far as to suggest that insect responses might be 'idiosyncratic', meaning that the response depends upon the particular plant species and the particular insect species and does not easily generalize beyond that (Bezemer *et al.*, 1999; Hunter, 2001; Newman *et al.*, 2003). See Box 5.1 for further consideration of insect population responses to climatic change.

## Pathogens

Despite their importance to crop production, surprisingly little experimental work has been done on plant pathogen responses to climatic change, particularly for plant viruses, where very little is known beyond speculation based on the impacts of climatic change on virus vectors (i.e. the organism, often an insect, that spreads the virus from plant to plant; Jones, 2009). Despite this lack of experimental work, there is a surprisingly large number of review articles on this topic. These articles are largely of the form of 'informed speculation' about the possible or likely impacts of climatic change on disease incidence and severity, based on insights from general biology and reasonable extrapolation.

As in Chapter 5, it seems a safe speculation that diseases will have different geographic distributions as their bioclimatic envelopes, or those of their obligate vectors and host plants (see Chapter 3), move due to climatic change. Also, pathogens that rely on vectors will depend in large part on the success or failure of the vector population. For diseases that are vectored by aphids, there is some evidence that these insects defy the trend and may actually increase under elevated $CO_2$ (Bezemer and Jones, 1998), implying that these diseases may become more significant in the future. However, the response is, once again, fairly complex with all possible outcomes having been observed in experiments and generated in models (Newman *et al.*, 2003).

Manning and von Tiedemann (1995), after finding the older literature sparse, made some reasonable speculations based on basic plant pathology and what we know about plant responses to elevated $CO_2$. This speculation provides a good framework for our consideration of the more recent literature. Increased biomass, in general, might simply provide more substrate and resource availability for plant pathogens. Larger crop canopies, if they are accompanied by increased canopy humidity, are likely to be favourable for the growth of rusts, powdery mildews, leaf spots and blights (Coakley *et al.*, 1999). Many pathogens rely on plant litter for overwinter survival. With more plant litter expected (due to increased biomass production) and warmer winter temperatures, climatic change may result in greater overwinter survival and hence greater initial inoculum at the start of the growing season (Coakley *et al.*, 1999), although decreased snow cover in some areas may counteract this.

However, as we have seen elsewhere, plants are not simply bigger: they are metabolomically different. Increased C:N ratios and higher concentrations of stored sugars would, all other things being equal, also favour pathogens like powdery mildews and rusts because they are sugar-dependent (Manning and von Tiedemann, 1995). On the other hand, as we saw in Chapter 4, plants tend to reduce their stomatal openings or indeed reduce their stomatal density in the face of increased $CO_2$ concentrations (Woodward, 1987). Both of these responses suggest reduced incidence and severity of powdery mildews and rusts because these pathogens enter the plant through the stomata (Lake and Wade, 2009), as do many bacterial pathogens (Melotto *et al.*, 2008). Despite the logic of this argument, Lake and Wade (2009) investigated the role of elevated $CO_2$ on infection by powdery mildew,

*Erysiphe cichoracearum*, and found that this pathogen was more common under elevated $CO_2$ (Fig. 11.11).

There have been a few attempts to model the epidemiological impacts of climatic change on disease incidence and severity. Salinari *et al.* (2006) modelled the epidemiology of downy mildew (*Plasmopra viticola*) on grapes in Italy, using both the Goddard Institute for Space Studies and Hadley Centre climate model outputs (see Chapter 2) from the A2 scenario (Chapter 2). For this scenario, both climate models predicted increased temperature and decreased rainfall for this region during the growing season. Their simulation results suggest that, all else being equal, increased temperature would strongly increase the incidence and severity of the disease, while decreased precipitation would decrease the incidence and severity. When combined, the results suggest that the negative effects of temperature more than outweigh the positive effects of precipitation, resulting in more frequent epidemics of downy mildew. In order to adapt to this change, the model suggests that two more fungicidal sprays per year

would need to be applied to the crop to maintain yield (Salinari *et al.*, 2006).

## Weeds

Again, like insect pests, we can reasonably expect a change in the distribution and abundance of crop weeds as their bioclimatic envelopes move due to climatic change. McDonald *et al.* (2009) recently conducted a study that exemplifies this approach for weeds of maize crops in the USA. They used an average of the outputs from the GFDL CM2.1, HadCM3 and PCM1 models (see Chapter 2) for the A1FI scenario (see Chapter 2). They come to two major conclusions. First, some states will see a nearly complete change in the composition of weed species associated with maize (Fig. 11.12). Second, for the two major species of weeds, *Abutilon theophrasti* and *Sorghumhale pense*, that are currently dominant in the northern and southern parts of the USA, respectively, both species' envelopes move north, indicating that *A. theopharsti* may cease to be a problem for most US states while *S. pense* will become a more widely distributed and important weed for nearly all of the eastern USA.

As with so many examples from community and population responses (see Chapters 5 and 6), weeds and crops may well respond differently to the changes in climate. Tungate *et al.* (2007) demonstrate this experimentally with soybeans and two of their important competitors, sickle-pod (*Senna obtusifolia*) and prickly sida (*Sida spinosa*). Figure 11.13 shows that the two weed species achieve maximum growth at higher temperatures than soybeans, indicating that these weeds will benefit more from increased temperatures than soybeans and thus perhaps be more problematic for control and management in the future.

Note that neither of the above examples considered the impacts of rising $CO_2$ in its assessment. The situation with weeds and crops in elevated $CO_2$ has often been generalized as follows. Since most crops use the $C_3$ photosynthetic pathway and most weeds use the $C_4$ pathway, and since $C_3$ plants ought to respond more to elevated $CO_2$ than $C_4$ plants, we ought to expect that crops will outgrow weeds; this expectation has some empirical support (Ziska, 2003b).

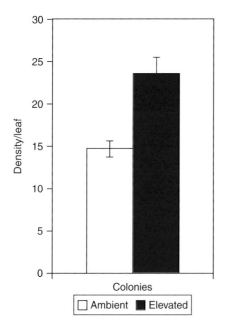

**Fig. 11.11.** The number of colonies of *Erysiphe cichoracearum* (powdery mildew) on *Arabidopsis thaliana* plants grown under elevated and ambient $CO_2$. (Redrawn and adapted from Lake and Wade, 2009.)

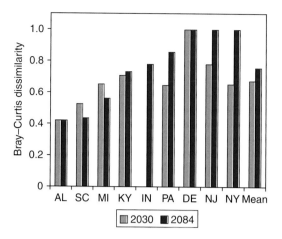

**Fig. 11.12.** The degree of similarity between the weed community, associated with maize crops, currently found in each US state and that expected to be found in 2030 or 2084. A Bray–Curtis value of 1 means that the future weed community is completely different from that found in contemporary climates. The plot shows that these states will all likely see large changes in the composition of the weed community, and that several states will see near complete substitutions of the weed community. (Redrawn and adapted from McDonald *et al.*, 2009.)

However, as Ziska (2003b) points out, this generalization misses the points that: (i) many weeds and crops share the same photosynthetic pathway; (ii) several important crops (e.g. sorghum, maize, millet and sugarcane) are $C_4$ plants; and (iii) biomass is not always the same as crop yield. Bunce and Ziska (2000) reviewed the literature on $C_3$ crop–$C_3$ weed and $C_4$ crop–$C_4$ weed interactions and concluded that increases in the vegetative growth of weeds was greater than that of the crop under elevated $CO_2$. Ziska (2003b) investigated the impacts of elevated $CO_2$ on the competition between sorghum (*Sorghum bicolor*, a $C_4$ crop) and velvetleaf (*Abutilon theophrasti*, a $C_3$ weed) or redroot pigweed (*Amaranthus retroflexus*, a $C_4$ weed). In ambient $CO_2$ neither weed resulted in significant crop yield loss, even though redroot pigweed resulted in a significant loss of biomass in the sorghum. However, at elevated $CO_2$, both weeds caused a loss of crop yield (17% on average for the velvetleaf and 23% on average for the redroot pigweed; Ziska, 2003b). This study

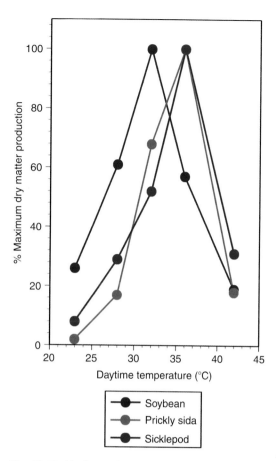

**Fig. 11.13.** Maximum dry matter production, as a function of temperature, for soybean and two of its important competitors, sicklepod (*Senna obtusifolia*) and prickly sida (*Sida spinosa*). Note that maximum dry mass production is achieved at higher temperatures for the two weeds than for the crop. (Redrawn and adapted from Tungate *et al.*, 2007.)

demonstrates nicely how results based on biomass production may lead to different answers from results based on yield.

## 11.5 Conclusions

The impacts of climatic change on agriculture have been reasonably well studied, but there is still much work to do. At this point, when taken as a whole, it looks like climatic change will result in small to moderate decreases in crop yields. These may be offset in whole or in part by conventional adaptation strategies and/or the production of

---

crops with new or improved traits that allow them to cope with or exploit climatic change. Much work is still needed to incorporate the impacts of climate change on crop protection into long-range predictions of climate change impacts and possible adaptation strategies. Lastly, it is important to understand the distinction between the impacts of climate change on agricultural yield and the impacts of the economic conditions that bring about climatic change on food security.

# 12 Impacts on Biodiversity

So far in this book we have examined how individuals, populations, communities and ecosystems respond to climate change. In this chapter we examine the consequences of these responses for biodiversity. Biodiversity has been defined as the 'variation of life at all levels of biological organization' (Gaston and Spicer, 2004). It includes diversity at the individual, population, community and landscape levels. Climate change can have a variety of effects – from changing individual growth rates to species niche expansion – which can in turn affect biodiversity patterns. Ecologists have hypothesized that:

> cross-scale interactions, likely those between the local demographics of species at the fine scale and the landscape configuration of patches at the broad scale . . . have significant implications for the conservation of biodiversity.
>
> (Willig *et al.*, 2007)

Thus, all of the information that we presented in previous chapters examining evolutionary, population and community responses are important to keep in mind in terms of their impacts on biodiversity. As we shall see, the level of organization that is most often talked about with regard to climate change is the number of species. However, genetic and landscape diversity are also very important to consider. Furthermore, biodiversity definitions are linked to spatial scale. Local responses (at small spatial scales, and usually defined at the community level) may differ from those examined at landscape, regional or even global scales. It is important to consider the responses of all of these scales of biodiversity to climate change. While still a controversial subject, the loss of biodiversity from any given ecosystem is generally thought to be detrimental because of the variation that its definition implies. Variation is thought to be useful for helping individuals, populations, communities and ecosystems adapt to changing environments.

## 12.1 Consequences of Biodiversity Loss

Ecosystems provide many services to us, from the tangible, like primary production (the basis of the food we eat), to the intangible and aesthetic (e.g. the enjoyment of wilderness). It is generally thought that loss of biodiversity leads to reduced ecosystem functioning and stability, and to fewer or less available 'ecological goods' such as food, fuel, fibre and medicines (Mooney *et al.*, 2009). Figure 12.1 schematically depicts the relationships between human well-being, ecosystem services and biodiversity (Millennium Ecosystem Assessment, 2005). From Fig. 12.1 we can see that climate change is but one of several factors that can alter biodiversity. Changes in biodiversity in turn can alter the various ways in which ecosystems function.

Why might biodiversity affect ecosystem functioning? Biodiversity is thought to increase ecosystem functioning via at least two different mechanisms. Since species, generally speaking, do different things (e.g. they use different resources), the more species present the more things get done (e.g. the more thoroughly the resources get used) and so the greater the provision of services within the system. This mechanism is referred to as 'niche complementarity'. The other mechanism is that species may differ in the *magnitude* of their effect on ecosystem functioning. For example, consider NPP (Chapters 7, 9 and 10). Some species will be highly productive, others not so

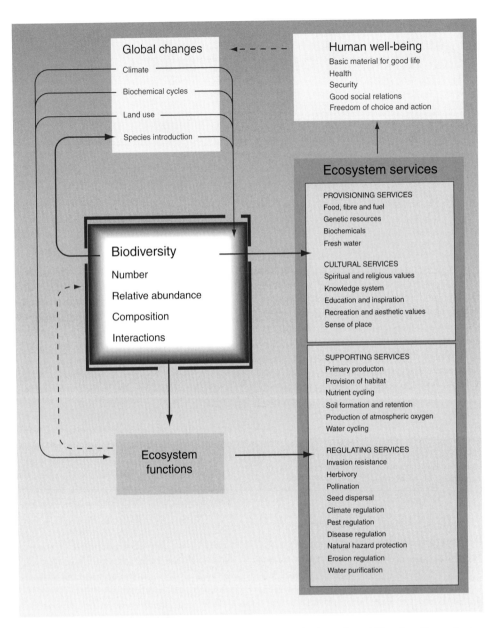

**Fig. 12.1.** The relationships between biodiversity and global change. (Redrawn from Millennium Ecosystem Assessment, 2005.)

much. With higher biodiversity in a system we increase the chance of including one or more of these 'high-functioning' species, just by chance alone. This is called the 'selection mechanism' (also called the 'sampling effect'). So reductions in biodiversity could decrease the degree of niche complementarity and/or the likelihood of a high-functioning species being present (see e.g. Bell

*et al.*, 2005; Fig. 12.2). Greater biodiversity might also lead to greater ecosystem *stability* because the more species there are, the greater the likelihood that a few will be resistant or be able to adapt to climate change or other stressors (Loreau *et al.*, 2001). This is called the 'insurance effect'. In addition, even if biodiversity per se is not important to ecosystem functioning

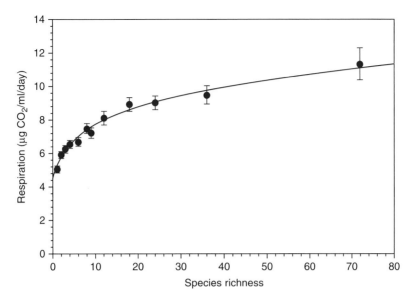

**Fig. 12.2.** The relationship between bacterial biodiversity and ecosystem functioning (leaf decomposition). The mean decomposition rate for each level of species richness is shown ± one standard error of the mean. (Redrawn from Bell *et al.*, 2005.)

because, for example, functioning depends mainly on the particular species present and not their diversity, another argument for preserving biodiversity is that we do not know which species will be critical to functioning and/or stability. The relationship between biodiversity and ecosystem functioning is not yet fully understood, nor is the relevance of much of this research to conservation entirely clear (Srivastava and Vellend, 2005). Nevertheless, it seems safe to say that biodiversity *will change* as a result of climatic change, even if the relevance of that change for ecosystem functioning is still a subject of much research (see e.g. Hooper *et al.*, 2005).

### 12.2 Measures of Biodiversity

Assessing the impacts of climatic change on biodiversity depends on which aspect of diversity is examined and, most importantly, on how biodiversity is measured. Other textbooks are available to provide a more comprehensive overview on measurements of biodiversity (see e.g. Magurran, 2004). Here we just mention a few of the more commonly used methods.

Most studies that have documented or predicted biodiversity impacts do so using simple measures of diversity such as species richness, i.e. the total number of species observed in a given area. This

measure of biodiversity is simple and relatively quick to assess, however it misses out on some important aspects of diversity. Shannon diversity is another commonly used method for measuring diversity. This measure is calculated by taking into account not only the number of species, but also their relative abundance. Shannon diversity ($H$) is calculated as:

$$H = -\sum_{i=1}^{N} p_i \log p_i \qquad (12.1)$$

where $p_i$ is the relative proportion (e.g. number of individuals or biomass) of species $i$ in a given area and $N$ is the total number of species (species richness). With this formula, when one species dominates a community in terms of relative abundance (e.g. in a four-species community, species 1 covers 97% of the area and species 2, 3 and 4 cover only 1% each), then the Shannon diversity is low. When all species are present in equal relative proportion (e.g. 25% each in a four-species community) then Shannon diversity is highest for that community. This reflects the fact that not just the presence of species is important, but also their relative abundance. Other variations of diversity measures include 'evenness', which is Shannon diversity divided by maximum diversity, or just a measure of the relative abundance of species in a community.

Another commonly used measure of diversity is Simpson's diversity index. This index gives greater weight to abundant species than Shannon diversity, but is generally less often used in studies on diversity response to climatic changes. It is important to note that there are many diversity indices but they do not all rank communities in the same order, and that, perhaps counter-intuitively, with some of the indices 'diversity' can actually increase with the loss of rare species (because the evenness increases; Hurlbert, 1971).

Most studies about biodiversity refer to diversity at the species level owing to the fact that it is often easiest to examine this level of diversity. However, there is no real reason why we should be concerned only with species-level diversity. Biodiversity responses to climatic change can be quite different when looking at effects of species diversity, phylogenetic diversity or functional diversity. Recently, diversity in ecosystem function, rather than species, has been identified to be as important, if not more important, to document. Functional diversity refers to the number of different 'functional groups'. Functional groups contain species that share similar ecosystem function, such as trophic level, growth form or physiology. Phylogenetic and taxonomic diversity refer to the number of distinct phylogenetic or taxonomic groups and are also important components of diversity, particularly from an evolutionary perspective. These measures of diversity are closer to measures of genetic diversity, and recent studies have suggested that the preservation of these kinds of diversity could be more important than preserving just species diversity, particularly for maintaining 'evolutionary potential' (see Chapter 8; Forest et al., 2007; Kramer et al., 2010).

It may be important to examine as many definitions (and thus measures) of diversity as possible, because each measure could show different responses to climatic change. For example, Hanley et al. (2004) found that species-level effects were more significant than functional group-level effects of elevated $CO_2$ in experimentally manipulated chalk grassland ecosystems in the UK. Recent studies also suggest that there may be phylogenetic consequences of climatic change because entire taxonomic groups may be at higher risk of extinction (Willis et al., 2008). Reusch et al. (2005) found that genetic diversity was very important in the ecosystem recovery of a species-poor seagrass community from extreme events associated with climatic change. They suggest that in some ecosystems, genetic diversity may replace the role of species

diversity in providing a buffer against climatic changes. It should be clear that it will continue to be important to examine genetic, functional and species-level diversity in predicting biological responses to climate change, because each level will give different kinds of information about impacts and also on potential adaptation or intervention strategies that we might take to adapt to climatic change.

## 12.3 Biodiversity Patterns at Local, Regional and Global Spatial Scales

The different spatial scales at which biodiversity is measured has given rise to a particular terminology that is used to describe diversity responses. 'Alpha diversity' is *within-habitat* species diversity (local, small-scale diversity), 'beta diversity' is *between-habitat* diversity (and thus reflects the variation of richness and composition in a region) and 'gamma diversity' is the species diversity of an *entire landscape*.

Generally speaking, local diversity is governed by processes such as predation, competitive exclusion and local abiotic conditions, and regional diversity is governed by larger-scale processes such as dispersal and landscape heterogeneity. However, local and regional diversity are linked; thus loss of species at one of these levels will affect the other. Reponses of diversity to climatic changes at the local level have been shown in many different ecosystems. These studies detect local extinctions (or extirpation, see Section 12.4) and/or changes in community composition. For example, Burgmer et al. (2007) found that increasing temperatures in lakes have changed the diversity and composition of freshwater macroinvertebrate communities. We saw many other examples of effects of climatic change on diversity and composition of communities in Chapter 6.

Finding a relationship between species diversity and local and regional spatial scales can help to determine how climatic change can affect diversity responses. Perhaps the best-known example is the species–area relationship (SAR). This relationship predicts that as area ($A$) increases, the number of species ($S$) will also increase according to:

$$S = cA^z \qquad (12.2)$$

where $c$ and $z$ are constants. The species–area relationship has been observed in many different ecosystems and is now considered to be one of the best-documented relationships in ecology. Many ecosystems seem to exhibit a relationship in which, on average, $z=0.25$. If a relationship between number of species and area exists, it can be used to predict the loss of species attributable to changes in habitat availability due to climatic change (see Section 12.4). However, these methods do have limitations and may either under- or overestimate extinction loss, since predictions depend on consistent and accurate estimations of $z$ (Lewis, 2006).

At local and even regional scales biodiversity loss may not be of great concern, particularly if there are sources for species recolonization from nearby communities. However, some areas contain an unusually high number of endemic species. These 'biodiversity hotspots' harbour very high numbers of endemic species, or species that occur only in certain regions and nowhere else in the world (Fig. 12.3; Myers et al., 2000). Biodiversity hotspots can be defined in many other ways. For example, we could alternatively consider the number of rare species. Rarity and endemism are not necessarily correlated. Malcolm et al. (2006) predicted how climatic change might affect biodiversity hotspots as compared with other biomes and found that these hotspots did not differ from other regions in many cases. However, in some places (Cape Florisic region, Indo-Burma, Mediterranean Basin, Southwest Australia and Tropical Andes), the effects of climatic change on hotspots were severe and predicted extinctions exceeded 2000 species per hotspot. Furthermore, they predict that extinctions due to global warming will be more important than deforestation activities in tropical areas as a cause of extinction.

At the global scale, biodiversity levels are also related to latitudinal position (see e.g. Gaston, 2000). Generally speaking, biodiversity is highest at low latitudes and declines with increasing latitude. The causes for this are still hotly debated. A recent study (Willig et al., 2003) suggested over 30 hypotheses to explain the latitudinal gradient, ranging from differences in evolutionary rates, energy available, productivity and geographic area available. Climatic factors obviously contribute to some of the explanation for the latitudinal gradient. We have seen in previous chapters that the effects of climatic change are thought to occur first in high-latitude, temperature-limited ecosystems. Nevertheless, many biodiversity hotspots occur at lower latitudes (e.g. tropics) and thus the effects of climatic change on biodiversity could be more severe in these areas. These same arguments can also apply to the case of elevational gradients in biodiversity, where, in general, higher diversity occurs at low elevations. For example, Forister et al. (2010) found that increasing temperatures have led to decreases in species richness of butterflies in the Sierra Nevada mountains of California. This mostly occurred in sites at lower elevations and this pattern has also been observed in other mountain regions (Wilson et al., 2007).

## 12.4 Biodiversity Decline and Climatic Change

Some have called the current and predicted losses in biodiversity due to climatic change and other global ecological changes the 'sixth mass extinction event'. To understand the magnitude of contemporary observed and predicted extinction rates, it is useful to examine past extinction events in the history of biota on Earth. There have been five mass extinction events in the past, as evidenced in the fossil record, but it is also important to remember that ~95% of all extinctions have occurred outside these 'events' (Raup, 1986, 1992; see also Box 12.1). Extinctions can occur at different spatial and temporal scales. In response to contemporary climatic change, sometimes these can occur over small scales. Local extinctions, where a species is lost in a given study area but can still be found in other habitats or regions, is referred to as 'extirpation'. As noted in previous chapters, extinctions can occur from various factors such an extreme reduction in species range or altered trophic interactions, including a reduction in the availability of food. In the following sections we examine more closely the rate and extent of these extinctions due to contemporary climatic change. While individual and population responses to climatic change can be diverse, we see that the sum of these responses usually results in an overall loss of biodiversity.

### Observed extirpations and extinctions within the past century

Risk factors for extinction include rarity (in time or space), small population sizes and specialized habitat requirements, among other characteristics. Climatic change, when it leads to declines in population sizes and/or range contractions (see Chapter 5), changes in plant and animal community compositions (see Chapter 6) and changes in ecosystem functioning (see Chapter 7), can lead to the endangerment of species (although note that certainly some species will benefit from climatic change).

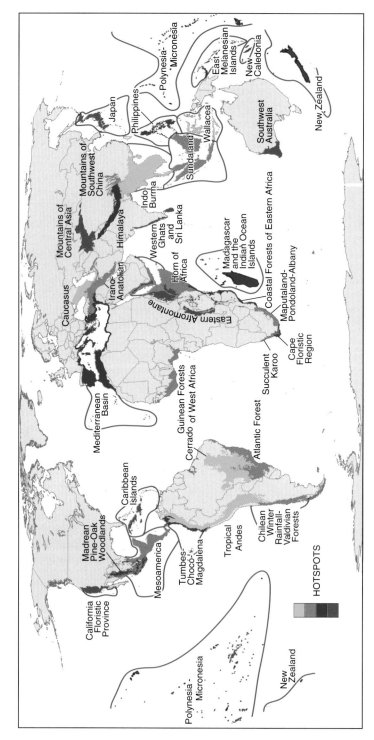

**Fig. 12.3.** Global hotspots of diversity where high levels of endemism occur and high amounts of habitat loss. (© Conservation International, www.conservation. org. Used with permission.)

Chapter 12

**Box 12.1. Palaeological extinction rates and climatic change.**

Palaeoecology uses the fossil record to reconstruct historic ecosystems (see also Chapter 1). Based on palaeoecological records, the background rate of extinctions on Earth is estimated to be about two to five taxonomic families of marine invertebrates and vertebrates every million years (Raup and Sepkoski, 1982). Since life evolved, several major (>50% animal species lost) mass extinctions have greatly exceeded the background extinction rate. In the past 540 million years, there have been five mass extinction events: End Ordovician, Late Devonian, End Permian, End Triassic and End Cretaceous. Many theories have been proposed about their causes. Mayhew *et al.* (2008) suggest that four out of five mass extinctions in palaeoecological history can be linked to climatic change. Many palaeoecological studies concentrate on the Holocene epoch (last 11,000 years) or the last glacial stage of the Pleistocene epoch (from 50,000 to 10,000 years ago), because older environments are less well represented in the fossil timeline of evolution. Such studies are useful for understanding the dynamics of ecosystem change and for studying pre-industrialized ecosystems.

Few topics in palaeoecological research match the late Pleistocene megafauna extinction debate, with most authorities championing either a human or climate change cause. Over 50,000 years ago, more than 150 genera of megafauna (>44 kg) populated the continents and by 10,000 years ago, at least 97 of those genera had gone extinct (Virtanen *et al.*, 1996).

The extinction pulse during the late Pleistocene exhibited the following special features: (i) extinction likelihood was strongly dependent on body size; (ii) the severity of the extinctions varied markedly between continents; and (iii) extinctions were associated in time with the appearance of humans, and also with a period of climate and hence habitat change (Owen-Smith, 1987). Evidence from palaeontology, climatology, archaeology and ecology supports the idea that humans contributed to extinction on some continents (see e.g. Alroy, 2001; Roberts *et al.*, 2001), but human hunting and land use were not solely responsible for the patterns of extinction everywhere. Instead, evidence suggests that the intersection of human impacts with pronounced climatic change drove the precise timing and geography of extinction in the northern hemisphere (Martin and Klein, 1984; Kutzbach *et al.*, 1998; Fiedel and Haynes, 2004). Much of what happened in the southern hemisphere is still unknown. New evidence from Australia supports the view that humans helped caused extinctions there (Prideaux *et al.*, 2007), but the correlation with climate is weak and contested. However, without Pleistocene climatic change, it is likely that species such as horses in Alaska and mammoths and giant Irish deer in central Eurasia would have survived longer, despite the presence of humans (Barnosky *et al.*, 2004). Firmer chronologies, realistic ecological models and regional palaeoecological insights are still needed to understand details of worldwide extinction patterns of the species involved.

---

It can be difficult to determine the cause of any observed extinction, since many factors contribute to produce this result. As a consequence, the examples we discuss in this section identify extinctions or extirpations (local extinctions) where climatic change is thought to have played a significant role, rather then being the sole cause of these events.

There are many examples of climate change-related extinctions or extirpations in the literature. Here we provide a small sample to give a flavour for what has been studied to date. Declines in plant species richness have been observed at high elevations in the Tibetan plateau (Klein *et al.*, 2004) and in both diversity and evenness at high latitudes in the tundra via experimental warming studies (Walker *et al.*, 2006). Reductions in diversity have also been observed via experimental warming studies in an alpine biodiversity hotspot in Norway (Klanderud and Totland, 2005). The extinction of two populations of checkerspot butterfly (*Euphydryas editha*)

was hastened by climatic change (as a result of increasing variability in precipitation; McLaughlin *et al.*, 2002). Pounds *et al.* (2006) have documented widespread extinctions in amphibians. Beever *et al.* (2003) found that 28% of pika (*Ochotona princeps*, a montane mammal) populations have been extirpated over the past century and that this could be attributed in part to climatic change. Numerous studies have linked extirpation to climate-related disturbance events such as extreme drought or low snowpack years (Singer and Thomas, 1996; Thomas *et al.*, 1996; although the connection of such events to climatic change per se is difficult to establish). From the 1970s to the 1990s, several previously common animal species went locally extinct in concert with higher winter precipitation in an arid ecosystem (Brown *et al.*, 1997).

The case of marine coral species has perhaps been the 'poster child' for climate change-driven extinction risk over the past decade (e.g. Carpenter *et al.*,

2008). Approximately a third of all studied species were in categories with elevated risk of extinction. 'Coral bleaching' occurs when stress, often caused by increased water temperature, causes the coral to expel their symbiotic zooxanthellae. The loss of these pigmented flagellate protozoa leads to the characteristic 'whitening' and hence the term 'bleaching'. These bleaching events often lead to the death of the coral. Whether or not corals go extinct this century will depend on the continued severity of climatic change, the extent of other environmental disturbances and the ability of corals to adapt (Veron, 2008). If bleaching events become frequent, many species may be unable to re-establish breeding populations before subsequent bleaching causes potentially irreversible declines, perhaps mimicking conditions that led to historic coral extinctions (Kiessling and Baron-Szabo, 2004).

### Predicted extinctions

Predictions of species extinction rates due to climatic change involve examination of the likelihood of species losing habitat. These are done using several different methods, such as the species–area relationship mentioned above. This method has been used to estimate biodiversity loss due specifically to coastal area lost from rising sea level. These so-called 'marine intrusions' are estimated to directly result in 157 to 381 species extinctions, depending on the scenario, just through the loss of suitable coastal habitat (Menon *et al.*, 2010).

Another common method is to use bioclimatic envelope models, sometimes in conjunction with the species–area curve. The general idea is that if habitat becomes unsuitable climatically, then most species will not be able to persist. For example, one study estimates that 15–37% of taxa will be threatened with extinction by 2050, depending on which climate-warming scenario is used in the analysis (Thomas *et al.*, 2004). Another study has suggested that more than half of 1350 European plant species studied could be lost by the end of the century (Thuiller *et al.*, 2005). This study suggested that mountainous areas would be most sensitive (which correlates well with the observation of reduced plant diversity in mountainous regions that we previously mentioned) and that plant diversity loss in the boreal region could be compensated by immigration from the south. Levinsky *et al.* (2007) have predicted that 1% (assuming unlimited migration) or 5–9% (assuming no migration) of European mammals risk extinc-

tion, while 32–46% (assuming unlimited migration) or 70–78% (assuming no migration) may be severely threatened (lose >30% of their current distribution). In an analysis of South African animal taxa, Erasmus *et al.* (2002) predicted that 17% of species will expand their ranges, 78% will undergo range contraction (4–98%), 3% will show no response and 2% will become locally extinct. While this does not seem too bad, these responses may lead to a loss of up to 66% of species in Kruger National Park, a protected area. We discuss the implications of climatic change on protected areas later in the chapter. Biodiversity losses have also been predicted for marine ecosystems, with the highest impacts anticipated at sub-polar regions, the tropics and semi-enclosed seas (Fig. 12.4).

## 12.5 Diversity–Productivity, Diversity–Stability and Diversity–Disturbance Relationships

As we have seen, one of the reasons why the study of biodiversity response is so important is its positive relationship to ecosystem productivity. The idea behind this is that productivity, or the amount of biomass produced per unit of time, is a good indicator of ecosystem function and is also an important consideration for ecosystem management. It is generally believed that higher species richness results in greater productivity (e.g. Naeem *et al.*, 1996; Tilman *et al.*, 1996); however, some studies have shown that more diverse ecosystems do not always lead to higher productivity, particularly in grasslands (Smith *et al.*, 2007), and it has been hard to demonstrate experimentally. The relationship between diversity and productivity may depend on climatic conditions. For example, an experimental study on bryophyte communities found that there was no relationship between diversity and productivity under constant climatic conditions, whereas biomass increased with richness during periods of drought (Mulder *et al.*, 2001). In another study it was found that the positive relationship between species diversity and productivity was enhanced by increased $CO_2$ concentrations in an experimental plant community (He *et al.*, 2002; but see Bell *et al.*, 2009 for a counter example).

Greater diversity is also generally thought to lead to greater stability. While there are many different definitions of stability, the most commonly applied ones in experimental studies are invasibility, variability, resistance and return rates (Ives and Carpenter, 2007). Very little is known about how

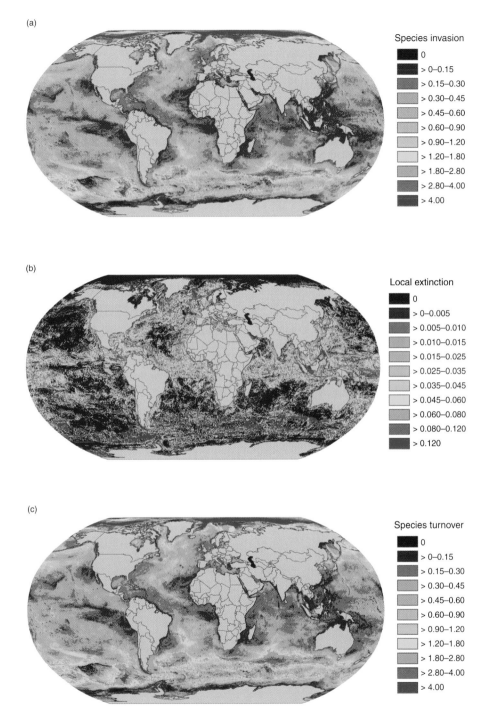

**Fig. 12.4.** Predicted extinctions due to climate change in marine metazoans. Impacts on biodiversity are expressed in terms of (a) species invasion intensity, (b) local extinction intensity and (c) species turnover for 1066 species of fish and invertebrates by 2050. Intensity is expressed as proportional to the initial species richness in each cell. (From Cheung *et al.*, 2009.)

this relationship may hold under stresses induced by climatic change. One study suggests that the relationship between diversity and stability may be dependent on environmental conditions and could be changed by global warming (Klanderud and Totland, 2008).

A well-established idea in ecology is that diversity (usually for this purpose measured as species richness) will be highest at intermediate levels of disturbance. This has been termed the 'intermediate disturbance hypothesis' (Connell, 1978). It is based on the assumption that, at very low levels of disturbance, some species in a community may outcompete others. At high levels of disturbance, populations do not have sufficient time to recover between disturbances and the risk of extinction increases. At intermediate levels of disturbance, populations of many species will remain below their carrying capacities and so competitive exclusion is less likely to occur, leading to higher levels of diversity (Fig. 12.5).

The intermediate disturbance hypothesis remains contentious, as there are certainly counter examples. Svensson et al. (2009) concluded that:

> tests and general syntheses of models of disturbance–diversity patterns would benefit from more explicit definitions of the components of disturbance, as well as a stronger focus on the importance of variation in inherent properties of natural assemblages.

These sources of variation could include climate change. As we have seen in other chapters, some components of climate change can affect disturbance regimes in ecosystems (e.g. fire, drought, pests). Some studies have suggested, however, that the intermediate disturbance hypothesis may not apply in highly stressed areas (see e.g. Piou et al., 2008; Letters, 2009; Sasaki et al., 2009). Mechanisms that allow ecosystems to recover from both (short-term) disturbance and (long-term) stress are complex (Trubina, 2009). Thus climatic change could have effects on diversity that do not fall into well-known responses of ecosystems to other kinds of perturbations. We examine further what is known about the interactions of climatic change and other perturbations in Chapter 13.

## 12.6  Invasive Species and Climate Change

Invasive species are often ranked as one of the most serious threats to biodiversity although, for plants at least, this conclusion is debatable (Sax and Gaines, 2003, 2008). Invasive species are typically introduced species that cause negative impacts on the environment that they invade. Invasions are considered distinct from other types of colonization events such as natural range expansion (Lee, 2002). In

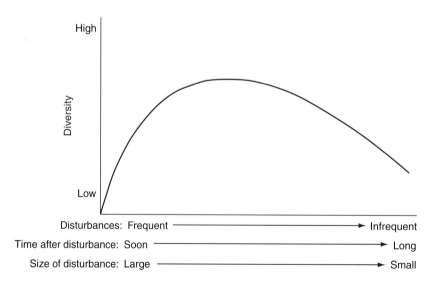

**Fig. 12.5.** The intermediate disturbance hypothesis and diversity. (Redrawn from Connell, 1978.)

most cases, invasive species are also non-native (exotic) species. However, sometimes a species native to one part of a region or country can become invasive when introduced to another part of a region/country. Invasive species can alter natural ecosystems by displacing native flora and fauna and thus alter ecosystem structure and function (Sutherst, 2000). Many invasive species have characteristics that are different from non-invasive species. For example, invasive species are typically successful and abundant whereas non-invasive species are rare and generally considered to be inferior competitors (Schweiger *et al.*, 2010). In addition, many invasive plants have broad climatic tolerances and large geographic ranges (Rejmanek and Richardson, 1996; Goodwin *et al.*, 1999), ultimately affecting responses to climatic change. Invasive plant species in particular often have characteristics that facilitate rapid range shifts, such as low seed mass and short time to maturity (Rejmanek and Richardson, 1996).

It is generally thought that climatic change will enhance the success of invasive species (Dukes and Mooney, 1999; Thuiller *et al.*, 2007). Some experimental evidence suggests, for example, that increasing $CO_2$ could favour existing invasive plant species (Ziska, 2003a). In addition, studies suggest that climatic change will affect interactions among native and alien animal species. For example, increased temperatures might benefit the Argentine ant (*Linepithema humile*), an invader of Mediterranean-climate regions, to the detriment of native ant species. In northern California, most native ant species reduce or cease foraging during the hottest hours of summer days, but Argentine ants remain active (Human and Gordon, 1996). If increased temperatures decrease the foraging time of native ants without affecting Argentine ant colonies, the ongoing displacement of native ant colonies by Argentine ants could accelerate.

Models have also been used to predict future ranges of many invasive species. Sutherst *et al.* (1996) forecast an increase of the potential range of the introduced cane toad (*Bufo marinus*) in a warmer Australia. By comparing current geographic ranges and estimated thermal requirements of 57 fish species with future climate scenarios, Eaton and Scheller (1996) found that climatic change will decrease habitat for cold-, cool- and even some warm-water fish species in North America. A few warm-water species that might be able to expand distributions are native to areas east of the North American Continental Divide, such as largemouth bass (*Micropterus salmoides*), green sunfish (*Lepomis cyanellus*) and bluegill (*Lepomis macrochirus*). These species have already attained dominance in many western North American watersheds, where they were introduced for sportfishing. Warmer climate might increase dominance and, if stocking continues, extend it across more of the western watersheds.

A recent review found that climatic change will likely remove climate-defined barriers for invasive species, ultimately giving them the ability to establish in various aquatic systems (Fig. 12.6; Rahel and Olden, 2008). Ecosystem management of invasive species can become complex in the face of climatic change because control of invasive species may interfere with conservation of native flora and fauna (Hellmann *et al.*, 2008). Finally, many exotic species that are not currently invasive may become so when introduced into new areas due to migrations expected from climatic change. Exotic species are indeed being introduced beyond their current biogeographical boundaries. Over a 32-year time period, Battisti *et al.* (2005) documented that the range of the pine processionary moth (*Thaumetopoea pityocampa*) had expanded 87 km at its northern boundary in France and 110–230 m at its upper altitudinal boundary in Italy. Laboratory and field experiments have linked the feeding behaviour and survival of this moth to minimum night-time temperatures and its expansion has been associated with warmer temperatures. Exotic species may disrupt existing ecological interactions in their new environments. Schweiger *et al.* (2010) focus on impacts of exotic species from the perspective of pollination and found that climatic change is expected to impact native plant pollinators by 'altering or disrupting temporal, spatial, behavioural, morphological or energetic matching and changing conditions of competition'.

## 12.7 Protected Areas and Other Management Practices in a Changing Climate

The management of protected areas will require new strategies due to climatic changes. One strategy may be to protect areas that are particularly resistant to climatic change. These areas are termed 'refugia', or localized areas where plants and animals survived historic climate change events. Refugia can be important at large regional scales and at small

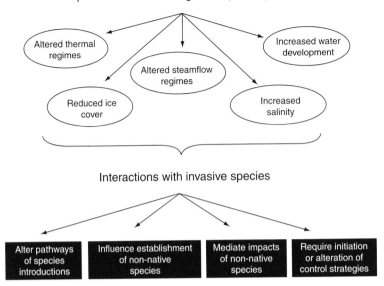

Impacts of climate change on aquatic systems

Altered thermal regimes

Altered steamflow regimes

Increased water development

Reduced ice cover

Increased salinity

Interactions with invasive species

| Alter pathways of species introductions | Influence establishment of non-native species | Mediate impacts of non-native species | Require initiation or alteration of control strategies |

**Fig. 12.6.** How climate change may affect invasive species. (Redrawn from Rahel and Olden, 2008.)

spatial scales. Some examples of regional refugia in North America include the southern Appalachian Mountains and the valleys of major rivers in the south-eastern coastal plains of the USA. In Europe, examples include Iberia, Italy and the Caucasus. Some studies have advocated that particular attention be paid to 'ecotones', transitional areas between two ecosystems at the local scale or two biomes at the continental scale. Ecotones usually represent a transition between two different climatic zones at broader spatial scales. Populations of species within and near ecotones may have genetic traits distinct from core populations, pre-adapting them to the physiological stress of climatic change (Killeen and Solórzano, 2008).

Some species poised to expand their ranges may be able to take advantage of new niches created by climatic change. The role of new niches has been studied for several invasive species. For example, Broennimann *et al.* (2007) and Broennimann and Guisan (2008) found that an invasive herbaceous plant, spotted knapweed (*Centaurea maculosa*), in western Canada has shown a shift in its distribution due to climatic change. Similarly Beaumont *et al.* (2009) found invasive plant species in the genus *Hieracium* to be able to occupy new niches and projected that current estimates of biological invasions could be underestimated. On the other hand, climatic

change could also lead to the decline of various invasive species in some areas. Linking invasion and restoration ecology, Bradley *et al.* (2009) find that:

retreat of once-intractable invasive species could create restoration opportunities across millions of hectares. Identifying and establishing native or novel species in places where invasive species contract will pose a considerable challenge for ecologists and land managers.

'Restoration ecology' focuses on returning ecosystems to previous states of ecological function. However, a return to historic states may not be possible due to the effects of climatic change on biodiversity (Harris *et al.*, 2006; Millar *et al.*, 2007). Thus, ecological restoration may need to take into account potential future range shifts and changing niches due to climatic change. 'Assisted migration', the idea of moving species northward of their current ranges in anticipation of climate change effects, may be coupled with restoration ecology efforts to increase the likelihood of survival of restored species. We explore these ideas further in Box 12.2.

## 12.8 Conservation Priorities for the Future

Conservation biology as a field has traditionally dealt with things such as deciding on the best size, shape and location of reserves for biodiversity conservation. With climatic change, priorities for conservation may change. Climate change poses obvious challenges to determining optimal strategies, because it creates moving targets of these things due to species range shifts, changes in species abundance and the variation of climatic changes across geographic regions (see Chapter 5 and Fig. 12.7; Peters and Darling, 1985; Hannah *et al.*, 2002). Thus, in addition to those species facing new threats, species already identified as threatened or endangered could face additional unforeseen threats if reserves happen to be located in areas subject to pronounced climate change effects such as species range shifts. For example, Araujo *et al.* (2004) found, based on a modelling study, that 6–11% of plant species in Europe could be lost from selected reserves within a 50-year period of climatic change.

Decisions about reserve design amount to optimizing cost-effectiveness, but this is a difficult concept to define when it comes to biodiversity. Some have suggested that reserves be placed in areas of high endemism and others in areas of overall highest diversity, such as in biodiversity hotspots. Incorporating climate change into these decisions may include location of reserves at species range edges or consideration of the likelihood of climate changing a reserve area's ecology. One study showed that butterfly populations survive longer when exposed to climatic fluctuations if they can inhabit several slopes facing different directions (thus having different moisture characteristics), suggesting that environmental heterogeneity should be taken into account in reserve design as a means of 'insurance'.

Some conservation priorities will likely remain unchanged. For example, we may want to continue to protect current areas of high endemism (biodiversity hotspots, as described above) or relatively undisturbed areas. However, some conservation strategies could differ wildly under future climatic conditions. Pearson and Dawson (2005) have suggested that some conventional strategies for reserve design should be entirely reversed (Fig. 12.8). Conservation organizations in the developed countries have now considered focusing not only on domestic conservation efforts, but also on protecting biodiversity hotspots in other parts of the world. As we have seen, tropical forests in particular harbour the vast majority of biodiversity on Earth, thus conservation of these forests must remain a priority (Williams *et al.*, 2003). Protecting these forests also

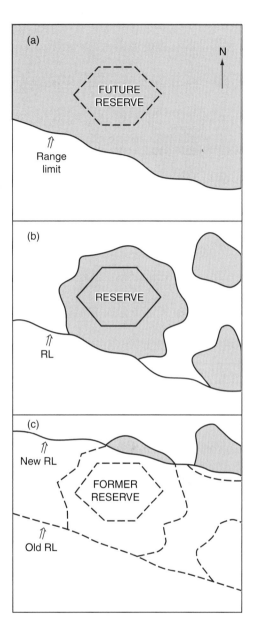

**Fig. 12.7.** The loss of reserves due to climate warming. Shaded areas indicate: (a) species distribution before human habitation; range limit (RL) indicates the southern limit; (b) fragmented species distribution after human habitation; (c) species distribution after warming. (Redrawn from Peters and Darling, 1985.)

has implications for climatic change mitigation (O'Connor, 2008), as tropical forest degradation and deforestation are direct sources of greenhouse gas emissions (see Chapter 10).

A recent comprehensive review (Heller and Zavaleta, 2009) suggests that the top recommendation for climate change adaptation strategy is to increase connectivity of existing ecosystems (design corridors, remove barriers for dispersal, locate reserves close to each other, reforestation); however, several other recommendations have also been suggested (McLachlan *et al.*, 2007, Table 12.1).

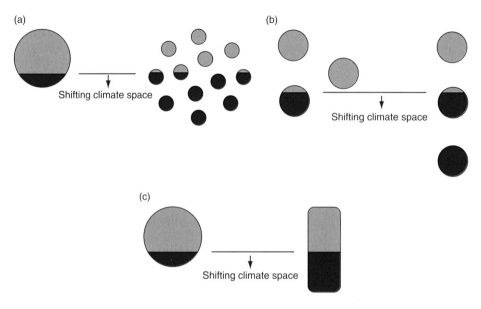

**Fig. 12.8.** Reserve design in a changing climate. The reserve configurations on the left are generally thought to be superior to those on the right, in the absence of climate change. Both left and right designs have the same total reserve area but that area is distributed differently. The light blue denotes area that is suitable under the current climate. The dark blue denotes area that is suitable in both current and future climates. We see that the reserve designs in the right result in more suitable areas in the face of climate change than those formerly preferred designs on the left. (Redrawn from Pearson and Dawson, 2005.)

**Table 12.1.** Some common recommendations for conservation priorities in response to climatic change as determined from literature review. (Adapted from Heller and Zavaleta, 2009.)

| Rank | Recommendation | No. of articles |
|---|---|---|
| 1 | Increase connectivity (design corridors, remove barriers for dispersal, locate reserves close to each other, reforestation) | 24 |
| 2 | Integrate climatic change into planning exercises (reserve, pest outbreaks, harvest schedules, grazing limits, incentive programmes) | 19 |
| 3 | Mitigate other threats (invasive species, fragmentation, pollution) | 17 |
| 4 | Study response of species to climatic change (physiological, behavioural, demographic) | 15 |
| | Practise intensive management to secure populations | 15 |
| | Translocate species | 15 |
| 5 | Increase number of reserves | 13 |
| 6 | Address scale problems (match modelling, management and experimental spatial scales for improved predictive capacity) | 12 |
| | Improve inter-agency regional coordination | 12 |
| 7 | Increase and maintain basic monitoring programmes | 11 |
| | Practise adaptive management | 11 |
| | Protect large areas, increase reserve size | 11 |

## 12.9 Conclusions

Climate change will undoubtedly result in the loss of some species. Some will be unable to migrate in concert with the changing climate, but equally will be unable to adapt to their new climate. Additionally, the movement of species that are able to shift their ranges may threaten other species in these new species assemblages. The extinction/extirpation of some species may have remarkable effects on the functioning of local ecosystems while the loss of other species my have little impact on such functioning. These changes in extinction patterns, and probably extinction rates, pose problems for conservationists and habitat managers. Climate change may need to be considered in the planning of future conservation priorities and projects, and in management decisions for such current projects.

### Box 12.2. Is 'assisted colonization' an appropriate conservation strategy?

Many species, particularly plant species, will not be able to migrate fast enough in response to climatic changes (see Chapter 5). As a result, some conservationists are suggesting that humans intervene by helping species to migrate. This phenomenon has been termed 'assisted migration' or 'assisted colonization' (see e.g. Hoegh-Guldberg *et al.*, 2008). One group called for the use of assisted migration to save the Florida torreya (*Torreya taxifolia*), a coniferous tree of which only 1000 individuals now remain within a very small geographic area (Barlow and Martin, 2004). Warming is predicted to reduce its suitable habitat and force it to migrate. This has become one of the most well-known cases of assisted migration in progress. Other species that are currently being considered for assisted migration include two species of regionally extinct butterflies in Britain (Carroll *et al.*, 2009). Early studies suggest that these kinds of assisted migrations can be successful (Willis *et al.*, 2009). Assisted migration is, however, a contentious intervention strategy because it puts different conservation goals at odds with one another. This is because the long-term consequence of introduced species into areas is not known. Indeed, introductions can lead to invasions or generally negative impacts on local species (Mueller and Hellmann, 2008). Some ecologists are arguing against the use of assisted migration for these reasons (see e.g. Ricciardi and Simberloff, 2009a,b).

More research on assisted migration is called for by scientists. For example, McLachlan *et al.* (2007) suggested we need to answer the following three questions:

> Is there a demographic threshold that should trigger the implementation of assisted migration? What suite of species should be prioritized as candidates for translocation? How should populations be introduced to minimize adverse ecological effects?

Some recommendations as to which cases should be prioritized for assisted migration have been suggested (Hoegh-Guldberg *et al.*, 2008) based on the spatial distribution of species and a decision framework (Fig. 12.9). Other ecologists are suggesting that the risks of extinction due to climate change are too great to not at least try to implement some careful assisted migration studies (Sax *et al.*, 2009). However, even if we conclude that assisted migration is a good option, current legislation may actually prohibit this option for many endangered species and thus legislation may need to change (Shirey and Lamberti, 2009). One study (Van der Veken *et al.*, 2008) found that commercial (most garden variety) versions of 357 European native plants are already located on average 1000 km north of their 'natural' ranges, suggesting that assisted migration has already been occurring, whether we like it or not. Finally, assisted migration may not just be applicable to the conservation of species, but also to help guide natural resource management, such as forestry, in the face of climatic change (Smulders *et al.*, 2009). For example, in reforesting harvested areas or in afforestation projects, managers might consider the future climate in their choice of tree species to plant.

Current management practices in response to climatic change may be contributing to biodiversity loss as opposed to actually helping to preserve it. This sometimes happens when economic interests come before ecological interests. For example, the construction of sea defences, flood management and fire exclusion may not be optimal management strategies to preserve biodiversity. Sea defences, for example, are often constructed at the expense of conserving coastal saltwater marsh habitat, which harbours many rare plants and migratory birds. Other areas will suffer prolonged drought and an increased risk of fire. Fire exclusion is a management technique being used in these areas; however, the role of human-induced fire on the long-term dynamics of ecosystems is still not fully understood. Adaptive strategies that take into account long-term dynamics and the interaction of human interventions and climatic change will need to be incorporated, as current management practices may not be sustainable (Hulme, 2005).

(a)

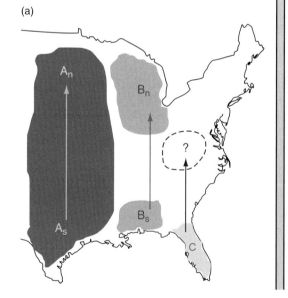

(b)

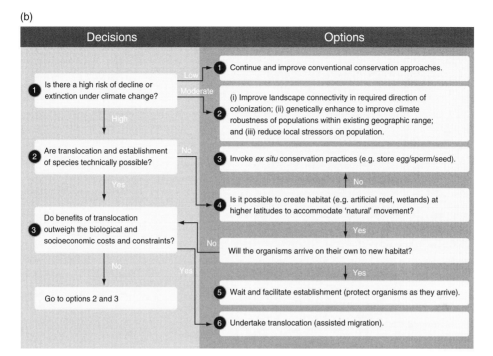

**Fig. 12.9.** (a) Different scenarios for requiring assisted migration. Species A has a continuous range from north to south. The $A_s$ populations are probably locally adapted to warmer climates. Translocating some of these individuals to populations further north may increase the chances that the recipient populations also become locally adapted to warmer temperatures as the climate changes. Species B is a more extreme version of this case since the disjunct nature of its range means that gene flow from south to north is far less likely than with species A. Species C is currently restricted to the south. Translocating individuals from C to locations further north where the climate will become suitable in the future may be necessary to prevent extinction. (Redrawn from Örstan, 2009). (b) General guidelines for assisted migration. (Redrawn from Hoegh-Guldberg *et al.*, 2008.)

# PART IV
# Final Considerations

BioCON is an ecological experiment designed to study the ways in which plant communities will respond to simultaneous decreases in biodiversity and increases in $CO_2$ concentrations and nitrogen deposition. The experiment was established in 1997 at University of Minnesota's Cedar Creek Ecosystem Science Reserve. The lead principal investigator is Peter Reich of the University of Minnesota. For more information see http://www.biocon.umn.edu/index.html. Experiments like this one will be discussed in Chapter 13. (Photo by D. Tilman, Department of Ecology, Evolution and Behavior, University of Minnesota.)

## Overview

We end this book with a final section over two chapters. In Chapter 13, we consider research that attempts to 'bring it all together' so to speak. In much of Parts II and III, the research focus has been on the individual impacts of changes in $CO_2$, temperature and precipitation even though these impacts will occur simultaneously and may well interact with each other, in some cases exacerbating the impacts of each other and in other cases cancelling the individual impacts. Bringing together multiple impacts of climatic change and other anthropogenic and natural stressors is arguably the most important type of impacts research, but it is also the most complex to carry out and to interpret. Such integrated research is not yet very common and has so far not yielded many very clear insights. Nevertheless, it is the future of impacts research and is worth considering even at these early stages.

Finally, we close the book with a chapter that focuses on the limits to our abilities to answer the kinds of research questions we posed throughout the book. Chapter 14 looks at both practical and philosophical limitations in our abilities to predict the biological impacts of climatic change. This chapter provides some important perspective for interpreting the results of climatic change impact studies and thinking about what those results mean for forming policy about climate adaptation and mitigation strategies. This might be the first contact with these kinds of ideas for many science students. It is hoped not; but in any case we hope this will not be their last contact with these ideas. They require a great deal of thought, so students shouldn't worry if their understanding doesn't come quickly.

# 13 Multiple Stressors

We have already identified the fact that climate change comprises many different components and factors; we have focused on increasing temperature, altered precipitation and elevated $CO_2$ concentrations. But climatic change is taking place in the context of other global ecological changes, both human-induced and 'natural'. Some of these changes may compound the negative effects of climatic change and others may diminish or cancel its effects. Potential interactions among climate factors and other factors have been a recurring theme throughout many chapters in this book. In this chapter we look more deliberately at a few multiple stressors and how they might interact to affect various levels of biological response.

## 13.1 Global Ecological Changes and Climatic Change

Climatic change is just one part of a number of different phenomena occurring both locally and globally that are affecting biological responses. Global ecological changes include habitat fragmentation and loss, acid deposition and pollution (nitrogen, ground-level ozone, mercury), increased UV-B radiation and invasion by exotic species and diseases (Fig. 13.1). Multiple anthropogenic stressors have already led to ecological surprises in terrestrial and aquatic systems (Christensen et al., 2006). Whether or not interaction will be synergistic or antagonistic will depend on which factors are involved and the organisms and ecosystems themselves. This is in part due to the fact that not all aspects of climate change are stressors to all species. For instance, $CO_2$ increases may well benefit several taxa through, for example, increases in photosynthesis

and WUE. Similarly, some aspects of pollution, such as nitrogen deposition, may also increase the productivity of species that are nitrogen-limited and thus may not be seen as a stressor. Other parts of climate change will however be stressors to some organisms or some ecosystems and thus may exacerbate other global ecological changes. In this chapter we examine our current knowledge about how some of the various components of climatic change and other natural and human-induced disturbances and stressors interact to produce complex biological responses at various levels of ecological organization.

## 13.2 Interactive Effects and Positive Feedbacks

In order to understand why multiple stressors merit a special discussion, indeed their own chapter, we need to understand the difference between additive effects and interactive effects. When stressor effects are additive, then:

$$F(x, y) = F(x) + F(y) \qquad (13.1)$$

where $F(\cdot)$ is an effect on some biological system of stressor $x$ or $y$. When stressors are interactive, then:

$$F(x, y) \neq F(x) + F(y) \qquad (13.2)$$

Take a simple example: let's say $F(\cdot)$ represents biomass of a plant species, $x$ represents increased temperature and $y$ represents increased precipitation. Say that an increase in temperature and the increase in precipitation both independently lead to an increase in biomass of this species, of say 25%. Then, when the plant is exposed to both increased temperature and increased precipitation simultaneously, the effect on biomass is simply a sum of the independent effects; i.e. a 50% increase. However, not all responses in biology are additive. In fact, most of the time they are not

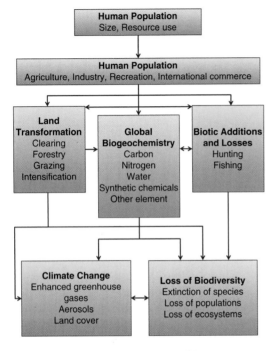

**Fig. 13.1.** Direct and indirect effects of humans causing multiple stressors on global earth systems, climate change and loss of biodiversity. (Redrawn from Vitousek *et al.*, 1997.)

additive: they are interactive or synergistic. Simply put, that means that the whole is more (or less) than the sum of its parts. In this example it should be easy to see that increased temperature increases growth when there is ample precipitation. However, when there is little precipitation (drought), temperature cannot continue to increase growth. Thus we say that the variables temperature and precipitation interact to produce a biological response. It does so, however, not because temperature might increase drought (although it does), but because drought effectively stops growth. Complicating matters, there might be ranges of the two variables over which they combine additively, a bit outside these ranges they interact.

In the presence of interactions we cannot easily predict the effects of several variables, unless we know the exact way that all of the variables interact. Figure 13.2 shows another example of interactive effects, this time with elevated $CO_2$ and temperature (Norby and Luo, 2004). Although there are well-established statistical methods to detect main effects and interactive effects, if several variables are considered we can see how a complete understanding of all the interactions may become very difficult. Several kinds of interaction are commonly observed in biological systems including multiplicative increases, suppressive effects and limitation by a third factor (when both factors cause an increase, but the combined effect is equal to the main effects in isolation).

Another important concept when discussing multiple stressors is feedback. Positive feedbacks occur when the outcome of a process accelerates the process

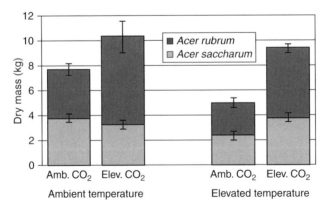

**Fig. 13.2.** The combined effects of $CO_2$ and temperature on tree biomass. Notice that the difference between tree biomass at elevated and ambient $CO_2$ depends on whether they are compared at elevated or ambient temperature. If these two factors combined additively then the size of the difference between the biomasses at the two $CO_2$ concentrations would be the same regardless of the temperature at which they are compared. That these differences are not the same indicates that temperature and $CO_2$ interact to determine tree biomass. (Redrawn from Norby and Luo, 2004.)

itself. Positive feedback stimulates change whereas negative feedback resists change. A well-known positive feedback in climate science is the fact that when snow melts, it exposes more dark ground (lower albedo), which absorbs more heat and causes more snow to melt. Another example is the positive feedback in Arctic tundra ecosystems due to thawing of permafrost (Oechel *et al.*, 1993). Negative feedbacks have a dampening or suppressing effect: a change to the system prompts a return to the pre-perturbation condition (e.g. the thermostat on a boiler ensures that the boiler comes on when the temperature dips too low, warming the house to the desired temperature, and shuts off when the temperature gets too high, again returning the house to the desired conditions). An example of negative feedback is that when atmospheric $CO_2$ levels rise, certain plants are able to grow better and thus act to remove more $CO_2$ from the atmosphere. As we shall see, multiple stressors and climate change interactions with disturbance have been observed to lead to positive feedback that accelerates change.

## 13.3 Stress versus Disturbance

Both stress and disturbance are potential components of climate change. Let us first define what we mean by 'stress' and 'disturbance', since we shall be examining both in this chapter. Barrett *et al.* (1976) defined stress as something foreign applied continuously to a system at an excessive level and give examples such as radiation and pollution. For animals, stress causes behavioural or physiological changes in the organism that may be costly for metabolic processes (Buchanan, 2000). Plant stress is most commonly defined in terms of limiting the rate of dry matter (biomass) production of vegetation (Grime, 2002). We have examined many kinds of response to climate change by various taxa and levels of biological organization (individuals, populations, communities) and, in many cases, climatic change (increased temperature, altered precipitation) has been seen as a stressor. However, in some cases, climatic change cannot be seen as a stressor at all spatial and temporal scales, or from the perspective of all organisms. For example, increased temperature can favour the growth of many organisms. Similarly, warming can cause heat stress over summer but it can reduce freezing stress over winter.

In ecology, disturbance is defined by any relatively discrete event in time that disrupts ecosystem, community or population structure and changes resources, substrate availability or the physical environment (Pickett and White, 1985). In addition, disturbances are temporary changes in average environmental conditions that cause a pronounced change in an ecosystem. Examples of disturbances include drought, flooding, fire, windstorms and insect outbreaks. Climate change alone may increase the frequency of various types of stress and disturbance, resulting in profound immediate effects on ecosystems and natural communities (Dale *et al.*, 2001). Because of these and their impacts on populations, the ecological effects of disturbance can continue for long periods of time.

Some have argued that in fact it is extreme event-based disturbances, rather than chronic trends of stress, which could have the greatest effects on biological systems responding to climatic change (Jentsch *et al.*, 2007). A recent review found that disturbance was the primary mechanism for converting ecosystems from carbon sinks to carbon sources (Baldocchi, 2008). In another study, a recent analysis of wildfire events found that, because of the effects of increased disturbance (due to either climatic change or natural variation), forests will actually become a net carbon source (not the sink that was expected as warming temperatures extend growing seasons; Kurz *et al.*, 2008b). In addition, events like hurricanes could have dramatic effects. Hurricane Katrina, for example, killed over 320 million trees in the southeastern USA, causing this area to release more carbon than was captured by the remainder of US forests during that year (Chambers *et al.*, 2007).

The fact that episodic disturbances can have consequences for the longer term or larger scales at which global ecological changes occur is also of increasing importance in the study of species range shifts, both human-induced and natural. For example, niche-based models (also referred to as bioclimatic models, envelope models or species distribution models) of species range shifts can have quite different predictions from process-based models (which can include disturbance; Morin and Thuiller, 2009; see also Chapter 3). Episodic disturbance events, such as fires and insect epidemics as well as land-use changes, need to be incorporated into models of climate change effects (Running, 2008). Observational studies have also documented the importance of natural disturbance for accelerating range shifts (Girard *et al.*, 2008). Many disturbances can lead to other disturbances; for example, an insect infestation following fire. Often these kinds of cascading events lead to surprising outcomes in ecosystem dynamics (Paine *et al.*, 1998; Fisher *et al.*, 2007).

However, an understanding of how various stressors interact with natural disturbance regimes is still largely missing (Smith *et al.*, 2009).

In Chapter 10 we introduced several kinds of disturbance in the context of forest ecosystems. Here we briefly examine these in the broader context of how stress and disturbance may interact in various different ecosystems.

## 13.4 Climate Change and Disturbance Events

Climatic change may lead to many kinds of disturbance events, but here we consider just three: drought, insect outbreaks and altered fire regimes.

### Drought

Warmer temperatures and/or decreased precipitation could lead to an increasing severity of drought (but see Chapter 2). Drought events in the past few millennia, as documented in the palaeoecological record, have caused major changes in both terrestrial and aquatic ecosystems. For example, in Africa, there were several short periods of highly arid conditions in the late Pleistocene which led to extremely low forest cover and decreased rates of evolution of fish species (Cohen *et al.*, 2007). In addition, changes in global temperature during the Holocene, after the last glacial period, led to biogeographical shifts in vegetation at continental scales (Williams *et al.*, 2004). While most of these periods of drought have been related to forest expansion in many parts of the world, some forests remained stable and some receded in favour of drier grassland ecosystems (Dumig *et al.*, 2008). Effects of drought on terrestrial ecosystems probably depend on the underlying stress levels of these ecosystems. We might reasonably expect different results in xeric, mesic and hydric systems (Fig. 13.3; Knapp *et al.*, 2008).

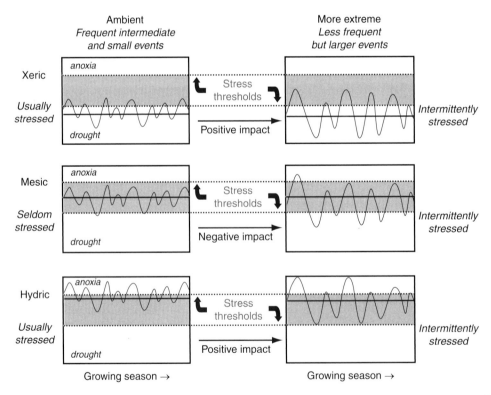

**Fig. 13.3.** The effects of more extreme precipitation regimes on ecosystems. Rectangles represent a single 'soil water bucket' in xeric (top), mesic (middle) and hydric (bottom) ecosystems. Within soil compartments, the heavy solid blue lines represent soil water; the thin solid blue lines represent temporal fluctuations in soil water; the dashed lines represent stress thresholds for water limitation (lower) and anoxia (upper); and the shaded area between denotes soil water levels that do not strongly limit ecological processes. (From Knapp *et al.*, 2008.)

A recent study found that temperature increases could shorten the time to drought-induced mortality in pinon pine (*Pinus edulis*) in the south-western USA and predicted that the interaction between stress and drought could lead to severe forest dieback (Adams *et al.*, 2009). Similarly, trees have had decreased growth rates, despite increasing $CO_2$ concentrations, that have been linked to drought events caused by climatic change in western USA (Van Mantgem *et al.*, 2009). The Amazon forest is very sensitive to drought events, becoming, at least temporarily, a carbon source rather than sink. This in turn could lead to positive feedback, accelerating climatic change (Phillips *et al.*, 2009). Tree growth in the tropics is negatively correlated with El Niño severity and positively related to precipitation, and the recent increasingly frequent and stronger El Niño droughts in this region have affected tropical rainforests at the population and community levels (Limsakul and Goes, 2008; Phillips *et al.*, 2009; Vincent *et al.*, 2009). In addition, a decline in precipitation is predicted in regions of the Amazon that are currently under strong additional stress due to crop production and ranching activities, and thus could lead to even greater deforestation in these regions (Fig. 13.4; Malhi *et al.*, 2008). We discuss interactions between climate change and land use in more detail below.

In several cases, not even increasing WUE due to atmospheric $CO_2$ fertilization effects (see Chapter 4) can compensate for these deleterious drought-related disturbances/stresses. For example, Clark *et al.* (2010) recently found that the diameter growth of six canopy tree species in Costa Rica was very sensitive to changes in water stress and to variation in mean annual temperature, but not to increasing atmospheric $CO_2$. Furthermore, permanent inventory plots in Panama and Malaysia also suggest that tropical tree growth is declining (Feeley *et al.*, 2007). In Brazil, it has been shown that the combination of warmer and drier conditions can lead to growth decline of *Araucaria angustifolia*, a keystone tree species often associated with ecotonal dynamics and forest expansion (Silva *et al.*, 2009). These negative responses are not limited to the tropics. Tree growth decline in European temperate beech forests (Peñuelas *et al.*, 2008) is another example of warming-induced drought stress. In Canadian boreal and mixed temperate forests, growth declines have also been detected, which in some cases can be related to temperature and precipitation changes over the same period (Silva *et al.*, 2010).

## Insect infestations

Forest insects generally have more limited geographic distributions than their host plants and have high mobility, so their distributions could change very rapidly in response to warmer climatic conditions (Ayres and Lombardero, 2000; Bale *et al.*, 2002). This response has been seen in previous

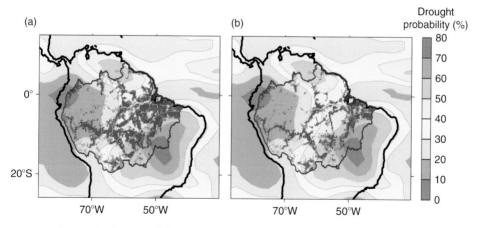

**Fig. 13.4.** The potential overlap between deforestation and drought due to climate change. Potential loss in forest cover (red) by 2050 under (a) business-as-usual forest management and (b) increased governance in forest management, superimposed on the probability of substantial drought, which is defined as a >20% reduction in dry-season rainfall by the late 21st century. The dry season is December to February south of the equator and June to August north of the equator. Precipitation scenarios represent mid-range (A1B) global greenhouse gas emissions scenarios from 21 climate models employed in the IPCC AR4. (From Malhi *et al.*, 2008. Reprinted with permission from AAAS.)

episodes of climate warming in the Palaeocene–Eocene period (Wilf and Labandeira, 1999). Because warmer temperatures can increase overwinter survival, climatic change may lead to larger or more frequent outbreaks of many different insects and this has been documented over the past century, as well as predicted for the future (Perry and Borchers, 1990).

Logan *et al.* (2010) found that mountain pine beetle infestations in the Greater Yellowstone ecosystem were more frequent and more widespread due to shifts in thermal habitat. Indeed, the contemporary outbreaks of mountain pine beetle in many areas in western North America have been attributed largely to warming, which has reduced overwinter mortality and provided the beetles access to previously inhospitable habitats (but see Box 10.1). Similarly, climatic warming may increase outbreaks in the northern and/or alpine regions of insect herbivore distributions, but may decrease outbreaks in the southern regions (Williams and Liebhold, 1995; Virtanen *et al.*, 1996). Table 13.1 shows some of the interacting effects of climate change on insect-related disturbances, which include greater crop pest problems and invasive species (Trumble and Butler, 2009). In a recent study, Kurz *et al.* (2008a) suggested a positive feedback between mountain pine beetle infestations and forest carbon losses, suggesting that they could exacerbate climatic changes. Some studies (e.g. Classen *et al.*, 2005) have also proposed that insect infestations could

lead to microclimate shifts (e.g. increased soil temperature) on the same order of magnitude as the global changes anticipated for this century.

Of additional concern is the fact that insects serve as vectors for many infectious diseases: as insects expand their range, so too might the diseases they carry. While ecological niche models are being used to predict the interaction between climate change and infectious diseases (Gonzalez *et al.*, 2010), the implications are still not fully understood (Lafferty, 2009).

## Fire

Palaeoecological evidence suggests that the indirect effects of global climatic change towards warmer, wetter environments reduced the occurrence of fire-related disturbances in some parts of the world, allowing forests to expand over the past few thousand years (Behling *et al.*, 2005). Changes in fire regimes over the past century have been observed with climatic change due to increased temperatures and increased drought. Fire regimes have several components, including frequency, size, intensity, seasonality, type and severity. Studies using GCMs predict that there could be an increase in fire severity of 10–50% across the USA by 2060 (Flannigan *et al.*, 2000). However, some fire history studies have suggested that fire frequency can decrease in some areas of the boreal forest despite increased temperature because of associated increases in precipitation (Bergeron and Archambault, 1993).

**Table 13.1.** Interacting effects between climate change and insect infestation. (See Trumble and Butler, 2009 for references.)

| Increasing atmospheric carbon dioxide leads to | Reference(s) |
|---|---|
| *Increasing...* | |
| Northward migration | Parmesan (2006) |
| Migration up elevation gradients | Epstein *et al.* (1998) |
| Insect developmental rates and oviposition | Regniere (1983) |
| Potential for insect outbreaks | Bale *et al.* (2002) |
| Invasive species introductions | Dukes and Mooney (1999) |
| Insect extinctions | Thomas *et al.* (2004) |
| Occurrence of human and animal diseases | Patz *et al.* (2003); Juliano and Lounibos (2005) |
| *Decreasing...* | |
| Effectiveness of insect biocontrol by fungi | Stacy and Fellowes (2002) |
| Reliability of economic threshold levels | Trumble and Butler (2009) |
| Insect diversity in ecosystems | Erasmus *et al.* (2002) |
| Parasitism | Fleming and Volney (1995); Hance *et al.* (2007) |

Because trees are long-lived and forests will respond slowly to environmental changes, it is likely that the short-term effects of climatic changes on fire will differ from their long-term effects (Gedalof, 2010).

The predicted future increases in area burned exceed 40% (Flannigan and Wagner, 1991; Rupp et al., 2001) and recent observations have confirmed these predictions, as the average area burned in western Canada has doubled in the last 20 years (Kasischke et al., 1999; Rupp et al., 2002). However, because there is a strong vegetation–fire feedback, vegetation shifts from conifer to deciduous forest in response to increased fire would likely reduce future fire occurrences. This assumption is supported by past changes in boreal vegetation, which occurred rapidly in response to warmer climates and were followed by significant changes in fire regimes (Chapin et al., 2006). In interior Alaska, fire frequency increased substantially with the arrival of black spruce (4000–6000 years BP) to the landscape despite the trend towards a cooler wetter climate, indicating strong vegetation effects on fire regimes.

More frequent fires in association with warmer and drier conditions could favour the expansion of open vegetation physiognomies (e.g. savannahs) at the expense of tree-dense vegetation forms. In turn, because grass-dominated landscapes are more flammable than wooded ecosystems, wildfire frequency and intensity are expected to be increased further (Hoffmann et al., 2002, 2003). Increasing fire disturbances would further reduce woody plant density within these open ecosystems and also towards forest–open vegetation boundaries, with unknown consequences for biodiversity and carbon sequestration.

## 13.5 $CO_2$ Elevation, Temperature and Precipitation

As we have seen in previous chapters, increased atmospheric $CO_2$ could have either negative or positive effects on ecosystems. However, when combined with other aspects of climate change, effects could differ. A meta-analysis study on plant herbivores found that predicted negative effects of increased $CO_2$ could be mitigated by temperature increases in some herbivores and suggested that conclusions of $CO_2$ elevation studies should not be extrapolated to climate change effects in general (Zvereva and Kozlov, 2006). While the effects of elevated atmospheric $CO_2$ on plants and herbivorous insects have been extensively studied (see

Box 5.1), relatively few studies have addressed the simultaneous increase in temperature and atmospheric $CO_2$ on other taxa. Here we focus a bit more on terrestrial systems.

In Chapter 4 we described the so-called 'CO$_2$ fertilization effect' whereby some plants benefit from higher atmospheric $CO_2$ concentrations. The enhancement of primary production in terrestrial ecosystems due to increasing atmospheric $CO_2$ levels is based on theoretical scaling up of the principles of leaf physiology to ecosystem-level processes, but it has also been observed in several ecosystem-scale experiments. In temperate and boreal forests, the synergistic effects of recent changes in climate and rising atmospheric $CO_2$ are expected to produce positive responses in primary productivity at larger scales (Bonan, 2008; but see Chapter 10). Generally speaking, high-altitude and high-latitude ecosystems should benefit most from increasing $CO_2$ and associated global warming because these ecosystems are assumed to be mainly temperature-limited. However, even tropical forests have been showing increasing growth rates due to greater atmospheric $CO_2$ concentrations (Lewis et al., 2009; Phillips et al., 2009).

This $CO_2$ fertilization effect on primary productivity may not always occur, due to interactions with other components of climate change or other stressors. Within the tropics, higher atmospheric $CO_2$ concentrations favour the growth of $C_3$ woody plants at the expense of the $C_4$ grasses (Ehleringer et al., 1997). Reductions in grass biomass, in turn, may lead to steep declines in fire frequency, which likely contributes to the expansion of forests as mentioned previously. In most cases, accurately predicting the outcomes of atmosphere–biosphere interactions remains a challenge (Huang et al., 2007).

Results from 16 FACE sites (see Chapter 3) representing four different global vegetation types (bog, forest, grassland, desert) indicate that expected effects of increasing $CO_2$ concentration on plant and ecosystem processes are well supported. Predictions for leaf $CO_2$ assimilation generally fit our understanding of limitations to photosynthesis, but at the ecosystem level predictions have been only partially supported. Moreover, since FACE studies are restricted to temperate systems, results may not hold for boreal and tropical forests (Hickler et al., 2008). In attempts to explain discrepancies within results for temperate forests, species identity and resource availability were found to be important factors influencing the response of ecosystems to elevated $CO_2$ (Nowak et al., 2004).

Few studies that are examining effects of various components of climate change are examining multiple factors (Rustad, 2008). However, this kind of examination is critical since most multi-factor experimental studies to date have shown interactive, often unpredicted effects of multiple components of climate change. For example, warming can result in reductions in soil water availability and thus limit the benefits of elevated $CO_2$ in grasslands (Niklaus and Körner, 2004). In other studies, however, interactive effects between elevated $CO_2$, warming and/or increased precipitation have been relatively minor for response variables such as soil respiration (Niinisto et al., 2004; Zhou et al., 2006), suggesting that not all aspects of biological response will be affected by interactive multiple stressors. Increasingly complex experiments may be needed in order to obtain more realistic predictions of the effects of $CO_2$ and temperature on other terrestrial ecosystems.

A new Danish climate change field-scale experiment was established in a heath/grassland ecosystem which includes full factorial combinations of elevated $CO_2$, elevated temperature and prolonged summer drought (Fig. 13.5; Mikkelsen et al., 2008). Initial results from this study are showing that different climate components can have different effects on ecosystems and that the combined effects often counteract any single main effect. In particular, it was found that the combined effects could lead to nutrient limitation (Andresen et al., 2010). In a recent study in which $CO_2$, temperature and precipitation were all experimentally manipulated, Castro et al. (2010) found that climate change factors and their interactions affected abundances of bacteria and fungi, but precipitation had a greater impact on soil microbial diversity. This highlights yet again the importance of examining various levels of ecosystem response.

**Fig. 13.5.** The CLIMAITE field site in Denmark where a shrubland–grassland ecosystem is exposed to future climate change. Twelve octagonal experimental study plots are placed in the landscape and divided into four subplots, receiving different combinations of elevated $CO_2$, elevated temperature and extended summer droughts. The 48 study plots in total provide six replicates of each of the eight combinations of the treatment factors, including six untreated reference plots. The study plot in the front of the photo is covered 50% by a screen which automatically covers half of the plot (two sub-plots) during night-time to reduce the heat loss and create night-time warming. To the left a perpendicular screen is used to automatically cover half of the study plot (two sub-plots) during rain events to remove the precipitation and expose the plots to drought. The study plot receives ambient $CO_2$ and provides all four combinations of 'non-elevated $CO_2$' (warming, drought, warming + drought, untreated reference). A similar plot in the background receives elevated $CO_2$ and provides all combinations with elevated $CO_2$ ($CO_2$, warming + $CO_2$, drought + $CO_2$, warming + drought + $CO_2$). (Photo: C. Beier, National Laboratory for Sustainable Energy, Danmarks Tekniske Universitet.)

## 13.6 Climate Change and Nitrogen Deposition

Interactions between aspects of climate change and nitrogen availability have come up repeatedly throughout this book. This is one of the few well-studied examples of climate change in the context of other stressors.

Global anthropogenically fixed nitrogen, due to pollution from fossil fuel burning, fertilizer production and legume cropping, is now much more abundant than natural sources of nitrogen fixation and is predicted to increase in coming decades (Fig. 13.6; Galloway *et al.*, 2004). Human activities are increasingly dominating nitrogen budgets at both the global and most regional scales. Although the terrestrial and open ocean nitrogen budgets are essentially disconnected, fixed forms of nitrogen are accumulating in the environment. Atmospheric deposition currently accounts for

roughly 12% of the reactive nitrogen entering terrestrial and coastal marine ecosystems globally, but in some regions, atmospheric deposition is even higher (about 33% in the USA). More importantly, these rates of deposition are sometimes greater than the uptake capacity of these systems (Galloway *et al.*, 2004). Atmospheric nitrogen deposition results in the release of greenhouse gases such as $N_2O$, which in turn can further contribute to climatic change and lead to additional stressors such as eutrophication and acid rain (Vitousek *et al.*, 1997). Reduction in soil buffering capacity (Bowman *et al.*, 2008), acidification, eutrophication and their effects on declines in species richness and/or abundance in terrestrial and aquatic ecosystems (Cairns and Yan, 2009; Stevens *et al.*, 2010) may become increasingly common.

For terrestrial ecosystems, increased nitrogen deposition has a fertilizing effect because most of

(a)

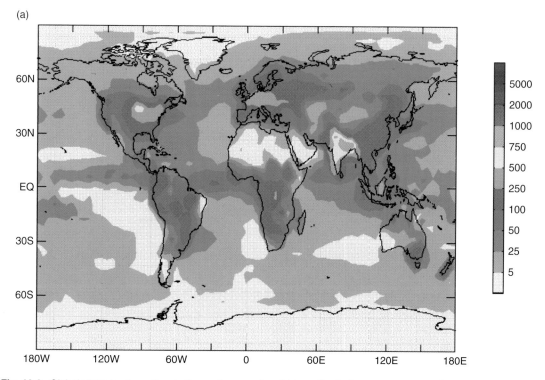

**Fig. 13.6.** Global nitrogen deposition in the past, present and future. Illustration represents spatial patterns of total inorganic nitrogen deposition in (a) 1980, (b) the early 1990s and (c) 2050, in mg N/m²/year. (From Galloway *et al.*, 2004.)

*Continued*

(b)

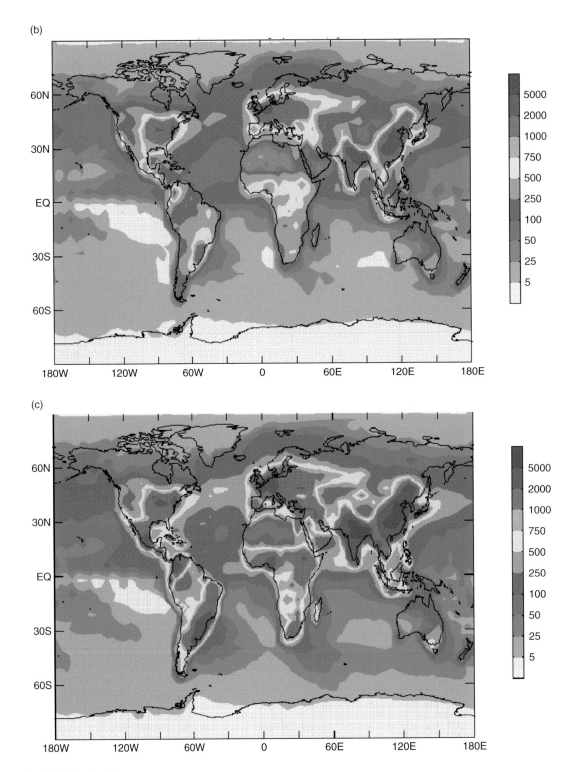

(c)

**Fig. 13.6.** Continued.

these systems are nitrogen-limited (Nadelhoffer *et al.*, 1999). The Jasper Ridge Global Change Experiment is another multi-factor study located in an annual grassland in the San Francisco Bay area of California, USA. This experiment has been incorporating various climate change factors with nitrogen deposition. This experiment shows the sometimes additive effects and at other times more subtle interacting effects of the multiple factors (elevated $CO_2$, nitrogen deposition, added precipitation and warming) on grassland diversity and species composition, and NPP (Zavaleta *et al.*, 2003; Norby *et al.*, 2007). Increased nitrate deposition could have stronger effects on these grasslands than climatic change (Dukes *et al.*, 2005).

More recently, however, researchers found for grasslands that responses to increased nitrogen deposition could depend strongly on future precipitation and that these relationships hold across grasslands of different compositional diversity (Harpole *et al.*, 2007; St Clair *et al.*, 2009). In temperate steppes in China, an experiment found non-additive effects on nitrogen and precipitation on net ecosystem carbon exchange (Niu *et al.*, 2010). Interactive effects were also significant for NPP among elevated $CO_2$, nitrogen deposition, added precipitation and warming; combinations of climate factors that included $CO_2$ elevation plus nitrogen deposition led to decreases in the overall increase in NPP in grassland ecosystems (Shaw *et al.*, 2002).

## 13.7  Climate Change and Ozone

Over the past few decades, increases in trace gases other than $CO_2$ have been observed in the atmosphere. One of these is ground-level ozone, which is toxic to plants and animals. Anthropogenic $NO_2$ emissions interact with sunlight to form ozone. An increase in ozone can be seen as another potential stressor to biological systems, particularly in the reduction of photosynthetic activity (Ashmore, 2005).

In forests, both models and experiments have shown that the $CO_2$ fertilization effect may be offset by temperature and rising ozone concentrations (Hanson *et al.*, 2005). The interactive effects of ozone and $CO_2$, for example, resulted in more than additive effects on soil respiration under silver birch trees (*Betula pendula*), but this

effect depended strongly on the genotype of the tree (Kasurinen *et al.*, 2004). When in high concentrations near the forest canopy, ozone may cause foliar injury and reduced growth rates (Matyssek and Sandermann, 2003). Chamber experiments have shown that high ozone levels can indeed negatively affect growth in young seedlings; however, tests using older seedlings have been inconclusive. There is no identifiable chemical or morphological marker for ozone effects on radial growth and extrapolation of results from seedlings in chambers to trees in forests is a stretch at best (Novak *et al.*, 2007). Moreover, even significant ozone-induced effects at the leaf level do not necessarily correspond to reduced tree-ring growth or WUE (Manning, 2005; Novak *et al.*, 2007). On the other hand, studies also suggest that if climatic change is associated with drought, the reduction in stomatal openings would tend to reduce the effects of ozone exposure (Dale *et al.*, 2001).

Increasing ozone may also affect animals, particularly insects. Climatic change may be exacerbating the disruptive effects of ozone on how insects perceive plant volatile organic compounds (VOCs). The proper recognition of VOCs is necessary for some plant–animal interactions, such as plant defence against insects, pollinator attraction, and other environmental stress adaptations (Yuan *et al.*, 2009). Higher atmospheric ozone concentration can degrade certain biogenic VOCs (Pinto *et al.*, 2007) and could reduce pollination efficiency and disturb trophic interactions. Increased ozone and $CO_2$ in future atmospheres could therefore have synergistically negative effects on VOC emissions (Calfapietra *et al.*, 2008).

## 13.8  Climate Change, Land Use and Habitat Loss

It is clear that land-use changes and their associated impacts (e.g. $N_2O$ emissions, as we discussed earlier) will have effects on biological systems and that if we want to understand effects of climate change, we may have to take these into consideration (Dale, 1997). Some have argued that climatic change and land-use changes are the two largest threats to biodiversity. In a comprehensive study on various global ecosystems, Sala *et al.* (2000) found that land-use change will probably have the largest effect on temperate ecosystems, followed by climatic change, nitrogen deposition, biotic exchange and elevated $CO_2$ concentration. In freshwater environments, biotic exchange will be most detrimental. In the Mediterranean, grassland ecosystems are expected to undergo the greatest proportional change in biodiversity; northern temperate systems are likely to experience the least biodiversity change due to the immense amount of land-use change that has already occurred. The interactions among the causes of biodiversity change represent one of the largest uncertainties in projections for future biodiversity (Sala *et al.*, 2000).

Modelling studies suggest that habitat loss and climatic change exacerbate each other's negative impacts on extinction and extirpation. This conclusion is supported by both empirical and observational studies. For example, Fig. 13.7 shows the interaction between climate change and habitat loss for projected populations of fairy shrimp (*Anostraca*) endemic to vernal pools in California, USA. Other studies on invertebrates and amphibians in this system demonstrate that the hydrologic changes due to climatic change may be compounded by cattle grazing (Pyke, 2004; Pyke and Marty, 2005). Furthermore, the extinction rates of

species are predicted to increase in response to climatic change when their habitat is fragmented. Travis (2003) refers to the combination of climatic change and habitat loss as 'a deadly anthropogenic cocktail'. Observational studies on butterflies and moths support this view that climatic change and fragmentation have synergistically negative effects (Ewers and Didham, 2005). For example, in a study of a 35-year dataset on 159 butterfly species along an elevational gradient, species richness decreased most at the lowest-elevation sites, i.e. those disproportionately impacted by habitat loss and climatic change. At the high-elevation sites, both species richness and species abundance increased, except for some alpine specialists that declined (Forister *et al.*, 2010).

Although studies on butterflies are disproportionately represented here, these interacting effects of climatic change and habitat loss are not restricted to butterflies. D'Andrea *et al.* (2009) used long-term (250 years) herbarium collections to study the spread of the weedy herbaceous plant species prickly lettuce (*Lactuca serriola*). This species is spread mainly by humans due to urbanization and agricultural activities. They found that from the 1900s to the 1970s prickly lettuce spread northward as the climate in Europe warmed.

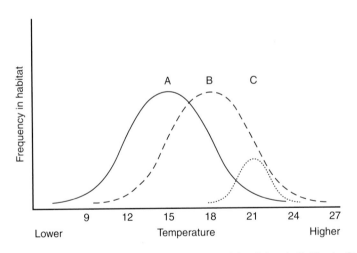

**Fig. 13.7.** The combined effects of habitat loss and climate change on fairy shrimp in California. Changes in the frequency of temperature conditions within a theoretical habitat experiencing climate and land-use change. Curve A represents the baseline distribution of habitat along an environmental gradient. Under climatic warming, temperatures across the habitat rise (curve B). Selective habitat loss removes cooler parts of the habitat and results in an additional shift towards warmer conditions, compounding climate-driven warming (curve C). (Redrawn from Pyke, 2005.)

Since the 1970s northward spread of this species seems to be unrelated to climate and more the result of habitat disturbance and other factors not related to climate. Of course care must be taken with observational studies. In some cases, effects that we think may be due to climatic change may in fact be caused by other factors. For example, Gehrig-Fasel *et al.* (2007) found that recent treeline shifts in the Swiss alps that were thought to be related to climatic change actually occurred due to human abandonment of grassland areas (see also Box 13.1).

Combining considerations of climatic change and land-use change, Jetz *et al.* (2007) found that there were geographic distinctions in the effects of climate change and land use on bird species, suggesting the importance of considering multiple stressors. They found that climate change effects at high latitudes are important. However, bird species most at risk were located mainly in narrow geographic ranges and endemic to the tropics, where range contractions are driven by anthropogenic land conversions. Similarly, at high elevations, land-use strategies could be particularly important since these activities may pose as a barrier to species attempting to migrate upslope (Feeley and Silman, 2010).

## 13.9 Conclusions

As we have seen in this and other chapters, climatic change actually refers to a number of things simultaneously. It refers to increasing $CO_2$, increasing temperatures and also changes in precipitation. We have seen examples here, and in previous chapters, of how some of these factors may interact with each other. However, so few studies have been done to date that our understanding of these interactions is incomplete at best. This makes prediction difficult. In addition to this basic fact, it is also clear that climatic changes are occurring in the context of other types of environmental change at the local or global level. Humans are altering air, land and water systems in a myriad of ways and many ecosystems also face natural disturbance regimes. We have attempted to explore a few examples where an understanding of how climate change factors may interact with other global ecological changes and/or disturbance regimes is beginning to emerge and it is important to know some of these examples. However, this is still a very young science. New papers on this subject are published daily. We explore further potential implications of these interactions in Chapter 14.

**Box 13.1. Human-induced versus climate-induced changes? Two vignettes.**

The Atlantic forest is a highly diverse but endangered ecosystem, restricted in distribution to parts of eastern and central Brazil. Conservation of this ecosystem requires knowledge of how climate change interacts with human disturbance and land-use pressures. In southern Brazil, the Atlantic forest occurs in patches on the landscape in a mosaic with natural grasslands used extensively for cattle ranching. Transitions between forest and non-forested open areas (e.g. grasslands, savannah) are usually associated with climatic factors and disturbance regimes. Fires tend to favour the maintenance of open vegetation physiognomies like grasslands instead of forests. Fires can be both climate-induced and anthropogenic in origin, where it is used as a method of controlling landscapes. Clearing of forests by humans for the establishment of rangeland is a major factor in the deforestation of these regions. Palaeoecological evidence suggests that forests have been expanding in southern Brazil since about 5000 years ago at the expense of grasslands (called 'campos') and that this also coincides with a reduction of fire frequency and intensity (Fig. 13.8; Behling and Pillar, 2007). Fire became frequent for the first time around 7400 years ago. This was probably in response to the arrival of Amerindians (Dillehay *et al.*, 1992), who may have used fire for hunting. This, in combination with a change in seasonal climatic conditions, may have led to accumulation of flammable biomass. Today, the Atlantic forest is under severe threat by multiple factors. A combination of warmer and drier conditions can lead to growth decline of *Araucaria angustifolia*. In addition, cleared areas are often planted with non-native pine species that have the potential to become invasive, further threatening the native forests. The future state of these forests thus requires an understanding of both climatic change and human-induced changes (Silva *et al.*, 2009).

A similar story has taken place in the coastal Douglas-fir (*Pseudotsuga menziesii*) forests of southern Vancouver Island, British Columbia, Canada. Dry and rocky sites in these forests are often dominated by Garry oak (*Quercus garryana*)-associated ecosystems. Pollen reconstructions of vegetation for this region indicate that oak appeared on the landscape approximately 6500 years ago (Brown and Hebda, 2002). What is remarkable about the appearance of oak in these ecosystems is that it occurred at a time when the regional climate was relatively cool and wet – precisely the sort of climatic conditions that favour more temperate species over oak. Curiously, the nearby Olympic Peninsula shows no evidence of oak despite having a similar climate and physiography. The most likely explanation for this disparity is provided within the sediment record itself: immediately preceding the appearance of oak pollen in the sedimentary record there is an increase in the abundance of macroscopic charcoal. This charcoal is attributable to prescribed fires set by the Coast Salish indigenous peoples as part of their land management practices (Turner, 1999). Fire was used for a wide range of purposes but, most importantly in a region where dietary carbohydrates were in very short supply, fire removed endemic vegetation and helped to promote the growth of camas lily (*Camassia quamash*) which was harvested for its starchy tuber. Oak and other associated species that were adapted to frequent low-intensity fire became established along with the camas lily. In the last 140 years, exclusion of fire has caused conifers to encroach many Garry oak-associated ecosystems (Gedalof *et al.*, 2006). Additional threats to these ecosystems have come from development, agriculture and species introductions. A wide range of bioclimatic envelope models suggest that Garry oak will expand its range in response to climatic change (e.g. Shafer *et al.*, 2001); however, there is little evidence that this has happened so far. Garry oak-associated ecosystems contain over 100 'species at risk', but less than 5% of their original habitat remains in a near-natural condition. Conserving these ecosystems will require a range of strategies – many of which have little to do with a changing climate.

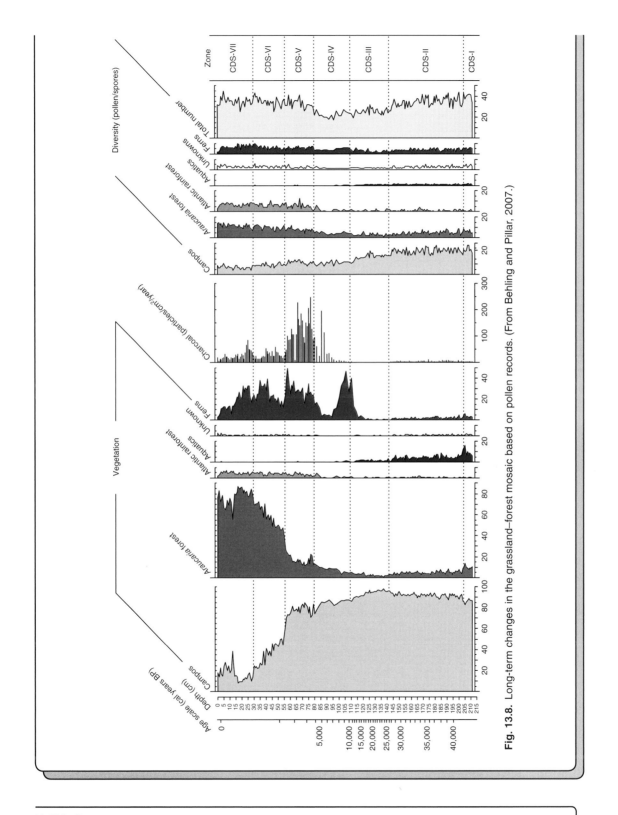

**Fig. 13.8.** Long-term changes in the grassland–forest mosaic based on pollen records. (From Behling and Pillar, 2007.)

# 14 The Limits of Science

## 14.1  Introduction

In this final chapter we consider some of the limits of science, and biology in particular, to answer society's most pressing questions about the impacts of climate change. Our aim in this chapter is to provide some perspective on things that are said about science and scientists in the media. To set the context for understanding the issues, we start by considering our task as scientists.

### The public wants and needs predictions

In 1993, Peter Kareiva and colleagues (Kareiva *et al.*, 1993: 6) stated in the introduction to their text *Biotic Interactions and Global Change*:

> they [fragmentation and climate change] represent crucial test cases for the science of ecology and evolutionary biology. . . . if we as community ecologists cannot predict the patterns of diversity change associated with these global onslaughts, one has to wonder just what community ecology can predict.

So how have we done so far with this challenge? To some extent the answer depends on how specific we expect the predictions to be. Forecasting that species will generally move towards the poles and toward higher altitudes (i.e. up mountainsides) is a prediction of sorts, but it is not very specific and probably not very 'policy-relevant'. Table 14.1 contains examples of questions that might reasonably be asked of biologists. The questions on the left are fairly specific in terms of ecosystem, species, place and time. The questions on the right are requests for generalizations. For now we just point out that we do not know the answers to the questions on the left. No one has (yet) done the research. We can take a stab

at some of the generalizations, and indeed we have at least presented evidence that would be used to do so throughout this book. But to understand why both types of question are so difficult, we need to first set them aside and take a look at some of the limits of science in providing such answers. We return to this topic in Section 14.4 below.

## 14.2  Factors Limiting Our Ability to Make Predictions

In Chapter 3 we considered some of the methodological difficulties and limitations for conducting research on the biological impacts of climatic change. Those limitations are important as they influence the inferences that can be supported from any single paper or group of papers. In this section we focus on bigger-picture, more philosophical limitations. We look at how the challenge of predicting impacts can be difficult due to commonly observed features of results in climate change experiments.

To motivate our discussion, let us start out by examining a real experiment, the Old Field Community Climate and Atmosphere Manipulation (OCCAM) experiment at the Oak Ridge National Environmental Research Park in Oak Ridge, Tennessee, USA. This experiment ran from 2002 until 2007 and examined the changes in a type of grassland community called an 'old field' in a full factorial design with two levels of $CO_2$ (ambient and +300 ppm), two air temperatures (ambient and +3°C) and two levels of precipitation (2 mm and 25 mm per week). Like some of the studies introduced in Chapter 13, this is one of the more comprehensive studies of climate change impacts in that it included three treatment variables in all combinations and ran for multiple years (Fig. 14.1).

**Table 14.1.** What are biologists being asked to predict?

| Examples of specific questions | Examples of generalizations |
|---|---|
| Will cassava (*Manihot esculenta*) production in Africa increase or decrease in the face of climatic change over the next 50 years? | Will our food and fibre supplies suffer? |
| Will purple loosestrife (*Lythrum salicaria*) be a bigger or smaller problem in the north-eastern USA in the next 40 years? | Will invasive species and/or human pathogens be helped or hindered by climate change? |
| Will Carolinian forest be the climax successional forest community in southern Ontario by the end of this century? | Will our landscapes be covered by different sets of species that look and function differently in the future? |
| Will the greater horseshoe bat (*Rhinolophus ferrumequinum*) become extinct in the UK by the end of this century? | Will species that we care about go extinct? |
| Is the New Zealand government planting the best species of trees in their afforestation scheme for the local climate in 2050? | Should we change our management practices in light of the coming changes? |
| Will carbon sequestration in Australian savannah soils increase or decrease over the course of this century? | Will we see substantial changes in our carbon and nutrient cycles? |

(a)

(b)

(c)

**Fig. 14.1.** The Old Field Community Climate and Atmosphere Manipulation experiment at the Oak Ridge National Environmental Research Park in Oak Ridge, Tennessee, USA. (a) Photo shows how the chambers were divided so that different soil moisture treatments could be implemented within the same chamber. (b) Photo shows the old field community that was constructed by hand within each chamber. (c) Photo shows the experiment underway. There were 12 chambers like this, three each for four combinations of $CO_2$ (ambient and +300 ppm) and temperature (ambient and +3°C). (Photos: R. Norby, Environmental Sciences Division, Oak Ridge National Laboratory.)

What will happen to this type of community under climatic change? To put this experiment in perspective, let us take a look at the projected climate for this location (25°54'N, 84°21'W). Figure 14.2a shows the anomaly in the daily minimum and maximum temperatures, compared with the +3°C used in the experiment. We see that +3°C is generally higher than any temperature projected for this location at any time during this century, at least for any of these three scenarios and two models, and by quite a lot. Furthermore, the projected anomaly is not constant with season, as it was in this experiment. Figure 14.2b shows the average daily precipitation anomaly for this location. The OCCAM treatment of 25 mm/week approximates ambient levels of precipitation and the experiment was designed to investigate the impacts of drought. However, as Figure 14.2b shows, projected drought is nowhere near as strong as that used in this experiment, at least for this location and these scenarios.

So our first observation is that this experiment is not simulating any specific projected climatic change for this location and the treatments, at least for temperature and precipitation, are considerably stronger

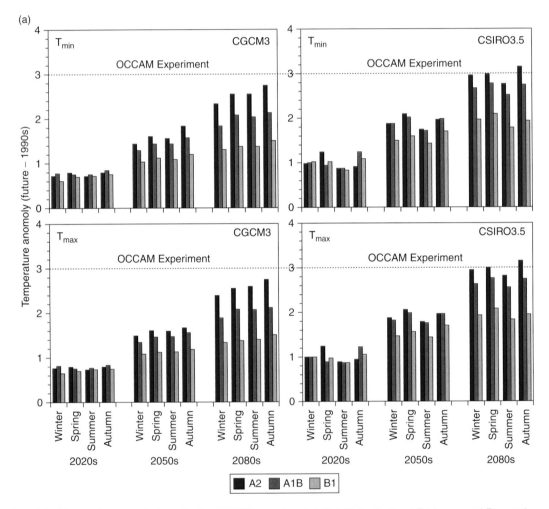

**Fig. 14.2.** Climate change projections for the OCCAM experiment at Oak Ridge National Environmental Research Park in Oak Ridge, Tennessee, USA. (a) The temperature anomaly (future temperature minus baseline) for the daily minimum and maximum air temperatures. 1971–2000 was used as the baseline period. The projections refer to three scenarios and two models (see Chapter 2). (b) The same information but for precipitation.

*Continued*

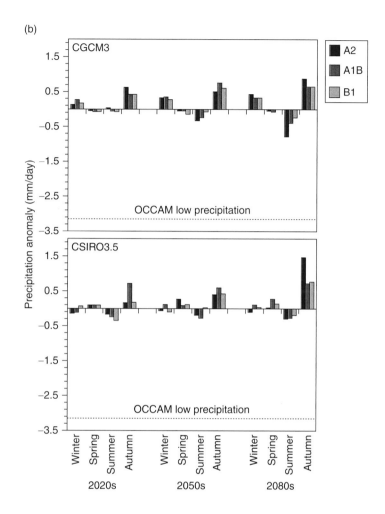

(b)

**Fig. 14.2.** Continued.

than the range of what is expected for this location. What did the researchers conclude from their experiment? Here is an excerpt from a recent publication from this experiment (Engel *et al.*, 2009):

*Will species vary in how foliar cover responds to [CO₂], warming and soil moisture?*

There were marked differences among species in how they responded to the various combinations of ambient or elevated [CO₂], ambient or elevated temperature and dry or moist soil. However, the responses also varied through time (both within years and among years), and the responses to [CO₂], warming or soil moisture often depended on the level of one of the other treatment factors. Hence, the species-specific responses of foliar cover to these atmospheric and climatic change factors cannot be simply characterized. Warming emerged as the dominant effect in 2003 and

2004, but species did not appear to respond to warming in 2005. In both 2003 and 2004, the response to warming varied within year, as evident in the temperature by date interactions. . . . Foliar cover responses to [CO₂] were surprisingly absent. The only species that responded to [CO₂] were *Plantago* and *Trifolium* in 2003, and *Solidago* in 2004 and 2005. C₃ plants are generally expected to respond to elevated [CO₂], albeit not necessarily through an increase in foliar cover, and physiological responses may not be reflected at the community level. We also detected no significant effects of [CO₂] on aboveground biomass (Garten *et al.* 2008). Previous research (Mooney *et al.* 1999) found that elevated [CO₂] increased aboveground biomass production an average of 14% in nine herbaceous communities, but production responses vary widely among ecosystems: elevated [CO₂] suppressed production in some ecosystems, but increased it by as much as 85% in others (Dukes *et al.* 2005). . . . In our system, the responses to the three treatment factors were not additive, rather, many interactions influenced foliar cover of individual species. The nature of these interactions varied over time, and increased in importance over the course of the experiment. . . . Prediction of species-specific responses of foliar cover will remain difficult, however, given the occurrence of interactions between the treatment factors and changes in the nature and strength of these interactions through time.

Engel *et al.* certainly discovered a number of interesting results, even if the mechanisms causing those results are not always clear. But the species-specific nature of the results and the presence of strong interactions, as Engel *et al.* suggest, mean that we are a long way from really understanding, and so being able to predict, what climatic change has in store for this system. And the lack of correspondence between the treatment levels and the actual projected climatic change for this location should suggest caution in making any strong conclusions about how *climatic change* (as opposed to these treatment levels) will affect this system. We elaborate on the problems of interactions and species-specific responses in the next two sections.

## Interactions hinder simple explanations and understanding

One of the conclusions of Engel *et al.*'s study is that the treatments interacted with each other and those interactions changed in time. The significance of this conclusion is subtle, and so easily missed. As we have seen in Chapter 13, when two variables interact, we cannot sensibly talk about the effects of either variable on its own. Although we know that both variables *matter* in that they affect the response variable, we really have little idea of *how* they matter because the interaction is not even stable through time. The most likely explanation is that some third variable, not part of the experiment but variable in time, is also influencing the response variable *and* interacting with $CO_2$, temperature, moisture or some combination of the three.

So interactions hamper our abilities to understand, and so to explain or predict, the impacts of our treatment variables on our response variable. And as we have also seen in Chapter 13, such interactions are not uncommon in climate change impact studies. It is important to recognize that such difficulties do not mean that we are *unable* to understand, and so to predict, the impacts of climatic

change; they only mean that our understanding will not come easily and that the predictions are unlikely to be simple ones. Our task is not just to document phenomena, but to work out mechanistic explanations that we can use to predict, even in the face of such complexity, the impacts of climatic change.

## Species-specific responses

Engel *et al.* describe so-called 'species-specific responses'. Like interactions, species-specific responses are not uncommon conclusions from climate change experiments. Indeed, we have mentioned a number of species-specific responses throughout this book (see particularly Box 5.1). Table 14.2 shows a small sampling of the most recent (at the time of writing) conclusions of species-specific responses to climatic change.

Again, like interactions, the significance of such a conclusion is quite important, but also quite subtle. Within similar groups of plants, animals and fungi we might expect, or at least hope, that there are general responses to climatic change. That is, perhaps we might, for example, reasonably expect that all temperate $C_3$ grasses would respond similarly to $CO_2$, temperature and precipitation. Perhaps we would expect that all folivorous insects that feed on oak trees would respond similarly to these environmental treatment factors, and so on. In Chapter 5 we examined some of the attempts of past studies to establish such 'functional group' responses to climatic change, and some trends are clearly emerging; however, we are still very far from knowing why, for example, some species (or groups of species) will adapt and survive and others will ultimately go extinct. A conclusion that results were 'species-specific' is often no more than an acknowledgment that no such general response was observed. On the one hand, these species-specific responses are part of what makes biology so interesting and so challenging, but on the other hand they are part of what makes meaningful prediction in biology difficult to achieve. At the extreme, if responses were really species-specific, then there would be no general conclusions to be drawn. Biology would be akin to stamp collecting in that every species by environment interaction would be unique and would require significant study before meaningful predictions could be made. With more than 1.5 million identified eukaryotic species, and possibly more than ten times that number undescribed, our job would be impossible

**Table 14.2.** A sampling of recent papers that invoke a 'species-specific' conclusion.

| Quote | Reference |
|---|---|
| 'These evolutionary correlations, coupled with the low levels of phylogenetic dependence we found, indicate that avian migration phenology adapts to climate change as a species-specific response.' | Vegvari *et al.* (2010) |
| 'Whilst significant correlations occurred between FADs [first arrival dates] of some of the species, there was considerable variability in these relationships indicating a species-specific response to rising temperatures.' | Askeyev *et al.* (2009) |
| 'The results from our study demonstrate how these species-specific responses make it challenging to predict the extent to which elevated $CO_2$ and $O_3$ will influence the biogeochemical cycling of C.' | Talhelm *et al.* (2009) |
| 'These trends support the hypothesis that the likelihood of being locally extinct is firmly related to species abundance at low to intermediate levels of drought, but it tends to become species-specific when conditions become more extreme or more prolonged as in our study.' | Prieto *et al.* (2009) |
| 'Despite these common patterns, the five most abundant species showed species-specific variation in climate responses regardless of their functional group.' | Matesanz *et al.* (2009) |

if there were no generality. Much of our challenge, both for biology in general and for climate change research in particular, is to find mechanisms by which we can see the generality. But until we can do so, our ability to predict the impacts of climate change is hampered significantly.

### Which climate change?

To answer either general or specific questions, we would ideally take some response variable that we understand in current climatic conditions, study the system (either experimentally or through models) under some set of future conditions and then make predictions about the response variable under those conditions. This caricature of the process is predicated on knowing what the future climate will look like (denoted as a point in the 'climate space' in Fig. 14.3a). The first problem is that climate scientists cannot tell us the exact location of that point. They can tell us the volume in that climate space where the point is probably located. So as biologists we really need to give bounds on the predicted response variable, bounds that are determined by the volume of the climate space (Fig. 14.3b). This volume (currently) comprises the results from the 18 climate models combined with the four to seven scenarios, interpreted by the six scenario models. Of course, climate scientists cannot give us clear sharp boundaries like those shown in Fig. 14.3b; the boundaries are fuzzy, as in Fig. 14.3c. Nevertheless, the philosophical problem is the same. Given that

future climate winds up somewhere in that space, what will happen to the response variable?

To answer that question we would really need to sample the responses to climate variable combinations that fall within that future climate space (Fig. 14.3b and c). Doing that is relatively easy for modelling studies, but it is nearly impossible for experimental studies. Think back to the Engel *et al.* study above. The combinations of treatments allowed them to compare two points in the climate space: (ambient $CO_2$ concentration, ambient temperature, precipitation 25 mm/week) versus (+300 ppm $CO_2$, +3°C, precipitation 2 mm/week). Yes, there are other comparisons embedded in that experiment, but they really just help to assess the strengths of the various main effects and interactions. The key comparison, as far as saying something about how this system may respond to future climate, is the one just described. And so while this experiment tells us something about the sensitivity of the system to these particular main effects and interactions, it tells us very little about the range of expected responses within the future climate space as predicted in Fig. 14.3c. Perhaps of equal importance, it tells us little about how the system may respond as it approaches this point in climate space, nor how it will respond when it passes that point. While the climate of any particular point in the future will have precise characteristics as shown in Fig. 14.3a, the Earth's climate is a constantly shifting target and interactions may differ as the climate system approaches or passes this point. To put it differently, not only

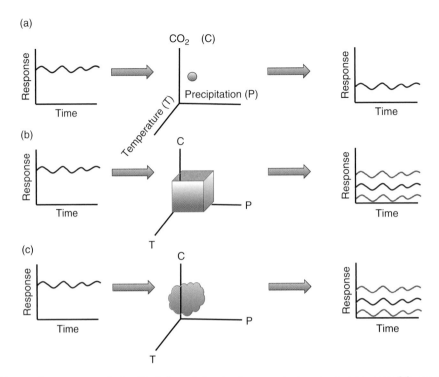

**Fig. 14.3.** Future climate spaces. A simplified view of climate change as being about changes in $CO_2$, temperature and precipitation; obviously climate change is more complex than this, but difficult to illustrate in a simple figure in three dimensions. (a) The simplest case, where we understand a response variable in the current climate conditions (the origin of the climate space graph) and then we investigate the changes in the response variable under the projected future climatic conditions (the point on the climate space graph). (b) The same situation except that we are unsure of exactly where the future climate will be, so we delimit the volume in which we think the point will reside. We then investigate the response variable throughout that future climate space. (c) The same situation but where we are more uncertain of exactly where the future climate volume will be located.

would we ideally like to know mean conditions at some point in the future, we would also like to know the trajectory the climate system will take in reaching that point. This is not a criticism of that experiment. It is simply an acknowledgment of the difficulties of doing this work at all, let alone comprehensively. That experiment took 5 years and cost several hundred thousand (US) dollars per year to conduct (R. Norby, personal communication). It was truly a Herculean effort that produced some excellent and intriguing science. It did not, however, yield a very clear idea of the impacts of future climatic change on that system, in that location.

So making meaningful predictions about the impacts of climate change is hampered by the presence of interactions, the commonly observed 'species-specific' nature of experimental results, and our abilities to replicate the experiment in enough different combin-

ations of climate variables to get beyond the specifics of the particular experiment. These practical, but fairly general, limitations should help put climate change biology research into perspective; yet we need to go a bit further to consider some even more general and more philosophical limitations on impact prediction. We take up these limits in the next section.

## 14.3 Persistent Uncertainty and the Limits of Science

In this section we consider the limits of science itself, rather than just the limits of climate change impacts research. This discussion is important in terms of putting some perspective on the task in front of us, our ability to complete that task and what 'accomplishment' would look like. We inevitably find ourselves engaged in conversation about climatic change with colleagues, friends and relatives. It may help these conversations to remind ourselves of what science is, what it is not, and what the limits of science are in general. Such a review is a big task (indeed, it is the subject of complete books), so we limit ourselves to issues that will likely be relevant in discussions of the impacts of climatic change. While this section of the book is about the limits of scientific knowledge, it is important to view the limits from the perspective of all that science has achieved. That science 'works' is undeniable: the evidence is in all of the achievements of science. And so we should remember that whatever the limitations of science, for discovering the workings of the natural world, we do not have any better tool than science:

> To exploit the fact that scientists cannot be certain about their claims (not counting trivial claims based on observations such as 'John is alive') simply because this uncertainty exists is intellectually dishonest. It's like blaming a rose for having thorns. Unlike roses, however, where we have alternatives, there is no alternative to science as a means of understanding the physical universe (including our minds but alas not our souls, whatever those things are).
>
> (Anonymous, http://tinyurl.com/ch14-quote)

## Process, deduction and induction

Science is a method, a process; it is not an accumulation of facts. To be a biologist is not to know the chemical formula for photosynthesis ($6H_2O + 6CO_2 \rightarrow C_6H_{12}O_6 + 6O_2$); it is to know how to go about discovering that formula. A bunch of facts is no more science, than a pile of bricks is a house.

As we said earlier in this book, science is not about absolutes, it is about very likely maybes. The terms 'scientific proof' and 'proven scientifically' really have no place in our conversations outside mathematics. Non-trivial scientific arguments are not generally deductive arguments (proofs), they are inductive arguments and, hence, 'weight of evidence' arguments. Let us illustrate this. A deductive argument takes the form:

Premise #1: All and only $C_3$ plants catalyse the primary uptake of $CO_2$ with the enzyme Rubisco.
Premise #2: Perennial ryegrass (*Lolium perenne*) catalyses the primary uptake of $CO_2$ with the enzyme Rubisco.
Conclusion: Therefore, perennial ryegrass is a $C_3$ plant.

Deductive arguments have two qualities that can be evaluated: 'validity' and 'soundness'. A valid argument is one where the conclusion must follow from the premises, if the premises are true. A sound argument is one that is valid, and all of the premises are true. The argument above is valid in that if both premises are true, there is no way for the conclusion to be false. It also happens to be a sound argument since premise #1 is trivially true, by definition, and premise #2 is true by observation. (Although whether premise #2 is always true, all the time, is not known and is possibly unknowable.) Now consider this argument:

Premise #1: Rates of photosynthesis in $C_3$ plants are $CO_2$-limited under current ambient $CO_2$ concentrations.
Premise #2: High rates of photosynthesis result in more carbon fixation and more biomass production.
Premise #3: Broomsedge (*Andropogon virginicus*) produces more biomass under elevated $CO_2$ conditions.
Conclusion: Therefore, broomsedge is a $C_3$ plant.

This argument is not valid because it is possible for a plant to accumulate more biomass under elevated $CO_2$ without increasing its rate of photosynthesis. For example, it can do so by increasing its WUE (see Chapter 4) and perhaps in other ways. So even if all three of the premises were true, the conclusion is not necessarily true. As it happens, even if this argument were valid, it would not be sound because premise #3 appears to be false, at least in some circumstances (Wray and Strain, 1986).

As we said earlier, non-trivial deductive arguments are rare in biology. Far more common are inductive arguments. Consider this argument:

Premise #1: $C_3$ plants tend to be $CO_2$-limited under current ambient $CO_2$ concentrations.
Premise #2: Soybeans (*Glycine max*) are $C_3$ plants.
Premise #3: Experiments in growth chambers, open-topped chambers and in FACE systems have all observed increased photosynthesis in soybeans under elevated $CO_2$ conditions.
Premise #4: Many crops are also nitrogen-limited.
Premise #5: Soybeans are legumes, and so are able to fix atmospheric nitrogen.
Premise #6: Crop growth can also be water-limited.
Premise #7: Elevated $CO_2$ tends to increase the WUE of plants by reducing transpiration and increasing root biomass.
⋮

Premise #X: Soybean yield is highly correlated with plant growth.
Conclusion: It therefore seems likely that soybean yield will increase in the future.

Validity and soundness are not properties of inductive arguments. It is entirely possible for all of the premises of an inductive argument to be true, but for the conclusion to still be false. It is also entirely possible for all of the premises of an inductive argument to be false and for the conclusion to be true nevertheless. Notice that even the form of the conclusion differs between inductive and deductive arguments. In deductive arguments the conclusion can be stated in the form 'Therefore . . .'. Conclusions to inductive arguments are stated in the form 'It therefore seems likely . . .', because we always acknowledge that with inductive arguments there can only be a likelihood, since there is neither validity nor soundness. The more premises (pieces of evidence) we can bring to bear, the more likely the conclusion may be true, but there is no sense in which such a conclusion is 'true' or 'right' or 'final' or 'conclusive'. It is simply our best current understanding, but it might be overturned by new evidence.

There are so many contingencies in biology that we can never test a hypothesis under every possible set of conditions that would be needed to make a definitive conclusion. Even if we could test the hypothesis under every set of conditions, we could not discount the possibility that we made an error in one or more test (type I or type II, see Chapter 3). In the end, we test a hypothesis under a few (hopefully realistic) conditions and then we make an induction.

### Theories, laws and scientific consensus

A commonly held, but incorrect, view of science is that hypotheses become theories that mature into laws (McComas, 1998). This view arises from an incorrect understanding of what these three elements of science are, how they may or may not be related, and the degree of faith we have in each. In this section we try to clear up these misunderstandings.

Let us start with laws. Scientific laws are statements about *patterns* in nature; they describe (predict) what will be observed in a given set of circumstances. Laws say nothing about *why* these observations would arise, only that they will. Theories, on the other hand, explain why we see what we see. Let us illustrate this with the law of gravity. The law of gravity can be stated as something like: the gravitational force between two objects is inversely proportional to the square of the distance between the two objects. The law

works well in most circumstances, but the law says nothing about *why* the force should be inversely proportional to the square of the distance. In fact there is no current, generally accepted *theory* of gravity (McComas, 1998). Laws and theories are different elements of science; they are separate but equal. Theories do not grow into laws and laws do not get downgraded into theories.

Scientific theories are often equated, in the minds of non-scientists, with the unsubstantiated musings of scientists. 'Global warming is only a theory' (search this in Google™ for some interesting reading) say some, as if that were a bad thing. They say this as if we should have a law of global warming rather than a theory. As far as understanding some natural phenomenon goes, there is *nothing* stronger in science than a mature and well-supported theory. A theory is a relatively mature, complex and wide-ranging causal explanation of some feature of the world (see e.g. Pigliucci, 2002). Sometimes (often) there are competing theories that attempt to provide causal explanations of the same natural feature. Scientists marshal evidence for and against these theories, tentatively concluding that the *weight of evidence* supports one theory more than another (i.e. they construct inductive arguments).

From time to time, well-established theories in science are overturned by new theories, because the new theories provide better, more wide-ranging explanations. That is the nature of science. It is never about absolute truths, it is always about very likely maybes. Similarly, it must be understood that scientific laws are just as tentative as scientific theories (McComas, 1998). Laws are no more absolute than theories. So even if we had such a thing as a (the) *law* of global warming, it would be no more secure than a (the) *theory* of global warming.

In Chapter 2 (Box 2.1) we mentioned the issue of 'consensus'. In that chapter we took a fairly pragmatic view of the concept. Here we briefly take a more philosophical view. When competing theories (causal explanations) exist for a feature of nature, through time one will tend to increase in favour among scientists, based on the weight of evidence. Eventually, an observer might conclude that a consensus has been reached that theory A is 'better' than theory B. The great philosopher of science, Thomas Kuhn (1970), referred to this situation as 'the dominant paradigm'. The dominant paradigm is not the whim of a few scientists; it is the conventional view of a discipline that arises through years of interaction and argument over ideas and the evidence in support of those ideas. There are no polls taken. The consensus, when it exists, is deduced from the writings and presentations of the scientists who work in that field.

It is a myth that new ideas are easily accepted in science; they are not. Kuhn's description of the process is that new ideas which are not terribly different from the current dominant paradigm, and particularly if they are proposed by someone working within the field, may at least get a hearing. Ideas that are too discordant with the dominant paradigm, particularly when they are proposed by people working outside that field, may not see the light of day, so to speak (McComas, 1998). Kuhn's idea was that science advances through a series of 'revolutions'. Most of the time scientists conduct 'normal science', science within the dominant paradigm, but sometimes we experience an upheaval, where the dominant paradigm is displaced by a new paradigm. Right now, the dominant paradigm favours the theory that greenhouse gas emissions, resulting from human activity, have contributed, and will continue to contribute, a significant and measurable amount of atmospheric warming. That is the 'consensus view'. No vote was taken. One can discern the consensus view, the dominant paradigm, from the scientific literature and public discourse of scientists. This does not mean that there are no scientists who work outside the dominant paradigm, nor that those scientists are necessarily wrong, but the dominant paradigm is the dominant paradigm for a reason. It is the argument that seems to best bear the weight of the evidence. As Carl Sagan once reportedly said, 'extraordinary claims require extraordinary evidence'. If you want to knock down the dominant paradigm, you need to bring some extraordinary evidence to light.

So, many critics of the consensus view are really misunderstanding how science works. Frequently these critics will say things like: 'That's not the way science works'. It certainly appears to be the way that science works. The real misunderstanding is that consensus views can be, and are occasionally, overturned. We experience a scientific revolution. However, simply pointing out that non-consensus views have, from time to time, turned out to be correct is not a convincing argument for ignoring the consensus view. We should not throw away every dominant paradigm because someone else has an alternative theory. If that alternative theory is better than the dominant paradigm, then the self-correcting nature of science will, eventually, lead to

a Kuhnian revolution and see the new theory replace the old one (see http://tinyurl.com/ch14-consensus for a nicely phrased expansion of this argument). The problem is that we have to make decisions now. We cannot wait 300 years to find out who was 'right'. We have a little more to say on this subject in the next section.

## 14.4 Conclusions

Throughout this book we have examined: what we know about climate change; how we know what we know; and then a variety of possible impacts of various components of climate change (mainly $CO_2$ concentrations, temperature and precipitation) on various ecological systems and at various levels of ecological organization. We have learned a good deal about the systems and questions that we have studied, and we have a number of predictions that have resulted, but they are limited in nature and specificity. What has not yet emerged is a coherent, reliable, comprehensive set of predictions.

What we are able to say about the biological impacts of climate change results from our experiments and our mathematical models. In Chapter 2 we examined some of the practical, *proximate* limits to gaining scientific knowledge about these impacts. We discussed the problems of adequate replication for avoiding type II errors of statistical inference. We discussed the problem of external validity versus experimental control. We discussed methodological and logistical limitations such as the instantaneous versus gradual application of experimental treatments such as $CO_2$ enrichment and experimental warming (see Box 3.1). Beyond these very practical limitations, we are hampered by the presence of treatment interactions and species-specific effects. We also do not know exactly what the future climate will be, and we have no realistic hope of conducting experiments over the entire range of plausible future climates. In and of themselves these limitations are non-trivial, but we have bigger philosophical issues to contend with: the problems of inference and induction.

We can understand the problem of inference as follows. Suppose we are interested in whether or not nitrogen stress affects the yield of maize. We might do an experiment, or set of experiments, over the course of a year or of several years. If nitrogen stress was likely to have arbitrary effects in different locations, then there would be no point in publishing the results of our experiments because they would not be helpful to anyone working anywhere except at our field station. But results of such experiments are published because we have a background of past success at generalizing from one field station to others, and to farms, particularly when those stations and farms are located relatively close to where the original experiment was conducted. We would not normally generalize from a field station in Canada to a farm in Australia because many interacting variables (e.g. climate and soil) will differ between the two locations (Newman *et al.*, 1997). The problem of inference then is the problem of recognizing how far we can generalize our results, beyond the individual experiment(s). The same considerations are true for the results of climate change impact studies, but arguably more so. In the first place we have vastly more experience of successful generalization of the results of nitrogen stress than we have for, say, elevated $CO_2$ experiments. Indeed we have barely a handful of FACE experiments worldwide (arguably our most externally valid experiments) from which to draw such generalizations. And second, we now have so much experimental research on the effects of nitrogen stress on plant growth, from so many different locations, under so many climatic conditions, that we have good, validated, mechanistic models of the impacts of nitrogen on plant growth. We are not yet anywhere near this level of understanding in our knowledge about the biological impacts of climatic change. And notice that this problem of inference is just as difficult and important for the specific questions as it is for the general questions in Table 14.1.

Finally, in the category of ultimate limitations, we have the problem of induction. This is a problem for all applications of science and is not unique to the subject matter of this book. There is no way around this problem. The problem of induction means that our knowledge will always remain tentative and uncertain; we have to admit that there might be a better explanation 'out there' for us to discover and so science is ultimately not capable of providing certain answers, only very likely maybes. This all might seem somewhat depressing but it behoves us to bear two points in mind. First, whatever the limitations of science, there is no better method for finding out about the natural world and science has accomplished great things regardless of these limitations. Second, all of the limitations discussed in this section become less serious as we continue to apply ourselves to answering the question(s) 'what are the likely biological impacts of climate change?'. As we develop more evidence for each of the relevant sub-questions, for more species, for more communities, in more ecosystems, each of the limitations we discussed becomes less worrisome. Meta-analyses help us overcome the problem of replication and help us to see which conclusions generalize over which areas and environments. Larger bodies of experimental work allow us to construct more mechanistic models and to validate those models. Such models further help us generalize our inferences across systems and environments. And finally, more evidence leads to stronger inductions. The more evidence we can bring to bear in an inductive argument, the more reasons we have for thinking our tentative conclusions might be correct. Climate change is a relatively new area of research. We expect significant advances over the next decade.

# References

Aberle, N., Lengfellner, K. and Sommer, U. (2007) Spring bloom succession, grazing impact and herbivore selectivity of ciliate communities in response to winter warming. *Oecologia* 150, 668–681.

Abrams, P.A. (2001) Describing and quantifying interspecific interactions: a commentary on recent approaches. *Oikos* 94, 209–218.

Ackerly, D.D. and Bazzaz, F.A. (1995) Plant-growth and reproduction along $CO_2$ gradients – nonlinear responses and implications for community change. *Global Change Biology* 1, 199–207.

Adams, H., Guardiola-Claramonte, M., Barron-Gafford, G., Villegas, J., Breshears, D., Zou, C., Troch, P. and Huxman, T. (2009) Temperature sensitivity of drought-induced tree mortality portends increased regional die-off under global-change-type drought. *Proceedings of the National Academy of Sciences USA* 106, 7063–7066.

Addington, R.N. and Seastedt, T.R. (1999) Activity of soil microarthropods beneath snowpack in alpine tundra and subalpine forest. *Pedobiologia* 43, 47–53.

Aerts, R. (2006) The freezer defrosting: global warming and litter decomposition rates in cold biomes. *Journal of Ecology* 94, 713–724.

Agosta, S. and Klemens, J. (2008) Ecological fitting by phenotypically flexible genotypes: implications for species associations, community assembly and evolution. *Ecology Letters* 11, 1123–1134.

Ainsworth, E.A. (2008) Rice production in a changing climate: a meta-analysis of responses to elevated carbon dioxide and elevated ozone concentration. *Global Change Biology* 14, 1642–1650.

Ainsworth, E.A. and Long, S.P. (2005) What have we learned from 15 years of free-air $CO_2$ enrichment (FACE)? A meta-analytic review of the responses of photosynthesis, canopy properties and plant production to rising $CO_2$. *New Phytologist* 165, 351–371.

Ainsworth, E.A. and Rogers, A. (2007) The response of photosynthesis and stomatal conductance to rising [$CO_2$]: mechanisms and environmental interactions. *Plant, Cell & Environment* 30, 258–270.

Ainsworth, E.A., Rogers, A. and Leakey, A.D.B. (2008) Targets for crop biotechnology in a future high-$CO_2$ and high-$O_3$ world. *Plant Physiology* 147, 13–19.

Allen, M.F., Klironomos, J.N., Treseder, K.K. and Oechel, W.C. (2005) Responses of soil biota to elevated $CO_2$ in a chaparral ecosystem. *Ecological Applications* 15, 1701–1711.

Alley, R.B. (2000) The Younger Dryas cold interval as viewed from central Greenland. *Quaternary Science Reviews* 19, 213–226.

Alley, R.B., Meese, D.A., Shuman, C.A., Gow, A.J., Taylor, K.C., Grootes, P.M., White, J.W.C., Ram, M., Waddington, E.D., Mayewski, P.A. and Zielinski, G.A. (1993) Abrupt increase in Greenland snow accumulation at the end of the Younger Dryas event. *Nature* 362, 527–529.

Allison, S.D. and Treseder, K.K. (2008) Warming and drying suppress microbial activity and carbon cycling in boreal forest soils. *Global Change Biology* 14, 2898–2909.

Alo, C.A. and Wang, G.L. (2008) Potential future changes of the terrestrial ecosystem based on climate projections by eight general circulation models. *Journal of Geophysical Research – Biogeosciences* 113, G01004; doi:10.1029/2007JG000528.

Alroy, J. (2001) A multispecies overkill simulation of the end-Pleistocene megafaunal mass extinction. *Science* 292, 1893–1896.

Amann, R.I., Ludwig, W. and Schleifer, K.H. (1995) Phylogenetic identification and *in situ* detection of individual microbial cells without cultivation. *Microbiological Reviews* 59, 143–169.

Amiro, B.D., Stocks, B.J., Alexander, M.E., Flannigan, M.D. and Wotton, B.M. (2001) Fire, climate change, carbon and fuel management in the Canadian boreal forest. *International Journal of Wildland Fire* 10, 405–413.

Anand, M., Gonzalez, A., Guichard, F.D.R., Kolasa, J. and Parrott, L. (2010) Ecological systems as complex systems: challenges for an emerging science. *Diversity* 2, 395–410.

Anderson, B., Akcakaya, H., Araujo, M., Fordham, D., Martinez-Meyer, E., Thuiller, W. and Brook, B. (2009) Dynamics of range margins for metapopulations under climate change. *Proceedings. Biological Sciences/The Royal Society* 276, 1415–1420.

Anderson, O.R. (2000) Abundance of terrestrial gymnamoebae at a northeastern US site: a four-year study, including the El Niño winter of 1997–1998. *Journal of Eukaryotic Microbiology* 47, 148–155.

Andresen, L., Michelsen, A., Jonasson, S., Schmidt, I., Mikkelsen, T., Ambus, P. and Beier, C. (2010) Plant nutrient mobilization in temperate heathland responds to elevated $CO_2$, temperature and drought. *Plant and Soil* 328, 1–16.

Andrew, C. and Lilleskov, E.A. (2009) Productivity and community structure of ectomycorrhizal fungal sporo-carps under increased atmospheric $CO_2$ and $O_3$. *Ecology Letters* 12, 813–822.

Aono, Y. and Kazui, K. (2008) Phenological data series of cherry tree flowering in Kyoto, Japan, and its applica-tion to reconstruction of springtime temperatures since the 9th century. *International Journal of Climatology* 28, 905–914.

Aranjuelo, I., Irigoyen, J.J., Nogues, S. and Sanchez-Diaz, M. (2009) Elevated $CO_2$ and water-availability effect on gas exchange and nodule development in $N_2$-fixing alfalfa plants. *Environmental and Experimental Botany* 65, 18–26.

Araujo, M. and Rahbek, C. (2006) How does climate change affect biodiversity? *Science* 313, 1396–1397.

Araujo, M., Cabeza, M., Thuiller, W., Hannah, L. and Williams, P. (2004) Would climate change drive spe-cies out of reserves? An assessment of existing reserve-selection methods. *Global Change Biology* 10, 1618–1626.

Arnold, S.S., Fernandez, I.J., Rustad, L.E. and Zibilske, L.M. (1999) Microbial response of an acid forest soil to experimental soil warming. *Biology and Fertility of Soils* 30, 239–244.

Arnone, J.A., Zaller, J.G., Ziegler, C., Zandt, H. and Korner, C. (1995) Leaf quality and insect herbivory in model tropical plant-communities after long-term exposure to elevated atmospheric $CO_2$. *Oecologia* 104, 72–78.

Arnott, S. and Ruxton, G. (2002) Sandeel recruitment in the North Sea: demographic, climatic and trophic effects. *Marine Ecology Progress Series* 238, 199–210.

Arp, W.J. (1991) Effects of source–sink relations on pho-tosynthetic acclimation to elevated $CO_2$. *Plant, Cell & Environment* 14, 869–875.

Ashmore, M.R. (2005) Assessing the future global impacts of ozone on vegetation. *Plant, Cell & Environment* 28, 949–964.

Askeyev, O.V., Sparks, T.H., Askeyev, I.V. and Tryjanowski, P. (2009) Spring migration timing of sylvia warblers in Tatarstan (Russia) 1957–2008. *Central European Journal of Biology* 4, 595–602.

Austin, E.E., Castro, H.F., Sides, K.E., Schadt, C.W. and Classen, A.T. (2009) Assessment of 10 years of $CO_2$ fumigation on soil microbial communities and func-tion in a sweetgum plantation. *Soil Biology & Biochemistry* 41, 514–520.

Ayres, E., Wall, D.H., Simmons, B.L., Field, C.B., Milchunas, D.G., Morgan, J.A. and Roy, J. (2008) Belowground nematode herbivores are resistant to elevated atmospheric $CO_2$ concentrations in grass-land ecosystems. *Soil Biology & Biochemistry* 40, 978–985.

Ayres, M.P. and Lombardero, M.J. (2000) Assessing the consequences of global change for forest disturbance from herbivores and pathogens. *Science of the Total Environment* 262, 263–286.

Babin-Fenske, J., Anand, M. and Alarie, Y. (2008) Rapid morphological change in stream beetle museum specimens correlates with climate change. *Ecological Entomology* 33, 646–651.

Bachelet, D., Neilson, R., Lenihan, J.M. and Drapek, R. (2001) Climate change effects on vegetation distribu-tion and carbon budget in the United States. *Ecosystems* 4, 164–185.

Bader, D.C., Covey, C., Gutowski, W.J. Jr, Held, I.M., Kunkel, K.E., Miller, R.L., Tokmakian, R.T. and Zhang, M.H. (2008) *Climate Models: An Assessment of Strengths and Limitations. A Report by the US Climate Change Science Program and the Subcommittee on Global Change Research.* Department of Energy, Office of Biological and Environmental Research, Washington, DC.

Balbontin, J., Møller, A., Hermosell, I., Marzal, A., Reviriego, M. and De Lope, F. (2009) Divergent pat-terns of impact of environmental conditions on life history traits in two populations of a long-distance migratory bird. *Oecologia* 159, 859–872.

Baldocchi, D.D. (2003) Assessing the eddy covariance technique for evaluating carbon dioxide exchange rates of ecosystems: past, present and future. *Global Change Biology* 9, 479–492.

Baldocchi, D. (2008) Turner review no. 15. Breathing of the terrestrial biosphere: lessons learned from a global network of carbon dioxide flux measurement systems. *Australian Journal of Botany* 56, 1–26.

Bale, J., Masters, G., Hodkinson, I., Awmack, C., Bezemer, T., Brown, V., Butterfield, J., Buse, A., Coulson, J. and Farrar, J. (2002) Herbivory in global climate change research: direct effects of rising temperature on insect herbivores. *Global Change Biology* 8, 1–16.

Baligar, V., Bunce, J., Machado, R. and Elson, M. (2008) Photosynthetic photon flux density, carbon dioxide concentration, and vapor pressure deficit effects on photosynthesis in cacao seedlings. *Photosynthetica* 46, 216–221.

Ball, A.S. (1997) Microbial decomposition at elevated $CO_2$ levels: effect of litter quality. *Global Change Biology* 3, 379–386.

Bally, M. and Garrabou, J. (2007) Thermodependent bacterial pathogens and mass mortalities in temper-ate benthic communities: a new case of emerging disease linked to climate change. *Global Change Biology* 13, 2078–2088.

Barlow, C. and Martin, P. (2004) Bring *Torreya taxifolia* north now. *Wild Earth* Winter/Spring, 52–56.

Barnola, J.M., Pimienta, P., Raynaud, D. and Korotkevich, Y.S. (1991) $CO_2$–climate relationship as deduced from the Vostok ice core: a re-examination based on new measurements and on a re-evaluation of the air dating. *Tellus* 43, 83–90.

Barnosky, A., Koch, P., Feranec, R., Wing, S. and Shabel, A. (2004) Assessing the causes of late Pleistocene extinctions on the continents. *Science* 306, 70–75.

Barrett, G., Van Dyne, G. and Odum, E. (1976) Stress ecology. *BioScience* 26, 192–194.

Barry, J.P., Baxter, C.H., Sagarin, R.D. and Gilman, S.E. (1995) Climate-related, long-term faunal changes in a California rocky intertidal community. *Science* 267, 672–675.

Barton, B., Beckerman, A. and Schmitz, O. (2009) Climate warming strengthens indirect interactions in an old-field food web. *Ecology* 90, 2346–2351.

Battarbee, R.W., Jones, V.J., Flower, R.J., Cameron, N.G., Bennion, H., Carvalho, L. and Juggins, S. (2001) Diatoms. In: Smol, J.P., Birks, H.J.B. and Last, W.M. (eds) *Tracking Environmental Change Using Lake Sediments*. Vol. 3. *Terrestrial, Algal and Siliceous Indicators*. Kluwer Academic Publishers, Dordrecht, The Netherlands, pp. 155–202.

Battisti, A., Stastny, M., Netherer, S., Robinet, C., Schopf, A., Roques, A. and Larsson, S. (2005) Expansion of geographic range in the pine processionary moth caused by increased winter temperatures. *Ecological Applications* 15, 2084–2096.

Battisti, D.S. and Naylor, R.L. (2009) Historical warnings of future food insecurity with unprecedented seasonal heat. *Science* 323, 240–244.

Battles, J.J., Robards, T., Das, A., Waring, K., Gilless, J.K., Biging, G. and Schurr, F. (2008) Climate change impacts on forest growth and tree mortality: a data-driven modeling study in the mixed-conifer forest of the Sierra Nevada, California. *Climatic Change* 87, Suppl. 1, S193–S213.

Bauer, W. (1972) Effect of water supply and seasonal distribution on spring wheat yields. *Bulletin 490*, Agricultural Experimental Station. North Dakota State University, Fargo, North Dakota.

Bazzaz, F.A. (1990) The response of natural ecosystems to the rising global $CO_2$ levels. *Annual Review of Ecology and Systematics* 21, 167–196.

Beaumont, L., Gallagher, R., Thuiller, W., Downey, P., Leishman, M. and Hughes, L. (2009) Different climatic envelopes among invasive populations may lead to underestimations of current and future biological invasions. *Diversity & Distributions* 15, 409–420.

Bebber, D., Brown, N. and Speight, M. (2002) Drought and root herbivory in understorey *Parashorea kurz* (Dipterocarpaceae) seedlings in Borneo. *Journal of Tropical Ecology* 18, 795–804.

Beckage, B., Osborne, B., Gavin, D., Pucko, C., Siccama, T. and Perkins, T. (2008) A rapid upward shift of a forest ecotone during 40 years of warming in the Green Mountains of Vermont. *Proceedings of the National Academy of Sciences USA* 105, 4197–4202.

Beedlow, P.A., Tingey, D.T., Phillips, D.L., Hogsett, W.E. and Olszyk, D.M. (2004) Rising atmospheric $CO_2$ and carbon sequestration in forests. *Frontiers in Ecology and the Environment* 2, 315–322.

Beever, E., Brussard, P. and Berger, J. (2003) Patterns of apparent extirpation among isolated populations of pikas (*Ochotona princeps*) in the Great Basin. *Journal of Mammalogy* 84, 37–54.

Bégin, C., Michaud, Y. and Filion, L. (1995) Dynamics of a Holocene cliff-top dune along mountain river, Northwest Territories, Canada. *Quaternary Research* 44, 392–404.

Behling, H. and Pillar, V. (2007) Late Quaternary vegetation, biodiversity and fire dynamics on the southern Brazilian highland and their implication for conservation and management of modern Araucaria forest and grassland ecosystems. *Philosophical Transactions of the Royal Society of London. Series B: Biological Sciences* 362, 243–251.

Behling, H., Pillar, V. and Bauermann, S. (2005) Late Quaternary grassland (campos), gallery forest, fire and climate dynamics, studied by pollen, charcoal and multivariate analysis of the São Francisco de Assis core in western Rio Grande do Sul (southern Brazil). *Review of Palaeobotany and Palynology* 133, 235–248.

Beijerinck, M.W. (1913) De infusies en de ontdekking der bakterien. In: Müller, J. (ed.) *Jaarboek van de Koninklijke Akademie van Wetenschappen.* Amsterdam, The Netherlands, pp. 1–28.

Bell, G. (2008) *Selection: The Mechanism of Evolution.* Oxford University Press, Oxford, UK.

Bell, G. and Collins, S. (2008) Adaptation, extinction and global change. *Evolutionary Applications* 1, 3–16.

Bell, T., Newman, J.A., Silverman, B.W., Turner, S.L. and Lilley, A.K. (2005) The contribution of species richness and composition to bacterial services. *Nature* 436, 1157–1160.

Bell, T., Lilley, A.K., Hector, A., Schmid, B., King, L. and Newman, J.A. (2009) A linear model method for biodiversity–ecosystem functioning experiments. *The American Naturalist* 174, 836–849.

Belnap, J., Phillips, S.L. and Miller, M.E. (2004) Response of desert biological soil crusts to alterations in precipitation frequency. *Oecologia* 141, 306–316.

Belnap, J., Phillips, S.L., Flint, S., Money, J. and Caldwell, M. (2008) Global change and biological soil crusts: effects of ultraviolet augmentation under altered precipitation regimes and nitrogen additions. *Global Change Biology* 14, 670–686.

Belote, R.T., Weltzin, J.F. and Norby, R.J. (2004) Response of an understory plant community to elevated [$CO_2$] depends on differential responses of dominant invasive species and is mediated by soil water availability. *New Phytologist* 161, 827–835.

Beltrami, H., Gosselin, C. and Mareschal, J.C. (2003) Ground surface temperatures in Canada: spatial and temporal variability. *Geophysical Research Letters* 30, 1499; doi:10.1029/2003GL017144.

Benning, T.L., Lapointe, D., Atkinson, C.T. and Vitousek, P.M. (2002) Interactions of climate change with biological invasions and land use in the Hawaiian islands: modeling the fate of endemic birds using a geographic information system. *Proceedings of the National Academy of Sciences USA* 99, 14246–14249.

Berg, M., Kiers, E., Driessen, G., Van Der Heijden, M., Kool, B., Kuenen, F., Liefting, M., Verhoef, H. and Ellers, J. (2010) Adapt or disperse: understanding species persistence in a changing world. *Global Change Biology* 16, 587–598.

Berger, A. (1988) Milankovitch theory and climate. *Reviews of Geophysics* 26, 624–657.

Bergeron, Y. and Archambault, S. (1993) Decreasing frequency of forest fires in the southern boreal zone of Quebec and its relation to global warming since the end of the 'Little Ice Age'. *The Holocene* 3, 255–259.

Berry, J. and Bjorkman, O. (1980) Photosynthetic response and adaptation to temperature in higher plants. *Annual Review of Plant Physiology* 31, 491–543.

Berryman, A. (2002) Population: a central concept for ecology? *Oikos* 97, 439–442.

Bertrand, A., Prevost, D., Bigras, F.J., Lalande, R., Tremblay, G.F., Castonguay, Y. and Belanger, G. (2007) Alfalfa response to elevated atmospheric $CO_2$ varies with the symbiotic rhizobial strain. *Plant and Soil* 301, 173–187.

Beyens, L., Ledeganck, P., Graae, B.J. and Nijs, I. (2009) Are soil biota buffered against climatic extremes? An experimental test on testate amoebae in Arctic tundra (Qeqertarsuaq, West Greenland). *Polar Biology* 32, 453–462.

Bezemer, T.M. and Jones, T.H. (1998) Plant–insect herbivore interactions in elevated atmospheric $CO_2$: quantitative analyses and guild effects. *Oikos* 82, 212–222.

Bezemer, T.M., Knight, K.J., Newington, J.E. and Jones, T.H. (1999) How general are aphid responses to elevated atmospheric $CO_2$? *Annals of the Entomological Society of America* 92, 724–730.

Biasi, C., Meyer, H., Rusalimova, O., Hammerle, R., Kaiser, C., Baranyi, C., Daims, H., Lashchinsky, N., Barsukov, P. and Richter, A. (2008) Initial effects of experimental warming on carbon exchange rates, plant growth and microbial dynamics of a lichen-rich dwarf shrub tundra in Siberia. *Plant and Soil* 307, 191–205.

Bigelow, S. and Canham, C. (2007) Nutrient limitation of juvenile trees in a northern hardwood forest: calcium and nitrate are preeminent. *Forest Ecology and Management* 243, 310–319.

Biknevicius, A., McFarlane, D. and MacPhee, R. (1993) Body size in *Amblyrhiza inundata* (Heptaxodontidae, Caviomorpha), an extinct megafaunal rodent from the Anguilla Bank, West Indies: estimates and implications. *American Museum Novitates* 3079, 1–25.

Billings, S.A. and Ziegler, S.E. (2005) Linking microbial activity and soil organic matter transformations in forest soils under elevated $CO_2$. *Global Change Biology* 11, 203–212.

Biro, P., Post, J. and Booth, D. (2007) Mechanisms for climate-induced mortality of fish populations in whole-lake experiments. *Proceedings of the National Academy of Sciences USA* 104, 9715–9719.

Bischoff, P.J. (2002) An analysis of the abundance, diversity and patchiness of terrestrial gymnamoebae in relation to soil depth and precipitation events following a drought in southeastern USA. *Acta Protozoologica* 41, 183–189.

Bjork, R.G., Majdi, H., Klemedtsson, L., Lewis-Jonsson, L. and Molau, U. (2007) Long-term warming effects on root morphology, root mass distribution, and microbial activity in two dry tundra plant communities in northern Sweden. *New Phytologist* 176, 862–873.

Black, B.A., Boehlert, G.W. and Yoklavich, M.M. (2008) Establishing climate–growth relationships for yellow-eye rockfish (*Sebastes ruberrimus*) in the Northeast Pacific using a dendrochronological approach. *Fisheries Oceanography* 17, 368–379.

Boag, P.T. and Grant, P.R. (1984) Darwin finches (geospiza) on Isla Daphne Major, Galapagos – breeding and feeding ecology in a climatically variable environment. *Ecological Monographs* 54, 463–489.

Bodri, L. and Čermák, V. (2007) *Borehole Climatology: A New Method on How to Reconstruct Climate*. Elsevier, Oxford, UK.

Bogeat-Triboulot, M.B., Bartoli, F., Garbaye, J., Marmeisse, R. and Tagu, D. (2004) Fungal ectomycorrhizal community and drought affect root hydraulic properties and soil adherence to roots of *Pinus pinaster* seedlings. *Plant and Soil* 267, 213–223.

Boisvenue, C. and Running, S.W. (2006) Impacts of climate change on natural forest productivity – evidence since the middle of the 20th century. *Global Change Biology* 12, 862–882.

Bokhorst, S., Huiskes, A., Convey, P., Van Bodegom, P.M. and Aerts, R. (2008) Climate change effects on soil arthropod communities from the Falkland Islands and the maritime Antarctic. *Soil Biology & Biochemistry* 40, 1547–1556.

Bokhorst, S.F., Bjerke, J.W., Tommervik, H., Callaghan, T.V. and Phoenix, G.K. (2009) Winter warming events damage sub-Arctic vegetation: consistent evidence from an experimental manipulation and a natural event. *Journal of Ecology* 97, 1408–1415.

Bonan, G.B. (2008) Forests and climate change: forcings, feedbacks, and the climate benefits of forests. *Science* 320, 1444–1449.

Bond-Lamberty, B., Wang, C. and Gower, S.T. (2004) Net primary production and net ecosystem production of a boreal black spruce wildfire chronosequence. *Global Change Biology* 10, 473–487.

Bond-Lamberty, B., Peckham, S.D., Ahl, D.E. and Gower, S.T. (2007) Fire as the dominant driver of central Canadian boreal forest carbon balance. *Nature* 450, 89–92.

Bouwman, L.A. and Zwart, K.B. (1994) The ecology of bacterivorous protozoans and nematodes in arable soil. *Agriculture, Ecosystems & Environment* 51, 145–160.

Bowatte, S., Asakawa, S., Okada, M., Kobayashi, K. and Kimura, M. (2007) Effect of elevated atmospheric $CO_2$ concentration on ammonia oxidizing bacteria communities inhabiting in rice roots. *Soil Science and Plant Nutrition* 53, 32–39.

Bowatte, S., Carran, R.A., Newton, P.C.D. and Theobald, P. (2008) Does atmospheric $CO_2$ concentration influence soil nitrifying bacteria and their activity? *Australian Journal of Soil Research* 46, 617–622.

Bowen, G. and Wilkinson, B. (2002) Spatial distribution of $\delta^{18}O$ in meteoric precipitation. *Geology* 30, 315–318.

Bowman, W., Cleveland, C., Hresko, J. and Baron, J. (2008) Negative impact of nitrogen deposition on soil buffering capacity. *Nature Geoscience* 1, 767–770.

Boyce, M., Haridas, C. and Lee, C. (2006) Demography in an increasingly variable world. *Trends in Ecology & Evolution* 21, 141–148.

Bradford, M.A., Davies, C.A., Frey, S.D., Maddox, T.R., Melillo, J.M., Mohan, J.E., Reynolds, J.F., Treseder, K.K. and Wallenstein, M.D. (2008) Thermal adaptation of soil microbial respiration to elevated temperature. *Ecology Letters* 11, 1316–1327.

Bradley, B., Oppenheimer, M. and Wilcove, D. (2009) Climate change and plant invasions: restoration opportunities ahead? *Global Change Biology* 15, 1511–1521.

Bradley, R.S. (1999) *Paleoclimatology – Reconstructing Climates of the Quaternary.* Academic Press, San Diego, California.

Bradshaw, A. (1991) The Croonian Lecture, 1991. Genostasis and the limits to evolution. *Philosophical Transactions: Biological Sciences* 333, 289–305.

Bradshaw, R.H.W. and Webb, T. III (1985) Relationships between contemporary pollen and vegetation data from Wisconsin and Michigan, US. *Ecology* 66, 721–737.

Brady, N.C. and Weil, R.R. (2004) *Elements of the Nature and Properties of Soils.* Pearson Education Inc., Upper Saddle River, New Jersey.

Brando, P., Nepstad, D., Davidson, E., Trumbore, S., Ray, D. and Camargo, P. (2008) Drought effects on litterfall, wood production and belowground carbon cycling in an Amazon forest: results of a throughfall reduction experiment. *Philosophical Transactions of the Royal Society of London. Series B: Biological Sciences* 363, 1839–1848.

Brasier, C.M. (1996) *Phytophthora cinnamomi* and oak decline in southern Europe. Environmental constraints including climate change. *Annales Des Sciences Forestieres* 53, 347–358.

Bredesen, E.L., Bos, D.G., Laird, K.R. and Cumming, B.F. (2002) A cladoceran-based paleolimnological assessment of the impact of forest harvesting on four lakes from the central interior of British Columbia, Canada. *Journal of Paleolimnology* 28, 389–402.

Briones, M.J.I., Ineson, P. and Piearce, T.G. (1997) Effects of climate change on soil fauna; responses of enchytraeids, diptera larvae and tardigrades in a transplant experiment. *Applied Soil Ecology* 6, 117–134.

Briones, M.J.I., Ostle, N.J., McNamara, N.R. and Poskitt, J. (2009) Functional shifts of grassland soil communities in response to soil warming. *Soil Biology & Biochemistry* 41, 315–322.

Broecker, W.S. (2006) Was the Younger Dryas triggered by a flood? *Science* 312, 1146–1148.

Broennimann, O. and Guisan, A. (2008) Predicting current and future biological invasions: both native and invaded ranges matter. *Biology Letters* 4, 585–589.

Broennimann, O., Treier, U., Muller-Scharer, H., Thuiller, W., Peterson, A. and Guisan, A. (2007) Evidence of climatic niche shift during biological invasion. *Ecology Letters* 10, 701–709.

Brommer, J. (2004) The range margins of northern birds shift polewards. *Annales Zoologici Fennici* 41, 391–397.

Bronson, D.R., Gower, S.T., Tanner, M., Linder, S. and Van Herk, I. (2008) Response of soil surface $CO_2$ flux in a boreal forest to ecosystem warming. *Global Change Biology* 14, 856–867.

Brouder, S.M. and Volenec, J.J. (2008) Impact of climate change on crop nutrient and water use efficiencies. *Physiologia Plantarum* 133, 705–724.

Brown, J., Valone, T. and Curtin, C. (1997) Reorganization of an arid ecosystem in response to recent climate change. *Proceedings of the National Academy of Sciences USA* 94, 9729–9733.

Brown, K.J. and Hebda, R.J. (2002) Origin, development, and dynamics of coastal temperate conifer rainforests of Southern Vancouver Island, Canada. *Canadian Journal of Forest Research* 32, 353–372.

Brubaker, L. (1988) Vegetation history and anticipating future vegetation change. In: Agee, J.K. and Johnson, D.R. (eds) *Ecosystem Management for Parks and Wilderness.* University of Washington Press, Seattle, Washington.

Brubaker, L., Anderson, P., Edwards, M. and Lozhkin, A. (2005) Beringia as a glacial refugium for boreal trees and shrubs: new perspectives from mapped pollen data. *Journal of Biogeography* 32, 833–848.

Bruce, K.D., Jones, T.H., Bezemer, T.M., Thompson, L.J. and Ritchie, D.A. (2000) The effect of elevated atmospheric carbon dioxide levels on soil bacterial communities. *Global Change Biology* 6, 427–434.

Buchanan, K. (2000) Stress and the evolution of condition-dependent signals. *Trends in Ecology & Evolution* 15, 156–160.

Bull, J. (1981) Sex ratio evolution when fitness varies. *Heredity* 46, 9–26.

Bunce, J. and Ziska, L. (2000) Crop ecosystem responses to climatic change: crop/weed interactions. In:

Reddy, K.R. and Hodges, H.F. (eds) *Climate Change and Global Crop Productivity*. CAB International, Wallingford, UK, pp. 333–352.

Bunn, A.G., Goetz, S.J. and Fiske, G.J. (2005) Observed and predicted responses of plant growth to climate across canada. *Geophysical Research Letters* 32, L16710; doi:10.1029/2005GL023646.

Burgess, T.L., Betancourt, J.L. and Busby, J.R. (1995) Geographic variation in plant species richness: lessons from the Sonoran Desert, USA and Mexico, and Northern Territory, Australia. In: Debano, L.F., Gottfried, G.J., Hamre, R.H., Edminster, C.B., Ffolliott, P.F. and Ortegarubio, A. (eds) *Biodiversity and Management of the Madrean Archipelago: The Sky Islands of Southwestern United States and Northwestern Mexico*. US Department of Agriculture, Forest Service Rocky Mountain Forest and Range Experimental Station, Fort Collins, Colorado, pp. 84–90.

Burgmer, T., Hillebrand, H. and Pfenninger, M. (2007) Effects of climate-driven temperature changes on the diversity of freshwater macroinvertebrates. *Oecologia* 151, 93–103.

Cable, J.M. and Huxman, T.E. (2004) Precipitation pulse size effects on Sonoran Desert soil microbial crusts. *Oecologia* 141, 317–324.

Cairns, A. and Yan, N. (2009) A review of the influence of low ambient calcium concentrations on freshwater daphniids, gammarids, and crayfish. *Environmental Reviews* 17, 67–79.

Calfapietra, C., Mugnozza, G., Karnosky, D., Loreto, F. and Sharkey, T. (2008) Isoprene emission rates under elevated $CO_2$. *New Phytologist* 179, 55–61.

Callaghan, T. (1994) Impact of climate change factors on the clonal sedge *Carex bigelowii*: implications for population growth and vegetative spread. *Ecography* 17, 321–330.

Callaway, R.M. (1997) Positive interactions in plant communities and the individualistic–continuum concept. *Oecologia* 112, 143–149.

Callaway, R.M. and Sabraw, C.S. (1994) Effects of variable precipitation on the structure and diversity of a California salt-marsh community. *Journal of Vegetation Science* 5, 433–438.

Campbell, B.D., Laing, W.A., Greer, D.H., Crush, J.R., Clark, H., Williamson, D.Y. and Given, M.D.J. (1995) Variation in grassland populations and species and the implications for community responses to elevated $CO_2$. *Journal of Biogeography* 22, 315–322.

Cane, M.A., Braconnot, P., Clement, A., Gildor, H., Joussaume, S., Kageyama, M., Khodri, M., Paillard, D., Tett, S. and Zorita, E. (2006) Progress in paleoclimate modeling. *Journal of Climate* 19, 5031–5057.

Cao, M.K. and Woodward, F.I. (1998) Net primary and ecosystem production and carbon stocks of terrestrial ecosystems and their responses to climate change. *Global Change Biology* 4, 185–198.

Caplat, P., Anand, M. and Bauch, C. (2008) Interactions between climate change, competition, dispersal, and disturbances in a tree migration model. *Theoretical Ecology* 1, 209–220.

Carney, K.M., Hungate, B.A., Drake, B.G. and Megonigal, J.P. (2007) Altered soil microbial community at elevated $CO_2$ leads to loss of soil carbon. *Proceedings of the National Academy of Sciences USA* 104, 4990–4995.

Carpenter, K., Abrar, M., Aeby, G., Aronson, R., Banks, S., Bruckner, A., Chiriboga, A., Cortes, J., Delbeek, J. and Devantier, L. (2008) One-third of reef-building corals face elevated extinction risk from climate change and local impacts. *Science* 321, 560–563.

Carroll, M., Anderson, B., Brereton, T., Knight, S., Kudrna, O. and Thomas, C. (2009) Climate change and translocations: the potential to re-establish two regionally-extinct butterfly species in Britain. *Biological Conservation* 142, 2114–2121.

Carroll, S., Hendry, A., Reznick, D. and Fox, C. (2007) Evolution on ecological time-scales. *Ecology* 21, 387–393.

Carthy, R.R., Foley, A.M. and Matsuzawa, Y. (2003) Incubation environment of loggerhead turtle nests. In: Bolten, A.B. and Witherington, B.E. (eds) *Loggerhead Sea Turtles*. Simthsonian Books, Washington, DC, pp. 144–153.

Castro, H.F., Classen, A.T., Austin, E.E., Norby, R.J. and Schadt, C.W. (2010) Soil microbial community responses to multiple experimental climate change drivers. *Applied and Environmental Microbiology* 76, 999–1007.

Catovsky, S. and Bazzaz, F.A. (1999) Elevated $CO_2$ influences the responses of two birch species to soil moisture: implications for forest community structure. *Global Change Biology* 5, 507–518.

Chambers, J., Fisher, J., Zeng, H., Chapman, E., Baker, D. and Hurtt, G. (2007) Hurricane Katrina's carbon footprint on US Gulf Coast forests. *Science* 318, 1107.

Chapin, F.S., Shaver, G.R., Giblin, A.E., Nadelhoffer, K.J. and Laundre, J.A. (1995) Responses of Arctic tundra to experimental and observed changes in climate. *Ecology* 76, 694–711.

Chapin, F.S., Matson, P.A. and Mooney, H.A. (2002) *Principles of Terrestrial Ecosystem Ecology*. Springer, New York.

Chapin, F.S., Oswood, M., Van Cleve, K., Viereck, L.A. and Verbyla, D.L. (2006) *Alaska's Changing Boreal Forest*. Oxford University Press, Oxford, UK.

Chapin, F.S., McFarland, J., McGuire, A.D., Euskirchen, E.S., Ruess, R.W. and Kielland, K. (2009) The changing global carbon cycle: linking plant–soil carbon dynamics to global consequences. *Journal of Ecology* 97, 840–850.

Charmantier, A., McCleery, R., Cole, L., Perrins, C., Kruuk, L. and Sheldon, B. (2008) Adaptive phenotypic plasticity in response to climate change in a wild bird population. *Science* 320, 800–803.

Chase, J.M. (2007) Drought mediates the importance of stochastic community assembly. *Proceedings of the National Academy of Sciences USA* 104, 17430–17434.

Chen, I.C., Shiu, H.J., Benedick, S., Holloway, J.D., Chey, V.K., Barlow, H.S., Hill, J.K. and Thomas, C.D. (2009) Elevation increases in moth assemblages over 42 years on a tropical mountain. *Proceedings of the National Academy of Sciences USA* 106, 1479–1483.

Chen, X., Tu, C., Burton, M.G., Watson, D.M., Burkey, K.O. and Hu, S.J. (2007) Plant nitrogen acquisition and interactions under elevated carbon dioxide: impact of endophytes and mycorrhizae. *Global Change Biology* 13, 1238–1249.

Chesson, P. (2000) General theory of competitive coexistence in spatially-varying environments. *Theoretical Population Biology* 58, 211–237.

Cheung, W., Lam, V., Sarmiento, J., Kearney, K., Watson, R. and Pauly, D. (2009) Projecting global marine biodiversity impacts under climate change scenarios. *Fish and Fisheries* 10, 235–251.

Chevallier, A.J.T., Lieffering, M., Carran, R.A. and Newton, P.C.D. (2006) Mineral nitrogen cycling through earthworm casts in a grazed pasture under elevated atmospheric $CO_2$. *Global Change Biology* 12, 56–60.

Chiang, J.M., Iverson, L.R., Prasad, A. and Brown, K.J. (2008) Effects of climate change and shifts in forest composition on forest net primary production. *Journal of Integrative Plant Biology* 50, 1426–1439.

Chikoski, J.M., Ferguson, S.H. and Meyer, L. (2006) Effects of water addition on soil arthropods and soil characteristics in a precipitation-limited environment. *Acta Oecologica–International Journal of Ecology* 30, 203–211.

Christensen, M., Graham, M., Vinebrooke, R., Findlay, D., Paterson, M. and Turner, M. (2006) Multiple anthropogenic stressors cause ecological surprises in boreal lakes. *Global Change Biology* 12, 2316–2322.

Chung, H.G., Zak, D.R., Reich, P.B. and Ellsworth, D.S. (2007) Plant species richness, elevated $CO_2$, and atmospheric nitrogen deposition alter soil microbial community composition and function. *Global Change Biology* 13, 980–989.

Clark, D., Clark, D. and Oberbauer, S. (2010) Annual wood production in a tropical rain forest in NE Costa Rica linked to climatic variation but not to increasing $CO_2$. *Global Change Biology* 16, 747–759.

Clark, D.A. (2004) Tropical forests and global warming: Slowing it down or speeding it up? *Frontiers in Ecology and the Environment* 2, 73–80.

Clark, J., Fastie, C., Hurtt, G., Jackson, S., Johnson, C., King, G., Lewis, M., Lynch, J., Pacala, S. and Prentice, C. (1998) Reid's paradox of rapid plant migration. *BioScience* 48, 13–24.

Clark, J., Campbell, J., Grizzle, H., Acosta-Martinez, V. and Zak, J. (2009) Soil microbial community response to drought and precipitation variability in the Chihuahuan Desert. *Microbial Ecology* 57, 248–260.

Classen, A., Hart, S., Whitman, T., Cobb, N. and Koch, G. (2005) Insect infestations linked to shifts in microclimate: important climate change implications. *Soil Science Society of America Journal* 69, 2049–2057.

Cleland, E.E., Peters, H.A., Mooney, H.A. and Field, C.B. (2006) Gastropod herbivory in response to elevated $CO_2$ and N addition impacts plant community composition. *Ecology* 87, 686–694.

Cleland, E.E., Chuine, I., Menzel, A., Mooney, H.A. and Schwartz, M.D. (2007) Shifting plant phenology in response to global change. *Trends in Ecology & Evolution* 22, 357–365.

Clements, F.E. (1916) *Plant Succession: An Analysis of the Development of Vegetation*. Carnegie Institute of Washington, Washington, DC.

Cline, W.R. (2007) *Global Warming and Agriculture: Impact Estimates by Country*. Center for Global Development and Peterson Institute for International Economics, Washington, DC.

Coakley, S.M., Scherm, H. and Chakraborty, S. (1999) Climate change and plant disease management. *Annual Review of Phytopathology* 37, 399–426.

Cohen, A., Stone, J., Beuning, K., Park, L., Reinthal, P., Dettman, D., Scholz, C., Johnson, T., King, J. and Talbot, M. (2007) Ecological consequences of early Late Pleistocene megadroughts in tropical Africa. *Proceedings of the National Academy of Sciences USA* 104, 16422–16427.

Cole, L., Bardgett, R.D., Ineson, P. and Hobbs, P.J. (2002) Enchytraeid worm (Oligochaeta) influences on microbial community structure, nutrient dynamics and plant growth in blanket peat subjected to warming. *Soil Biology & Biochemistry* 34, 83–92.

Coleman, D.C., Crossley, D.A. and Hendrix, P.F. (2004) *Fundamentals of Soil Ecology*. Elsevier Academic Press, San Diego, California.

Colinvaux, P., De Oliveira, P., Moreno, J., Miller, M. and Bush, M. (1996) A long pollen record from lowland Amazonia: forest and cooling in glacial times. *Science* 274, 85–88.

Colinvaux, P., Bush, M., Steinitz-Kannan, M. and Miller, M. (1997) Glacial and postglacial pollen records from the Ecuadorian Andes and Amazon. *Quaternary Research* 48, 69–78.

Collins, S. and Bell, G. (2004) Phenotypic consequences of 1,000 generations of selection at elevated $CO_2$ in a green alga. *Nature* 431, 566–569.

Collins, S.A., Sultemeyer, D. and Bell, G. (2006) Changes in C uptake in populations of *Chlamydomonas reinhardtii* selected at high $CO_2$. *Plant, Cell & Environment* 29, 1812–1819.

Conant, R.T., Steinweg, J.M., Haddix, M.L., Paul, E.A., Plante, A.F. and Six, J. (2008) Experimental warming shows that decomposition temperature sensitivity increases with soil organic matter recalcitrance. *Ecology* 89, 2384–2391.

Connell, J.H. (1978) Diversity in tropical rain forests and coral reefs. *Science* 199, 1302–1310.

Conover, D., Van Voorhees, D. and Ehtisham, A. (1992) Sex ratio selection and the evolution of environmental sex determination in laboratory populations of *Menidia menidia*. *Evolution* 46, 1722–1730.

Cook, E.R. and Kairiukstis, L.A. (eds) (1990) *Methods of Dendrochronology: Applications in the Environmental Sciences*. Kluwer Academic Publishers, Dordrecht, The Netherlands.

Cornic, G. (2000) Drought stress inhibits photosynthesis by decreasing stomatal aperture – not by affecting ATP synthesis. *Trends in Plant Science* 5, 187–188.

Cottenie, K. and De Meester, L. (2003) Comment to Oksanen (2001): reconciling Oksanen (2001) and Hurlbert (1984). *Oikos* 100, 394–396.

Coulson, S.J., Hodkinson, I.D., Block, W., Webb, N.R. and Worland, M.R. (1995) Low summer temperatures – a potential mortality factor for High Arctic soil microarthropods. *Journal of Insect Physiology* 41, 783–792.

Couteaux, M.M. and Bolger, T. (2000) Interactions between atmospheric $CO_2$ enrichment and soil fauna. *Plant and Soil* 224, 123–134.

Couteaux, M.M., Mousseau, M., Celerier, M.L. and Bottner, P. (1991) Increased atmospheric $CO_2$ and litter quality – decomposition of sweet chestnut leaf litter with animal food webs of different complexities. *Oikos* 61, 54–64.

Couteaux, M.M., Monrozier, L.J. and Bottner, P. (1996) Increased atmospheric $CO_2$: chemical changes in decomposing sweet chestnut (*Castanea sativa*) leaf litter incubated in microcosms under increasing food web complexity. *Oikos* 76, 553–563.

Coviella, C.E. and Trumble, J.T. (1999) Effects of elevated atmospheric carbon dioxide on insect–plant interactions. *Conservation Biology* 13, 700–712.

Cramer, W., Bondeau, A., Woodward, F.I., Prentice, I.C., Betts, R.A., Brovkin, V., Cox, P.M., Fisher, V., Foley, J.A., Friend, A.D., Kucharik, C., Lomas, M.R., Ramankutty, N., Sitch, S., Smith, B., White, A. and Young-Molling, C. (2001) Global response of terrestrial ecosystem structure and function to $CO_2$ and climate change: results from six dynamic global vegetation models. *Global Change Biology* 7, 357–373.

Croxall, J., Trathan, P. and Murphy, E. (2002) Environmental change and Antarctic seabird populations. *Science* 297, 1510–1514.

Dahlsten, D., Tait, S., Rowney, D. and Gingg, B. (1993) A monitoring system and development of ecologically sound treatments for elm leaf beetle. *Journal of Arboriculture* 19, 181–181.

Dakos, V., Scheffer, M., Van Nes, E., Brovkin, V., Petoukhov, V. and Held, H. (2008) Slowing down as an early warning signal for abrupt climate change.

*Proceedings of the National Academy of Sciences USA* 105, 14308–14312.

Dale, V. (1997) The relationship between land-use change and climate change. *Ecological Applications* 7, 753–769.

Dale, V., Joyce, L., McNulty, S., Neilson, R., Ayres, M., Flannigan, M., Hanson, P., Irland, L., Lugo, A. and Peterson, C. (2001) Climate change and forest disturbances. *BioScience* 51, 723–734.

Dalgleish, H.J., Koons, D.N. and Adler, P.B. (2010) Can life-history traits predict the response of forb populations to changes in climate variability? *Journal of Ecology* 98, 209–217.

D'Andrea, L., Broennimann, O., Kozlowski, G., Guisan, A., Morin, X., Keller-Senften, J. and Felber, F. (2009) Climate change, anthropogenic disturbance and the northward range expansion of *Lactuca serriola* (Asteraceae). *Journal of Biogeography* 36, 1573–1587.

Daniels, L.D. and Veblen, T.T. (2004) Spatiotemporal influences of climate on altitudinal treeline in northern Patagonia. *Ecology* 85, 1284–1296.

Darby, B.J., Housman, D.C., Zaki, A.M., Shamout, Y., Adl, S.M., Belnap, J. and Neher, D.A. (2006) Effects of altered temperature and precipitation on desert protozoa associated with biological soil crusts. *Journal of Eukaryotic Microbiology* 53, 507–514.

D'Arrigo, R., Jacoby, G., Buckley, B., Sakulich, J., Frank, D., Wilson, R., Curtis, A.V. and Anchukaitis, K. (2009) Tree growth and inferred temperature variability at the North American Arctic treeline. *Global and Planetary Change* 65, 71–82.

Daufresne, M. and Boet, P. (2007) Climate change impacts on structure and diversity of fish communities in rivers. *Global Change Biology* 13, 2467–2478.

Davi, H., Dufrene, E., Francois, C., Le Maire, G., Loustau, D., Bosc, A., Rambal, S., Granier, A. and Moors, E. (2006) Sensitivity of water and carbon fluxes to climate changes from 1960 to 2100 in European forest ecosystems. *Agricultural and Forest Meteorology* 141, 35–56.

Davis, M.A., Pritchard, S.G., Mitchell, R.J., Prior, S.A., Rogers, H.H. and Runion, G.B. (2002) Elevated atmospheric $CO_2$ affects structure of a model regenerating longleaf pine community. *Journal of Ecology* 90, 130–140.

Davis, M.B. (1981) Quaternary history and the stability of forest communities. In: West, D.C., Shugart, H.H. and Botkin, D.B. (eds) *Forest Succession: Concepts and Application*. Springer-Verlag, New York, pp. 132–153.

De Boeck, H.J., Lemmens, C., Zavalloni, C., Gielen, B., Malchair, S., Carnol, M., Merckx, R., Van Den Berge, J., Ceulemans, R. and Nijs, I. (2008) Biomass production in experimental grasslands of different species richness during three years of climate warming. *Biogeosciences* 5, 585–594.

De Graaff, M.A., Van Groenigen, K.J., Six, J., Hungate, B. and Van Kessel, C. (2006) Interactions between plant growth and soil nutrient cycling under elevated $CO_2$: a meta-analysis. *Global Change Biology* 12, 2077–2091.

Delucia, E., Sasek, T. and Strain, B. (1985) Photosynthetic inhibition after long-term exposure to elevated levels of atmospheric carbon dioxide. *Photosynthesis Research* 7, 175–184.

Deslippe, J.R., Egger, K.N. and Henry, G.H.R. (2005) Impacts of warming and fertilization on nitrogen-fixing microbial communities in the Canadian High Arctic. *FEMS Microbiology Ecology* 53, 41–50.

Desprez-Loustau, M.L., Marcais, B., Nageleisen, L.M., Piou, D. and Vannini, A. (2006) Interactive effects of drought and pathogens in forest trees. *Annals of Forest Science* 63, 597–612.

De Valpine, P. and Harte, J. (2001) Plant responses to experimental warming in a montane meadow. *Ecology* 82, 637–648.

Devictor, V., Julliard, R., Couvet, D. and Jiguet, F. (2008) Birds are tracking climate warming, but not fast enough. *Proceedings. Biological Sciences/The Royal Society* 275, 2743–2748.

Diamond, J.M. (1975) Assembly of species communities. In: Cody, M.L. and Diamond J.M. (eds) *Ecology and Evolution of Communities*. Harvard University Press, Cambridge, Massachusetts, pp. 342–445.

Diaz, S., Fraser, L.H., Grime, J.P. and Falczuk, V. (1998) The impact of elevated $CO_2$ on plant–herbivore interactions: experimental evidence of moderating effects at the community level. *Oecologia* 117, 177–186.

Dillehay, T., Calderûn, G. and Politis, G. (1992) Earliest hunters and gatherers of South America. *Journal of World Prehistory* 6, 145–204.

Doi, H. (2008) Delayed phenological timing of dragonfly emergence in Japan over five decades. *Biology Letters* 4, 388–391.

Domis, L.N.D., Mooij, W.M. and Huisman, J. (2007) Climate-induced shifts in an experimental phytoplankton community: a mechanistic approach. *Hydrobiologia* 584, 403–413.

Dorrepaal, E., Toet, S., Van Logtestijn, R.S.P., Swart, E., Van De Weg, M.J., Callaghan, T.V. and Aerts, R. (2009) Carbon respiration from subsurface peat accelerated by climate warming in the subarctic. *Nature* 460, 616–619.

Doughty, C.E. and Goulden, M.L. (2008) Are tropical forests near a high temperature threshold? *Journal of Geophysical Research – Biogeosciences* 113, G00B07; doi:10.1029/2007JG000632.

Drigo, B., Kowalchuk, G.A., Yergeau, E., Bezemer, T.M., Boschker, H.T.S. and Van Veen, J.A. (2007) Impact of elevated carbon dioxide on the rhizosphere communities of *Carex arenaria* and *Festuca rubra*. *Global Change Biology* 13, 2396–2410.

Drigo, B., Kowalchuk, G.A. and Van Veen, J.A. (2008) Climate change goes underground: effects of elevated atmospheric $CO_2$ on microbial community structure and activities in the rhizosphere. *Biology and Fertility of Soils* 44, 667–679.

Drigo, B., Van Veen, J.A. and Kowalchuk, G.A. (2009) Specific rhizosphere bacterial and fungal groups respond differently to elevated atmospheric $CO_2$. *ISME Journal* 3, 1204–1217.

Dukes, J.S. and Mooney, H.A. (1999) Does global change increase the success of biological invaders? *Trends in Ecology & Evolution* 14, 135–139.

Dukes, J.S., Chiariello, N.R., Cleland, E.E., Moore, L.A., Shaw, M.R., Thayer, S., Tobeck, T., Mooney, H.A. and Field, C.B. (2005) Responses of grassland production to single and multiple global environmental changes. *PloS Biology* 3, 1829–1837.

Dumig, A., Schad, P., Rumpel, C., Dignac, M. and Kögel-Knabner, I. (2008) Araucaria forest expansion on grassland in the Southern Brazilian highlands as revealed by [14]C and [13]C studies. *Geoderma* 145, 143–157.

Dunn, A.L., Barford, C.C., Wofsy, S.C., Goulden, M.L. and Daube, B.C. (2007) A long-term record of carbon exchange in a boreal black spruce forest: means, responses to interannual variability, and decadal trends. *Global Change Biology* 13, 577–590.

Easterling, W.E., Aggarwal, P.K., Batima, P., Brander, K.M., Erda, L., Howden, S.M., Kirilenko, A., Morton, J., Soussana, J.F., Schmidhuber, J. and Tubiello, F.N. (2007) Food, fibre and forest products. In: Parry, M.L., Canziani, O.F., Palutikof, J.P., Van Der Linden, P.J. and Hanson, C.E. (eds) *Climate Change 2007: Impacts, Adaptation and Vulnerability. Contribution of Working Group II to the Fourth Assessment Report of the Intergovernmental Panel on Climate Change*. Cambridge University Press, Cambridge, UK, pp. 273–313.

Eaton, J. and Scheller, R. (1996) Effects of climate warming on fish thermal habitat in streams of the United States. *Limnology and Oceanography* 41, 1109–1115.

Ehleringer, J., Cerling, T. and Helliker, B. (1997) $C_4$ photosynthesis, atmospheric $CO_2$, and climate. *Oecologia* 112, 285–299.

Elliott, J.A., Thackeray, S.J., Huntingford, C. and Jones, R.G. (2005) Combining a regional climate model with a phytoplankton community model to predict future changes in phytoplankton in lakes. *Freshwater Biology* 50, 1404–1411.

Ellis, R., Vogel, J. and Fuls, A. (1980) Photosynthetic pathways and the geographical distribution of grasses in South West Africa/Namibia. *South African Journal of Science* 76, 307–314.

Endler, J. (1986) *Natural Selection in the Wild*. Princeton University Press, Princeton, New Jersey.

Engel, E.C., Weltzin, J.F., Norby, R.J. and Classen, A.T. (2009) Responses of an old-field plant community to interacting factors of elevated [$CO_2$], warming, and soil moisture. *Journal of Plant Ecology* 2, 1–11.

Erasmus, B., Van Jaarsveld, A., Chown, S., Kshatriya, M. and Wessels, K. (2002) Vulnerability of South African animal taxa to climate change. *Global Change Biology* 8, 679–693.

Euskirchen, E.S., McGuire, A.D., Chapin, F.S., Yi, S. and Thompson, C.C. (2009) Changes in vegetation in Northern Alaska under scenarios of climate change, 2003–2100: implications for climate feedbacks. *Ecological Applications* 19, 1022–1043.

Evans, J.C. and Prepas, E.E. (1996) Potential effects of climate change on ion chemistry and phytoplankton communities in prairie saline lakes. *Limnology and Oceanography* 41, 1063–1076.

Ewers, R. and Didham, R. (2005) Confounding factors in the detection of species responses to habitat fragmentation. *Biological Reviews* 81, 117–142.

Fang, J.Y., Piao, S., Field, C.B., Pan, Y., Guo, Q.H., Zhou, L.M., Peng, C.H. and Tao, S. (2003) Increasing net primary production in china from 1982 to 1999. *Frontiers in Ecology and the Environment* 1, 293–297.

FAO (2009) *State of the World's Forests 2009*. Food and Agriculture Organization of the United Nations, Rome.

Feeley, K. and Silman, M. (2010) Land-use and climate change effects on population size and extinction risk of Andean plants. *Global Change Biology* 16, 3215–3222.

Feeley, K., Joseph Wright, S., Nur Supardi, M., Kassim, A. and Davies, S. (2007) Decelerating growth in tropical forest trees. *Ecology Letters* 10, 461–469.

Feuchtmayr, H., Moran, R., Hatton, K., Connor, L., Heyes, T., Moss, B., Harvey, I. and Atkinson, D. (2009) Global warming and eutrophication: effects on water chemistry and autotrophic communities in experimental hypertrophic shallow lake mesocosms. *Journal of Applied Ecology* 46, 713–723.

Fiedel, S. and Haynes, G. (2004) A premature burial: comments on Grayson and Meltzer's 'Requiem for overkill'. *Journal of Archaeological Science* 31, 121–131.

Finzi, A.C., Sinsabaugh, R.L., Long, T.M. and Osgood, M.P. (2006) Microbial community responses to atmospheric carbon dioxide enrichment in a warm-temperate forest. *Ecosystems* 9, 215–226.

Fischlin, A., Midgley, G.F., Price, J.T., Leemans, R., Gopal, B., Turley, C., Roundevell, M.D.A., Dube, O.P., Tarazona, J. and Velichko, A.A. (2007) Ecosystems, their properties, goods and services. In: Parry, M.L., Canziani, O.F., Palutikof, J.P., Van Der Linden, P.J. and Hanson, C.E. (eds) *Climate Change 2007: Impacts, Adaptation and Vulnerability. Contribution of Working Group II to the Fourth Assessment Report of the Intergovernmental Panel on Climate Change.* Cambridge University Press, Cambridge, UK, pp. 211–272.

Fisher, R.A., Williams, M., Da Costa, A.L., Malhi, Y., Da Costa, R.F., Almeida, S. and Meir, P. (2007) The response of an Eastern Amazonian rain forest to drought stress: results and modelling analyses from a throughfall exclusion experiment. *Global Change Biology* 13, 2361–2378.

Flannigan, M. and Wagner, C. (1991) Climate change and wildfire in Canada. *Canadian Journal of Forest Research* 21, 66–72.

Flannigan, M., Stocks, B. and Wotton, B. (2000) Climate change and forest fires. *Science of the Total Environment* 262, 221–229.

Flannigan, M., Krawchuk, M., De Groot, W., Wotton, B. and Gowman, L. (2009) Implications of changing climate for global wildland fire. *International Journal of Wildland Fire* 18, 483–507.

Fleming, R.A. (2000) Climate change and insect disturbance regimes in Canada's boreal forests. *World Resource Review* 12, 520–554.

Fleming, R.A. and Candau, J.-N. (1998) Influences of climatic change on some ecological processes of an insect outbreak system in Canada's boreal forests and the implications for biodiversity. *Environmental Monitoring and Assessment* 49, 235–249.

Fleurat-Lessard, F. (1990) Effect of modified atmospheres on insect and mites infesting stored products. In: Calderon, M. and Barkai-Golan, R. (eds) *Food Preservation by Modified Atmospheres*. CRC Press, Boca Raton, Florida, pp. 21–38.

Flexas, J. and Medrano, H. (2002) Drought-inhibition of photosynthesis in $C_3$ plants: stomatal and non-stomatal limitations revisited. *Annals of Botany* 89, 183–189.

Foley, J.A., Defries, R., Asner, G.P., Barford, C., Bonan, G., Carpenter, S.R., Chapin, F.S., Coe, M.T., Daily, G.C., Gibbs, H.K., Helkowski, J.H., Holloway, T., Howard, E.A., Kucharik, C.J., Monfreda, C., Patz, J.A., Prentice, I.C., Ramankutty, N. and Snyder, P.K. (2005) Global consequences of land use. *Science* 309, 570–574.

Fonty, E., Sarthou, C., Larpin, D. and Ponge, J.F. (2009) A 10-year decrease in plant species richness on a neotropical inselberg: detrimental effects of global warming? *Global Change Biology* 15, 2360–2374.

Forest, F., Grenyer, R., Rouget, M., Davies, T., Cowling, R., Faith, D., Balmford, A., Manning, J., Proches, S. and Van Der Bank, M. (2007) Preserving the evolutionary potential of floras in biodiversity hotspots. *Nature* 445, 757–760.

Forister, M., McCall, A., Sanders, N., Fordyce, J., Thorne, J., Oĭbrien, J., Waetjen, D. and Shapiro, A. (2010) Compounded effects of climate change and habitat alteration shift patterns of butterfly diversity. *Proceedings of the National Academy of Sciences USA* 107, 2088–2092.

Foukal, P., Fröhlich, C., Spruit, H. and Wigley, T.M.L. (2006) Variations in solar luminosity and their effect on the Earth's climate. *Nature* 443, 161–166.

Fransson, P.M.A., Taylor, A.F.S. and Finlay, R.D. (2001) Elevated atmospheric $CO_2$ alters root symbiont community structure in forest trees. *New Phytologist* 152, 431–442.

Frederiksen, H.B., Ronn, R. and Christensen, S. (2001) Effect of elevated atmospheric $CO_2$ and vegetation type on microbiota associated with decomposing straw. *Global Change Biology* 7, 313–321.

Frey, S.D., Drijber, R., Smith, H. and Melillo, J. (2008) Microbial biomass, functional capacity, and community structure after 12 years of soil warming. *Soil Biology & Biochemistry* 40, 2904–2907.

Friis, K., Damgaard, C. and Holmstrup, M. (2004) Sublethal soil copper concentrations increase mortality in the earthworm *Aporrectodea caliginosa* during drought. *Ecotoxicology and Environmental Safety* 57, 65–73.

Fritts, H.C. (1976) *Tree Rings and Climate*. Academic Press, London.

Frouz, J., Novakova, A. and Jones, T.H. (2002) The potential effect of high atmospheric $CO_2$ on soil fungi–invertebrate interactions. *Global Change Biology* 8, 339–344.

Galloway, J., Dentener, F., Capone, D., Boyer, E., Howarth, R., Seitzinger, S., Asner, G., Cleveland, C., Green, P. and Holland, E. (2004) Nitrogen cycles: past, present, and future. *Biogeochemistry* 70, 153–226.

Gaston, K. (2000) Global patterns in biodiversity. *Nature* 405, 220–227.

Gaston, K. and Blackburn, T. (2000) *Pattern and Process in Macroecology*. Blackwell Publishing, Oxford, UK.

Gaston, K. and Spicer, J. (2004) *Biodiversity: An Introduction*. Blackwell Publishing, Oxford, UK.

Gedalof, Z. (2010) Climate and spatial patterns of wildfire. In: McKenzie, D., Millar, C. and Falk, D.A. (eds) *The Landscape Ecology of Fire*. Springer, Berlin, pp. 89–116.

Gedalof, Z., Pellatt, M. and Smith, D. (2006) From prairie to forest: three centuries of environmental change at Rocky Point, Vancouver Island, BC. *Northwest Science* 80, 34–46.

Gehrig-Fasel, J., Guisan, A. and Zimmermann, N. (2007) Tree line shifts in the Swiss Alps: climate change or land abandonment? *Journal of Vegetation Science* 18, 571–582.

Gienapp, P., Teplitsky, C., Alho, J.S., Mills, J.A. and Meril, J. (2008) Climate change and evolution: disentangling environmental and genetic responses. *Molecular Ecology* 17, 167–178.

Gillett, N.P., Weaver, A.J., Zwiers, F.W. and Flannigan, M.D. (2004) Detecting the effect of climate change on Canadian forest fires. *Geophysical Research Letters* 31, L18211; doi:10.1029/2004GL020876.

Girard, F., Payette, S. and Gagnon, R. (2008) Rapid expansion of lichen woodlands within the closed-crown boreal forest zone over the last 50 years caused by stand disturbances in Eastern Canada. *Journal of Biogeography* 35, 529–537.

Gjerdrum, C., Vallee, A., St Clair, C., Bertram, D., Ryder, J. and Blackburn, G. (2003) Tufted puffin reproduction reveals ocean climate variability. *Proceedings of the National Academy of Sciences USA* 100, 9377–9382.

Gleason, H.A. (1926) The individualistic concept of the plant association. *Bulletin of the Torrey Botanical Club* 53, 7–26.

Goetz, S.J., Mack, M.C., Gurney, K.R., Randerson, J.T. and Houghton, R.A. (2007) Ecosystem responses to recent climate change and fire disturbance at northern high latitudes: observations and model results contrasting Northern Eurasia and North America. *Environmental Research Letters* 2, 045031; doi:10.1088/1748-9326/2/4/045031.

Golinski, M., Bauch, C. and Anand, M. (2008) The effects of endogenous ecological memory on population stability and resilience in a variable environment. *Ecological Modelling* 212, 334–341.

Gomulkiewicz, R. and Holt, R. (1995) When does evolution by natural selection prevent extinction? *Evolution* 49, 201–207.

Gonzalez, C., Wang, O., Strutz, S., Gonzalez-Salazar, C., Sanchez-Cordero, V. and Sarkar, S. (2010) Climate change and risk of leishmaniasis in North America: predictions from ecological niche models of vector and reservoir species. *PLoS Neglected Tropical Diseases* 4, e585.

Goodwin, B., McAllister, A. and Fahrig, L. (1999) Predicting invasiveness of plant species based on biological information. *Conservation Biology* 13, 422–426.

Gower, S.T. (2003) Patterns and mechanisms of the forest carbon cycle. *Annual Review of Environment and Resources* 28, 169–204.

Grace, J. and Van Gardingen, P.R. (1997) Sites of naturally elevated carbon dioxide. In: Raschi, A., Miglietta, F., Tognetti, R. and Van Gardingen, P.R. (eds) *Plant Responses to Elevated $CO_2$*. Cambridge University Press, Cambridge, UK, pp. 1–6.

Grant, B.R. and Grant, P.R. (1993) Evolution of Darwin's finches caused by a rare climatic event. *Proceedings of the Royal Society of London, B* 251, 111–117.

Grant, W.C. (1857) Description of Vancouver Island. *Journal of the Royal Geographical Society of London* 27, 268–320.

Gregory, P., Johnson, S., Newton, A. and Ingram, J. (2009) Integrating pests and pathogens into the climate change/food security debate. *Journal of Experimental Botany* 60, 2827–2838.

Grime, J.P. (1998) Benefits of plant diversity to ecosystems: immediate, filter and founder effects. *Journal of Ecology* 86, 902–910.

Grime, J.P. (2002) *Plant Strategies, Vegetation Processes, and Ecosystem Properties*. John Wiley and Sons, Chichester, UK.

Grimm, V. and Railsback, S.F. (2005) *Individual-Based Modeling and Ecology*. Princeton University Press, Princeton, New Jersey.

Groffman, P., Driscoll, C., Fahey, T., Hardy, J., Fitzhugh, R. and Tierney, G. (2001) Colder soils in a warmer world: a snow manipulation study in a northern hardwood forest ecosystem. *Biogeochemistry* 56, 135–150.

Grubb, P.J. (1977) The maintenance of species-richness in plant communities: the importance of the regeneration niche. *Biological Reviews* 52, 107–145.

Gruter, D., Schmid, B. and Brandl, H. (2006) Influence of plant diversity and elevated atmospheric carbon dioxide levels on belowground bacterial diversity. *BMC Microbiology* 6, 8.

Guiot, J. (1987) Late quaternary climatic change in France estimated from multivariate pollen time series. *Quaternary Research* 28, 100–118.

Hagvar, S. and Klanderud, K. (2009) Effect of simulated environmental change on alpine soil arthropods. *Global Change Biology* 15, 2972–2980.

Hairston, N., Ellner, S., Geber, M., Yoshida, T. and Fox, J. (2005) Rapid evolution and the convergence of ecological and evolutionary time. *Ecology Letters* 8, 1114–1127.

Hamilton, J.G., Delucia, E.H., George, K., Naidu, S.L., Finzi, A.C. and Schlesinger, W.H. (2002) Forest carbon balance under elevated $CO_2$. *Oecologia* 131, 250–260.

Hampe, A. (2004) Bioclimate envelope models: what they detect and what they hide. *Global Ecology and Biogeography* 13, 469–471.

Hampe, A. and Petit, R. (2005) Conserving biodiversity under climate change: the rear edge matters. *Ecology Letters* 8, 461–467.

Hanley, M., Trofimov, S. and Taylor, G. (2004) Species-level effects more important than functional group-level responses to elevated $CO_2$: evidence from simulated turves. *Ecology* 18, 304–313.

Hannah, L., Midgley, G., Lovejoy, T., Bond, W., Bush, M., Lovett, J., Scott, D. and Woodward, F. (2002) Conservation of biodiversity in a changing climate. *Conservation Biology* 16, 264–268.

Hanson, P., Wullschleger, S., Norby, R., Tschaplinski, T. and Gunderson, C. (2005) Importance of changing $CO_2$, temperature, precipitation, and ozone on carbon and water cycles of an upland-oak forest: incorporating experimental results into model simulations. *Global Change Biology* 11, 1402–1423.

Harpole, W., Potts, D. and Suding, K. (2007) Ecosystem responses to water and nitrogen amendment in a California grassland. *Global Change Biology* 13, 2341–2348.

Harris, J., Hobbs, R., Higgs, E. and Aronson, J. (2006) Ecological restoration and global climate change. *Restoration Ecology* 14, 170–176.

Harrison, R.G., Jones, C.D. and Hughes, J.K. (2008) Competing roles of rising $CO_2$ and climate change in the contemporary European carbon balance. *Biogeosciences* 5, 1–10.

Harshman, L. and Hoffmann, A. (2000) Laboratory selection experiments using drosophila: what do they really tell us? *Trends in Ecology & Evolution* 15, 32–36.

Hart, J. and Grissino-Mayer, H. (2009) Gap-scale disturbance processes in secondary hardwood stands on the Cumberland Plateau, Tennessee, USA. *Plant Ecology* 201, 131–146.

Harte, J. and Shaw, R. (1995) Shifting dominance within a montane vegetation community – results of a climate-warming experiment. *Science* 267, 876–880.

Hartl, D. and Clark, A. (1997) *Principles of Population Genetics*. Sinauer Associates, Sunderland, Massachusetts.

Hassall, C., Thompson, D.J., French, G.C. and Harvey, I.F. (2007) Historical changes in the phenology of British odonata are related to climate. *Global Change Biology* 13, 933–941.

Hawkes, L., Broderick, A., Godfrey, M. and Godley, B. (2007) Investigating the potential impacts of climate change on a marine turtle population. *Global Change Biology* 13, 923–932.

He, J.S., Bazzaz, F.A. and Schmid, B. (2002) Interactive effects of diversity, nutrients and elevated $CO_2$ on experimental plant communities. *Oikos* 97, 337–348.

Heijmans, M., Klees, H., De Visser, W. and Berendse, F. (2002) Response of a sphagnum bog plant community to elevated $CO_2$ and N supply. *Plant Ecology* 162, 123–134.

Heinemeyer, A., Ridgway, K.P., Edwards, E.J., Benham, D.G., Young, J.P.W. and Fitter, A.H. (2004) Impact of soil warming and shading on colonization and community structure of arbuscular mycorrhizal fungi in roots of a native grassland community. *Global Change Biology* 10, 52–64.

Heinemeyer, A., Hartley, I.P., Evans, S.P., De La Fuente, J.A.C. and Ineson, P. (2007) Forest soil $CO_2$ flux: uncovering the contribution and environmental responses of ectomycorrhizas. *Global Change Biology* 13, 1786–1797.

Heller, N. and Zavaleta, E. (2009) Biodiversity management in the face of climate change: a review of 22 years of recommendations. *Biological Conservation* 142, 14–32.

Hellmann, J. and Pineda-Krch, M. (2007) Constraints and reinforcement on adaptation under climate change: selection of genetically correlated traits. *Biological Conservation* 137, 599–609.

Hellmann, J., Byers, J., Bierwagen, B. and Dukes, J. (2008) Five potential consequences of climate change for invasive species. *Conservation Biology* 22, 534–543.

Hendry, A.P. and Kinnison, M.T. (1999) Perspective: The pace of modern life: measuring rates of contemporary microevolution. *Evolution* 53, 1637–1653.

Henry, H.A.L. (2007) Soil freeze–thaw cycle experiments: trends, methodological weaknesses and suggested improvements. *Soil Biology & Biochemistry* 39, 977–986.

Henry, H.A.L. (2008) Climate change and soil freezing dynamics: historical trends and projected changes. *Climatic Change* 87, 421–434.

Heusser, C.J. (1960) *Late-Pleistocene Environments of North Pacific North America. Special Publication No. 35.* American Geological Society, Boulder, Colorado.

Hickler, T., Smith, B., Prentice, I.C., Mjofors, K., Miller, P., Arneth, A. and Sykes, M.T. (2008) $CO_2$ fertilization in temperate FACE experiments not representative of boreal and tropical forests. *Global Change Biology* 14, 1531–1542.

Hickling, R., Roy, D.B., Hill, J.K., Fox, R. and Thomas, C.D. (2006) The distributions of a wide range of taxonomic groups are expanding polewards. *Global Change Biology* 12, 450–455.

Higgins, P.A.T. (2007) Biodiversity loss under existing land use and climate change: an illustration using northern South America. *Global Ecology and Biogeography* 16, 197–204.

Higgins, P.A.T. and Harte, J. (2006) Biophysical and biogeochemical responses to climate change depend on dispersal and migration. *BioScience* 56, 407–417.

Hillstrom, M.L. and Lindroth, R.L. (2008) Elevated atmospheric carbon dioxide and ozone alter forest insect abundance and community composition. *Insect Conservation and Diversity* 1, 233–241.

Hitch, A. and Leberg, P. (2007) Breeding distributions of North American bird species moving north as a result of climate change. *Conservation Biology* 21, 534–539.

Hobbie, S.E. and Chapin, F.S. (1998) Response of tundra plant biomass, aboveground production, nitrogen, and $CO_2$ flux to experimental warming. *Ecology* 79, 1526–1544.

Hobbs, R.J., Yates, S. and Mooney, H.A. (2007) Long-term data reveal complex dynamics in grassland in relation to climate and disturbance. *Ecological Monographs* 77, 545–568.

Hochkirch, A. and Damerau, M. (2009) Rapid range expansion of a wing-dimorphic bush-cricket after the 2003 climatic anomaly. *Biological Journal of the Linnean Society* 97, 118–127.

Hodkinson, I.D., Coulson, S.J., Webb, N.R. and Block, W. (1996) Can High Arctic soil microarthropods survive elevated summer temperatures? *Functional Ecology* 10, 314–321.

Hoegh-Guldberg, O. (1999) Climate change, coral bleaching and the future of the world's coral reefs. *Marine and Freshwater Research* 50, 839–866.

Hoegh-Guldberg, O., Hughes, L., McIntyre, S., Lindenmayer, D., Parmesan, C., Possingham, H. and Thomas, C. (2008) Assisted colonization and rapid climate change. *Science* 321, 345–346.

Hoeksema, J.D., Lussenhop, J. and Teeri, J.A. (2000) Soil nematodes indicate food web responses to elevated atmospheric $CO_2$. *Pedobiologia* 44, 725–735.

Hoffmann, W., Schroeder, W. and Jackson, R. (2002) Positive feedbacks of fire, climate, and vegetation and the conversion of tropical savanna. *Geophysical Research Letters* 29, 2051; doi:10.1029/2002GL015856.

Hoffmann, W., Orthen, B. and Do Nascimento, P. (2003) Comparative fire ecology of tropical savanna and forest trees. *Functional Ecology* 17, 720–726.

Hogg, E.H., Brandt, J.P. and Kochtubajda, B. (2002) Growth and dieback of aspen forests in northwestern Alberta, Canada, in relation to climate and insects. *Canadian Journal of Forest Research* 32, 823–832.

Holmstrup, M., Maraldo, K. and Krogh, P.H. (2007) Combined effect of copper and prolonged summer drought on soil microarthropods in the field. *Environmental Pollution* 146, 525–533.

Holzapfel, A.M. and Vinebrooke, R.D. (2005) Environmental warming increases invasion potential of alpine lake communities by imported species. *Global Change Biology* 11, 2009–2015.

Hooper, D., Chapin, F.S., Ewel, J., Hector, A., Inchausti, P., Lavorel, S., Lawton, J., Lodge, D., Loreau, M. and Naeem, S. (2005) Effects of biodiversity on ecosystem functioning: a consensus of current knowledge. *Ecological Monographs* 75, 3–35.

Houghton, J. (2004) Global *Warming: The Complete Briefing*. Cambridge University Press, Cambridge, UK.

Houghton, J., Ding, Y., Griggs, D., Noguer, M., Van Der Linden, P., Dai, X., Maskell, K. and Johnson, C. (2001) *Climate Change 2001: The Scientific Basis.* Cambridge University Press, Cambridge, UK.

Howden, M., Soussana, J., Tubiello, F., Chhetri, N., Dunlop, M. and Aggarwal, P. (2007) Adaptation strategies for climate change. *Proceedings of the National Academy of Sciences USA* 104, 19691–19698.

Huang, J., Bergeron, Y., Denneler, B., Berninger, F. and Tardif, J. (2007) Response of forest trees to increased atmospheric $CO_2$. *Critical Reviews in Plant Sciences* 26, 265–283.

Huang, S., Pollack, H.N. and Shen, P.-Y. (2000) Temperature trends over the past five centuries reconstructed from borehole temperatures. *Nature* 403, 756–758.

Hudson, J.M.G. and Henry, G.H.R. (2009) Increased plant biomass in a High Arctic heath community from 1981 to 2008. *Ecology* 90, 2657–2663.

Hughes, T.P., Baird, A.H., Bellwood, D.R., Card, M., Connolly, S.R., Folke, C., Grosberg, R., Hoegh-Guldberg, O., Jackson, J.B.C., Kleypas, J., Lough, J.M.,

Marshall, P., Nystrom, M., Palumbi, S.R., Pandolfi, J.M., Rosen, B. and Roughgarden, J. (2003) Climate change, human impacts, and the resilience of coral reefs. *Science* 301, 929–933.

Hulme, P. (2005) Adapting to climate change: is there scope for ecological management in the face of a global threat? *Ecology* 42, 784–794.

Human, K. and Gordon, D. (1996) Exploitation and inter-ference competition between the invasive Argentine ant, *Linepithema humile*, and native ant species. *Oecologia* 105, 405–412.

Hungate, B.A., Jaeger, C.H., Gamara, G., Chapin, F.S. and Field, C.B. (2000) Soil microbiota in two annual grasslands: responses to elevated atmospheric $CO_2$. *Oecologia* 124, 589–598.

Hungate, B.A., Dukes, J.S., Shaw, M.R., Luo, Y. and Field, C.B. (2003) Atmospheric science. Nitrogen and climate change. *Science* 302, 1512–1513.

Hunt, S., Newman, J. and Otis, G. (2006) *Threats and Impacts of Exotic Pests under Climate Change: Implications for Canada's Forest Ecosystems and Carbon Stocks. A BIOCAP Research Integration Program Synthesis Paper.* BIOCAP Canada, Kingston, Ontario.

Hunter, M. (2001) Effects of elevated atmospheric carbon dioxide on insect–plant interactions. *Agricultural and Forest Entomology* 3, 153–159.

Huntley, B., Spicer, R.A., Chaloner, W.G. and Jarzembowski, E.A. (1993) The use of climate response surfaces to reconstruct palaeoclimate from quaternary pollen and plant macrofossil data. *Philosophical Transactions: Biological Sciences* 341, 215–224.

Hurlbert, S.H. (1971) The nonconcept of species diver-sity: a critique and alternative parameters. *Ecology* 52, 577–586.

Hurlbert, S.H. (1984) Pseudoreplication and the design of ecological field experiments. *Ecological Monographs* 54, 187–211.

Hurlbert, S.H. (2004) On misinterpretations of pseudo-replication and related matters: a reply to Oksanen. *Oikos* 104, 591–597.

Huse, G. and Ellingsen, I. (2008) Capelin migrations and climate change – a modelling analysis. *Climatic Change* 87, 177–197.

Ibanez, I., Clark, J. and Dietze, M. (2008) Evaluating the sources of potential migrant species: implications under climate change. *Ecological Applications* 18, 1664–1678.

IPCC (2007) Climate Change 2007: Synthesis Report. In: Team, C.W., Pachauri, R.K. and Reisinger, A. (eds) *Contribution of Working Groups I, II and III to the Fourth Assessment Report of the Intergovernmental Panel on Climate Change.* Intergovernmental Panel on Climate Change, Geneva, Switzerland.

Ives, A. and Carpenter, S. (2007) Stability and diversity of ecosystems. *Science* 317, 58–62.

Jablonski, L., Wang, X. and Curtis, P. (2002) Plant repro-duction under elevated $CO_2$ conditions: a meta-analysis of reports on 79 crop and wild species. *New Phytologist* 156, 9–26.

Jacob, J., Greitner, C. and Drake, B. (1995) Acclimation of photosynthesis in relation to rubisco and non-structural carbohydrate contents and *in situ* carboxyl-ase activity in *Scirpus olneyi* grown at elevated $CO_2$ in the field. *Plant, Cell & Environment* 18, 875–884.

Jagerbrand, A.K., Alatalo, J.M., Chrimes, D. and Molau, U. (2009) Plant community responses to 5 years of simulated climate change in meadow and heath ecosystems at a subarctic-alpine site. *Oecologia* 161, 601–610.

Janecek, T.R. and Rea, D.K. (1985) Quaternary fluctua-tions in the northern hemisphere trade winds and westerlies. *Quaternary Research* 24, 150–163.

Jansen, E., Overpeck, J., Briffa, K.R., Duplessy, J.-C., Joos, F., Masson-Delmotte, V., Olago, D., Otto-Bliesner, B., Peltier, W.R., Rahmstorf, S., Ramesh, R., Raynaud, D., Rind, D., Solomina, O., Villalba, R. and Zhang, D. (2007) Palaeoclimate. In: Solomon, S., Qin, D., Manning, M., Chen, Z., Marquis, M., Averyt, K.B., Tignor, M. and Miller, H.L. (eds) *Climate Change 2007: The Physical Science Basis. Contribution of Working Group I to the Fourth Assessment Report of the Intergovernmental Panel on Climate Change.* Cambridge University Press, Cambridge, UK.

Janus, L.R., Angeloni, N.L., McCormack, J., Rier, S.T., Tuchman, N.C. and Kelly, J.J. (2005) Elevated atmo-spheric $CO_2$ alters soil microbial communities associ-ated with trembling aspen (*Populus tremuloides*) roots. *Microbial Ecology* 50, 102–109.

Janzen, D. (1985) On ecological fitting. *Oikos* 45, 308–310.

Jaramillo, J., Chabi-Olaye, A., Kamonjo, C., Jaramillo, A., Vega, F.E., Poehling, H.-M. and Borgemeister, C. (2009) Thermal tolerance of the coffee berry borer *Hypothenemus hampei*: predictions of climate change impact on a tropical insect pest. *PLoS One* 4, e6487.

Jennings, S. and Brander, K. (2010) Predicting the effects of climate change on marine communities and the consequences for fisheries. *Journal of Marine Systems* 79, 418–426.

Jenouvrier, S., Weimerskirch, H., Barbraud, C., Park, Y. and Cazelles, B. (2005) Evidence of a shift in the cyclicity of Antarctic seabird dynamics linked to cli-mate. *Proceedings. Biological Sciences/The Royal Society* 272, 887–895.

Jentsch, A., Kreyling, J. and Beierkuhnlein, C. (2007) A new generation of climate-change experiments: events, not trends. *Frontiers in Ecology and the Environment* 5, 365–374.

Jetz, W., Wilcove, D. and Dobson, A. (2007) Projected impacts of climate and land-use change on the global diversity of birds. *PLoS Biology* 5, 1211–1219.

Jiguet, F., Gadot, A., Julliard, R., Newson, S. and Couvet, D. (2007) Climate envelope, life history traits and the resilience of birds facing global change. *Global Change Biology* 13, 1672–1684.

Jin, V.L. and Evans, R.D. (2007) Elevated $CO_2$ increases microbial carbon substrate use and nitrogen cycling in Mojave desert soils. *Global Change Biology* 13, 452–465.

Jinks, J. and Connolly, V. (1973) Selection for specific and general response to environmental differences. *Heredity* 30, 33–40.

Jobbagy, E.G. and Jackson, R.B. (2000) The vertical distribution of soil organic carbon and its relation to climate and vegetation. *Ecological Applications* 10, 423–436.

Johnson, N.C., Wolf, J., Reyes, M.A., Panter, A., Koch, G.W. and Redman, A. (2005) Species of plants and associated arbuscular mycorrhizal fungi mediate mycorrhizal responses to $CO_2$ enrichment. *Global Change Biology* 11, 1156–1166.

Johnson, S.N. and McNicol, J.W. (2010) Elevated $CO_2$ and aboveground–belowground herbivory by the clover root weevil. *Oecologia* 162, 209–216.

Johnson, S.N., Gregory, P.J., McNicol, J.W., Oodally, Y., Zhang, X.X. and Murray, P.J. (2010) Effects of soil conditions and drought on egg hatching and larval survival of the clover root weevil (*Sitona lepidus*). *Applied Soil Ecology* 44, 75–79.

Johnson, W. and Webb, T. III (1989) The role of blue jays (*Cyanocitta cristata* L.) in the postglacial dispersal of fagaceous trees in eastern North America. *Journal of Biogeography* 16, 561–571.

Jonasson, S., Castro, J. and Michelsen, A. (2004) Litter, warming and plants affect respiration and allocation of soil microbial and plant C, N and P in Arctic mesocosms. *Soil Biology & Biochemistry* 36, 1129–1139.

Jones, C.G., Lawton, J.H. and Shachak, M. (1994) Organisms as ecosystem engineers. *Oikos* 69, 373–386.

Jones, P.D. (1999) Surface air temperature and its changes over the past 150 years. *Reviews of Geophysics* 37, 173–199.

Jones, R.A.C. (2009) Plant virus emergence and evolution: origins, new encounter scenarios, factors driving emergence, effects of changing world conditions, and prospects for control. *Virus Research* 141, 113–130.

Jónsdóttir, I.S., Magnússon, B., Gudmundsson, J., Elmarsdóttir, A. and Hjartarson, H. (2005) Variable sensitivity of plant communities in Iceland to experimental warming. *Global Change Biology* 11, 553–563.

Jordan, D. and Ogren, W. (1984) The $CO_2/O_2$ specificity of ribulose 1,5-bisphosphate carboxylase/oxygenase. *Planta* 161, 308–313.

Jump, A. and Peñuelas, J. (2005) Running to stand still: adaptation and the response of plants to rapid climate change. *Ecology Letters* 8, 1010–1020.

Kampichler, C., Kandeler, E., Bardgett, R.D., Jones, T.H. and Thompson, L.J. (1998) Impact of elevated atmospheric $CO_2$ concentration on soil microbial biomass and activity in a complex, weedy field model ecosystem. *Global Change Biology* 4, 335–346.

Kandeler, E., Tscherko, D., Bardgett, R.D., Hobbs, P.J., Kampichler, C. and Jones, T.H. (1998) The response of soil microorganisms and roots to elevated $CO_2$ and temperature in a terrestrial model ecosystem. *Plant and Soil* 202, 251–262.

Kanerva, T., Palojarvi, A., Ramo, K. and Manninen, S. (2008) Changes in soil microbial community structure under elevated tropospheric $O_3$ and $CO_2$. *Soil Biology & Biochemistry* 40, 2502–2510.

Kao-Kniffin, J. and Balser, T.C. (2007) Elevated $CO_2$ differentially alters belowground plant and soil microbial community structure in reed canary grass-invaded experimental wetlands. *Soil Biology & Biochemistry* 39, 517–525.

Kareiva, P., Kingsolver, J. and Huey, R. (1993) *Biotic Interactions and Global Change*. Sinauer Associates, Sunderland, Massachusetts.

Karowe, D., Seimens, D. and Mitchell-Olds, T. (1997) Species-specific response of glucosinolate content to elevated atmospheric $CO_2$. *Journal of Chemical Ecology* 23, 2569–2582.

Kasischke, E., Bergen, K., Fennimore, R., Sotelo, F., Stephens, G., Janetos, A. and Shugart, H. (1999) Satellite imagery gives clear picture of Russia's boreal forest fires. *Eos – Transactions of the American Geophysical Union* 80, 141–147.

Kasurinen, A., Kokko-Gonzales, P., Riikonen, J., Vapaavuori, E. and Holopainen, T. (2004) Soil $CO_2$ efflux of two silver birch clones exposed to elevated $CO_2$ and $O_3$ levels during three growing seasons. *Global Change Biology* 10, 1654–1665.

Kasurinen, A., Peltonen, P.A., Julkunen-Tiitto, R., Vapaavuori, E., Nuutinen, V., Holopainen, T. and Holopainen, J.K. (2007) Effects of elevated $CO_2$ and $O_3$ on leaf litter phenolics and subsequent performance of litter-feeding soil macrofauna. *Plant and Soil* 292, 25–43.

Keith, S.A., Newton, A.C., Herbert, R.J.H., Morecroft, M.D. and Bealey, C.E. (2009) Non-analogous community formation in response to climate change. *Journal for Nature Conservation* 17, 228–235.

Kelly, C., Chase, M., De Bruijn, A., Fay, M. and Woodward, F. (2003) Temperature-based population segregation in birch. *Ecology Letters* 6, 87–89.

Kiessling, W. and Baron-Szabo, R. (2004) Extinction and recovery patterns of scleractinian corals at the Cretaceous–Tertiary boundary. *Palaeogeography, Palaeoclimatology, Palaeoecology* 214, 195–223.

Killeen, T. and Solórzano, L. (2008) Conservation strategies to mitigate impacts from climate change in Amazonia. *Philosophical Transactions of the Royal Society of London. Series B: Biological Sciences* 363, 1881–1888.

Kim, S.Y., Lee, S.H., Freeman, C., Fenner, N. and Kang, H. (2008) Comparative analysis of soil microbial communities and their responses to the short-term drought in bog, fen, and riparian wetlands. *Soil Biology & Biochemistry* 40, 2874–2880.

Kimball, B.A., Mauney, J.R., Nakayama, F.S. and Idso, S.B. (1993) Effects of increasing atmospheric $CO_2$ on vegetation. *Vegetatio* 104/105, 65–75.

Kimball, B.A., Kobayashi, K. and Bindi, M. (2002) Responses of agricultural crops to free-air $CO_2$ enrichment. *Advances in Agronomy* 77, 293–368.

King, J.S., Thomas, R.B. and Strain, B.R. (1997) Morphology and tissue quality of seedling root systems of *Pinus taeda* and *Pinus ponderosa* as affected by varying $CO_2$, temperature, and nitrogen. *Plant and Soil* 195, 107–119.

Kirilenko, A.P. and Sedjo, R.A. (2007) Climate change impacts on forestry. *Proceedings of the National Academy of Sciences USA* 104, 19697–19702.

Kirschbaum, M.U.F. (2006) The temperature dependence of organic-matter decomposition – still a topic of debate. *Soil Biology & Biochemistry* 38, 2510–2518.

Klamer, M., Roberts, M.S., Levine, L.H., Drake, B.G. and Garland, J.L. (2002) Influence of elevated $CO_2$ on the fungal community in a coastal scrub oak forest soil investigated with terminal-restriction fragment length polymorphism analysis. *Applied and Environmental Microbiology* 68, 4370–4376.

Klanderud, K. and Totland, Ÿ. (2005) Simulated climate change altered dominance hierarchies and diversity of an alpine biodiversity hotspot. *Ecology* 86, 2047–2054.

Klanderud, K. and Totland, Ÿ. (2008) Diversity–stability relationships of an alpine plant community under simulated environmental change. *Arctic, Antarctic, and Alpine Research* 40, 679–684.

Klein, J., Harte, J. and Zhao, X. (2004) Experimental warming causes large and rapid species loss, dampened by simulated grazing, on the Tibetan Plateau. *Ecology Letters* 7, 1170–1179.

Klimes, L., Klimesova, J., Hendriks, R. and Van Groenendael, J. (1997) Clonal plant architecture: a comparative analysis of form and function. In: de Kroon, H. and Van Groenendael, J. (eds) *The Ecology and Evolution of Clonal Plants*. Backhuys Publishers, Leiden, The Netherlands, pp. 1–29.

Klironomos, J.N., Rillig, M.C. and Allen, M.F. (1996) Below-ground microbial and microfaunal responses to *Artemisia tridentata* grown under elevated atmospheric $CO_2$. *Functional Ecology* 10, 527–534.

Klironomos, J.N., Rillig, M.C., Allen, M.F., Zak, D.R., Kubiske, M. and Pregitzer, K.S. (1997) Soil fungal–arthropod responses to *Populus tremuloides* grown under enriched atmospheric $CO_2$ under field conditions. *Global Change Biology* 3, 473–478.

Klironomos, J.N., Allen, M.F., Rillig, M.C., Piotrowski, J., Makvandi-Nejad, S., Wolfe, B.E. and Powell, J.R. (2005) Abrupt rise in atmospheric $CO_2$ overestimates community response in a model plant–soil system. *Nature* 433, 621–624.

Knapp, A.K., Fay, P.A., Blair, J.M., Collins, S.L., Smith, M.D., Carlisle, J.D., Harper, C.W., Danner, B.T., Lett, M.S. and McCarron, J.K. (2002) Rainfall variability, carbon cycling, and plant species diversity in a mesic grassland. *Science* 298, 2202–2205.

Knapp, A.K., Beier, C., Briske, D.D., Classen, A.T., Luo, Y., Reichstein, M., Smith, M.D., Smith, S.D., Bell, J.E., Fay, P.A., Heisler, J.L., Leavitt, S.W., Sherry, R., Smith, B. and Weng, E. (2008) Consequences of more extreme precipitation regimes for terrestrial ecosystems. *BioScience* 58, 811–821.

Knepp, R.G., Hamilton, J.G., Mohan, J.E., Zangerl, A.R., Berenbaum, M.R. and Delucia, E.H. (2005) Elevated $CO_2$ reduces leaf damage by insect herbivores in a forest community. *New Phytologist* 167, 207–218.

Kokko, H. and Lopez-Sepulcre, A. (2006) From individual dispersal to species ranges: perspectives for a changing world. *Science* 313, 789–791.

Kolb, S., Carbrera, A., Kammann, C., Kampfer, P., Conrad, R. and Jackel, U. (2005) Quantitative impact of $CO_2$ enriched atmosphere on abundances of methanotrophic bacteria in a meadow soil. *Biology and Fertility of Soils* 41, 337–342.

Konvicka, M., Maradova, M., Benes, J., Fric, Z. and Kepka, P. (2003) Uphill shifts in distribution of butterflies in the Czech Republic: effects of changing climate detected on a regional scale. *Global Ecology and Biogeography* 12, 403–410.

Körner, C. (1995) Towards a better experimental basis for upscaling plant-responses to elevated $CO_2$ and climate warming. *Plant, Cell & Environment* 18, 1101–1110.

Körner, C., Asshoff, R., Bignucolo, O., Hattenschwiler, S., Keel, S.G., Pelaez-Riedl, S., Pepin, S., Siegwolf, R.T.W. and Zotz, G. (2005) Carbon flux and growth in mature deciduous forest trees exposed to elevated $CO_2$. *Science* 309, 1360–1362.

Kramer, K., Degen, B., Buschbom, J., Hickler, T., Thuiller, W., Sykes, M. and De Winter, W. (2010) Modelling exploration of the future of European beech (*Fagus sylvatica* L.) under climate change – range, abundance, genetic diversity and adaptive response. *Forest Ecology and Management* 259, 2213–2222.

Kuhn, T. (1970) *The Structure of Scientific Revolutions*. University of Chicago Press, Chicago, Illinois.

Kurukulasuriya, P. and Rosenthal, S. (2003) *Climate Change and Agriculture. Climate Change Series Paper No. 91*. Environment Department, World Bank, Washington, DC.

Kurz, W., Dymond, C., Stinson, G., Rampley, G., Neilson, E., Carroll, A., Ebata, T. and Safranyik, L. (2008a) Mountain pine beetle and forest carbon feedback to climate change. *Nature* 452, 987–990.

Kurz, W., Stinson, G., Rampley, G., Dymond, C. and Neilson, E. (2008b) Risk of natural disturbances makes future contribution of Canada's forests to the global carbon cycle highly uncertain. *Proceedings of the National Academy of Sciences USA* 105, 1551–1555.

Kurz, W.A., Stinson, G. and Rampley, G. (2008c) Could increased boreal forest ecosystem productivity offset carbon losses from increased disturbances? *Philosophical Transactions of the Royal Society of London. Series B: Biological Sciences* 363, 2261–2269.

Kutzbach, J.E., Gallimore, R.G., Harrison, S.P., Behling, P., Selin, R. and Laarif, F. (1998) Climate and biome simulations for the past 21,000 years. *Quarternary Science Review* 17, 473–506.

Kytoviita, M.M., Le Thiec, D. and Dizengremel, P. (2001) Elevated $CO_2$ and ozone reduce nitrogen acquisition by *Pinus halepensis* from its mycorrhizal symbiont. *Physiologia Plantarum* 111, 305–312.

LaDeau, S.L. and Clark, J.S. (2001) Rising $CO_2$ levels and the fecundity of forest trees. *Science* 292, 95–98.

Lafferty, K. (2009) The ecology of climate change and infectious diseases. *Ecology* 90, 888–900.

Lagergren, F., Grelle, A., Lankreijer, H., Molder, M. and Lindroth, A. (2006) Current carbon balance of the forested area in Sweden and its sensitivity to global change as simulated by biome-BGC. *Ecosystems* 9, 894–908.

Lagomarsino, A., Knapp, B.A., Moscatelli, M.C., De Angelis, P., Grego, S. and Insam, H. (2007) Structural and functional diversity of soil microbes is affected by elevated $[CO_2]$ and N addition in a poplar plantation. *Journal of Soils and Sediments* 7, 399–405.

Lake, J.A. and Wade, R.N. (2009) Plant–pathogen interactions and elevated $CO_2$: morphological changes in favour of pathogens. *Journal of Experimental Botany* 60, 3123–3131.

Lamarche, V.C., Graybill, D.A., Fritts, H.C. and Rose, M.R. (1984) Increasing atmospheric carbon dioxide: tree ring evidence for growth enhancement in natural vegetation. *Science* 225, 1019–1021.

Lamb, H.H. and Woodroffe, A. (1970) Atmospheric circulation during the last ice age. *Quaternary Research* 1, 29–58.

Lambers, H., Chapin, F.S.I. and Pons, T.L. (2008) *Plant Physiological Ecology*. Springer, Berlin.

Last, W., Smol, J. and Birks, H. (2001) *Tracking Environmental Change Using Lake Sediments: Physical and Geochemical Methods*. Springer, Berlin.

Last, W., Smol, J. and Birks, H. (2002) *Tracking Environmental Change Using Lake Sediments: Basin Analysis, Coring, and Chronological Techniques*. Kluwer Academic Publishers, Dordrecht, The Netherlands.

Laurance, W.F., Oliveira, A.A., Laurance, S.G., Condit, R., Nascimento, H.E.M., Sanchez-Thorin, A.C., Lovejoy, T.E., Andrade, A., D'Angelo, S., Ribeiro, J.E. and Dick, C.W. (2004) Pervasive alteration of tree communities in undisturbed Amazonian forests. *Nature* 428, 171–175.

Law, R. and Morton, R.D. (1993) Alternative permanent states of ecological communities. *Ecology* 74, 1347–1361.

Lawton, J.H. (1993) Range, population abundance and conservation. *Trends in Ecology & Evolution* 8, 409–413.

Lawton, J.H. (1998) Ecological experiments with model systems. In: Resetarits, W.J.J. and Bernardo, J. (eds) *Experimental Ecology: Issues and Perspectives*. Oxford University Press, Oxford, UK, pp. 170–182.

Leake, J.R., Ostle, N.J., Rangel-Castro, J.I. and Johnson, D. (2006) Carbon fluxes from plants through soil organisms determined by field $^{13}CO_2$ pulse-labelling in an upland grassland. *Applied Soil Ecology* 33, 152–175.

Leakey, A.D.B. (2009) Rising atmospheric carbon dioxide concentration and the future of $C_4$ crops for food and fuel. *Proceedings. Biological Sciences/The Royal Society* 276, 2333–2343.

Leakey, A.D.B., Ainsworth, E.A., Bernacchi, C.J., Rogers, A., Long, S.P. and Ort, D.R. (2009) Elevated $CO_2$ effects on plant carbon, nitrogen, and water relations: Six important lessons from FACE. *Journal of Experimental Botany* 60, 2859–2876.

Lean, J. (2010) Cycles and trends in solar irradiance and climate. *WIREs Climatic Change* 1, 111–122.

LeBauer, D. and Treseder, K. (2008) Nitrogen limitation of net primary productivity in terrestrial ecosystems is globally distributed. *Ecology* 89, 371–379.

Lee, C. (2002) Evolutionary genetics of invasive species. *Trends in Ecology & Evolution* 17, 386–391.

Leithead, M., Anand, M. and Silva, L. (2010) Northward migrating trees establish in treefall gaps at the northern limit of the temperate–boreal ecotone, Ontario, Canada. *Oecologia* 164, 1095–1106.

Lenihan, J.M., Bachelet, D., Neilson, R.P. and Drapek, R. (2008) Response of vegetation distribution, ecosystem productivity, and fire to climate change scenarios for California. *Climatic Change* 87, Suppl. 1, S215–S230.

Lenoir, J., Gègout, J., Marquet, P., De Ruffray, P. and Brisse, H. (2008) A significant upward shift in plant species optimum elevation during the 20th century. *Science* 320, 1768–1771.

Le Roux, P.C. and McGeoch, M.A. (2008) Rapid range expansion and community reorganization in response to warming. *Global Change Biology* 14, 2950–2962.

Letters, E. (2009) The intermediate disturbance hypothesis applies to tropical forests, but disturbance contributes little to tree diversity. *Ecology Letters* 12, 1–8.

Levinsky, I., Skov, F., Svenning, J. and Rahbek, C. (2007) Potential impacts of climate change on the distributions and diversity patterns of European mammals. *Biodiversity and Conservation* 16, 3803–3816.

Lewis, D.H. and Smith, D.J. (2004) Little ice age glacial activity in Strathcona Provincial Park, Vancouver

Island, British Columbia, Canada. *Canadian Journal of Earth Sciences* 41, 285–297.

Lewis, O. (2006) Climate change, species–area curves and the extinction crisis. *Philosophical Transactions of the Royal Society of London. Series B: Biological Sciences* 361, 163–171.

Lewis, S., Phillips, O., Baker, T., Lloyd, J., Malhi, Y., Almeida, S., Higuchi, N., Laurance, W., Neill, D. and Silva, J. (2004) Concerted changes in tropical forest structure and dynamics: evidence from 50 South American long-term plots. *Philosophical Transactions of the Royal Society of London. Series B: Biological Sciences* 359, 421–436.

Lewis, S., Lopez-Gonzalez, G., Sonkè, B., Affum-Baffoe, K., Baker, T., Ojo, L., Phillips, O., Reitsma, J., White, L. and Comiskey, J. (2009) Increasing carbon storage in intact African tropical forests. *Nature* 457, 1003–1006.

Lewontin, R. and Cohen, D. (1969) On population growth in a randomly varying environment. *Proceedings of the National Academy of Sciences USA* 62, 1056–1060.

Li, J., Zhou, J.M. and Duan, Z.Q. (2007) Effects of elevated $CO_2$ concentration on growth and water usage of tomato seedlings under different ammonium/nitrate ratios. *Journal of Environmental Sciences – China* 19, 1100–1107.

Li, Q., Liang, W.J., Jiang, Y., Shi, Y., Zhu, J.G. and Neher, D.A. (2007) Effect of elevated $CO_2$ and N fertilisation on soil nematode abundance and diversity in a wheat field. *Applied Soil Ecology* 36, 63–69.

Li, Q., Xu, C.G., Liang, W.J., Zhong, S., Zheng, X.H. and Zhu, J.G. (2009) Residue incorporation and N fertilization affect the response of soil nematodes to the elevated $CO_2$ in a Chinese wheat field. *Soil Biology & Biochemistry* 41, 1497–1503.

Li, T., Grant, R.F. and Flanagan, L.B. (2004) Climate impact on net ecosystem productivity of a semi-arid natural grassland: modeling and measurement. *Agricultural and Forest Meteorology* 126, 99–116.

Lichter, J., Billings, S.A., Ziegler, S.E., Gaindh, D., Ryals, R., Finzi, A.C., Jackson, R.B., Stemmler, E.A. and Schlesinger, W.H. (2008) Soil carbon sequestration in a pine forest after 9 years of atmospheric $CO_2$ enrichment. *Global Change Biology* 14, 2910–2922.

Limsakul, A. and Goes, J. (2008) Empirical evidence for interannual and longer period variability in Thailand surface air temperatures. *Atmospheric Research* 87, 89–102.

Lindberg, N., Engtsson, J.B. and Persson, T. (2002) Effects of experimental irrigation and drought on the composition and diversity of soil fauna in a coniferous stand. *Journal of Applied Ecology* 39, 924–936.

Lipson, D.A., Blair, M., Barron-Gafford, G., Grieve, K. and Murthy, R. (2006) Relationships between microbial community structure and soil processes under elevated atmospheric carbon dioxide. *Microbial Ecology* 51, 302–314.

Litton, C.M. and Giardina, C.P. (2008) Below-ground carbon flux and partitioning: global patterns and response to temperature. *Functional Ecology* 22, 941–954.

Liu, W.X., Zhang, Z. and Wan, S.Q. (2009) Predominant role of water in regulating soil and microbial respiration and their responses to climate change in a semi-arid grassland. *Global Change Biology* 15, 184–195.

Lloret, F., Peñuelas, J. and Estiarte, M. (2004) Experimental evidence of reduced diversity of seedlings due to climate modification in a Mediterranean-type community. *Global Change Biology* 10, 248–258.

Logan, J., McFarlane, W. and Willcox, L. (2010) Whitebark pine vulnerability to climate change induced mountain pine beetle disturbance in the greater Yellowstone ecosystem. *Ecological Applications* 20, 895–902.

Logan, J.A. and Powell, J.A. (2001) Ghost forests, global warming, and the mountain pine beetle (Coleoptera: Scolytidae). *American Entomologist* 47, 160–172.

Logan, J.A., Régnière, J. and Powell, J.A. (2003) Assessing the impacts of global warming on forest pest dynamics. *Frontiers in Ecology and the Environment* 1, 130–137.

Long, S.P. (1991) Modification of the response of photosynthetic productivity to rising temperature by atmospheric $CO_2$ concentrations: has its importance been underestimated? *Plant, Cell & Environment* 14, 729–739.

Long, S.P. and Bernacchi, C.J. (2003) Gas exchange measurements, what can they tell us about the underlying limitations to photosynthesis? Procedures and sources of error. *Journal of Experimental Botany* 54, 2393–2401.

Long, S.P., Ainsworth, E.A., Leakey, A.D. and Morgan, P.B. (2005) Global food insecurity. Treatment of major food crops with elevated carbon dioxide or ozone under large-scale fully open-air conditions suggests recent models may have overestimated future yields. *Philosophical Transactions of the Royal Society of London. Series B: Biological Sciences* 360, 2011–2020.

Long, S.P., Ainsworth, E.A., Leakey, A.D., Nösberger, J. and Ort, D.R. (2006) Food for thought: lower-than-expected crop yield stimulation with rising $CO_2$ concentrations. *Science* 312, 1918–1921.

Loreau, M., Naeem, S., Inchausti, P., Bengtsson, J., Grime, J., Hector, A., Hooper, D., Huston, M., Raffaelli, D. and Schmid, B. (2001) Biodiversity and ecosystem functioning: current knowledge and future challenges. *Science* 294, 804–808.

Lorius, C., Jouzel, J., Raynaud, D., Hansen, J. and Le Treut, H. (1990) The ice-core record: climate sensitivity and future greenhouse warming. *Nature* 347, 139–145.

Lovelock, J. (1987) *Geophysiology: A New Look at Earth Science*. John Wiley, New York.

Luckman, B.H. (1995) Calendar-dated, early 'Little Ice Age' glacier advance at Robson Glacier, British Columbia, Canada. *The Holocene* 5, 149–159.

Luckman, B.H. (2000) The Little Ice Age in the Canadian Rockies. *Geomorphology* 32, 357–384.

Lugato, E. and Berti, A. (2008) Potential carbon sequestration in a cultivated soil under different climate change scenarios: a modelling approach for evaluating promising management practices in northeast Italy. *Agriculture, Ecosystems & Environment* 128, 97–103.

Luo, Y., Su, B., Currie, W.S., Dukes, J.S., Finzi, A., Hartwig, U., Hungate, B., McMurtrie, R.E., Oren, R., Parton, W.J., Pataki, D.E., Shaw, M.R., Zak, D.R. and Field, C.B. (2004) Progressive nitrogen limitation of ecosystem responses to rising atmospheric carbon dioxide. *BioScience* 54, 731–739.

Luo, Y., Hui, D. and Zhang, D. (2006) Elevated $CO_2$ stimulates net accumulations of carbon and nitrogen in land ecosystems: a meta-analysis. *Ecology* 87, 53–63.

Luo, Y.Q. (2007) Terrestrial carbon-cycle feedback to climate warming. *Annual Review of Ecology Evolution and Systematics* 38, 683–712.

Luo, Y.Q., Gerten, D., Le Maire, G., Parton, W.J., Weng, E.S., Zhou, X.H., Keough, C., Beier, C., Ciais, P., Cramer, W., Dukes, J.S., Emmett, B., Hanson, P.J., Knapp, A., Linder, S., Nepstad, D. and Rustad, L. (2008) Modeled interactive effects of precipitation, temperature, and $[CO_2]$ on ecosystem carbon and water dynamics in different climatic zones. *Global Change Biology* 14, 1986–1999.

Lussenhop, J., Treonis, A., Curtis, P.S., Teeri, J.A. and Vogel, C.S. (1998) Response of soil biota to elevated atmospheric $CO_2$ in poplar model systems. *Oecologia* 113, 247–251.

MacInnes, C., Dunn, E., Rusch, D., Cooke, F. and Cooch, F. (1990) Advancement of goose nesting dates in the Hudson Bay region, 1951–1986. *Canadian Field Naturalist (Ottawa ON)* 104, 295–297.

Mack, M.C., Schuur, E.A.G., Bret-Harte, M.S., Shaver, G.R. and Chapin, F.S. (2004) Ecosystem carbon storage in Arctic tundra reduced by long-term nutrient fertilization. *Nature* 431, 440–443.

Mack, M.C., Treseder, K.K., Manies, K.L., Harden, J.W., Schuur, E.A.G., Vogel, J.G., Randerson, J.T. and Chapin, F.S. (2008) Recovery of aboveground plant biomass and productivity after fire in mesic and dry black spruce forests of interior Alaska. *Ecosystems* 11, 209–225.

MacKenzie, B.R., Gislason, H., Mollmann, C. and Koster, F.W. (2007) Impact of 21st century climate change on the Baltic Sea fish community and fisheries. *Global Change Biology* 13, 1348–1367.

MacLean, D.A., Gray, D., Porter, K.B., MacKinnon, W., Budd, M., Beaton, K., Fleming, R.A. and Volney, W.J.A. (2002) Climate change effects on insect outbreaks and management opportunities for carbon sequestration. In: Shaw, C.H. and Apps, M.J. (eds) *The Role of Boreal Forests and Forestry in the Global Carbon Budget*, pp. 119–128. Proceedings of the International Boreal Forest Research Association 2000 Conference, 8–12 May, 2000, Edmonton, Alberta. Canadian Forest Service, Edmonton, Canada.

MacLeod, C.D., Bannon, S.M., Pierce, G.J., Schweder, C., Learmonth, J.A., Herman, J.S. and Reid, R.J. (2005) Climate change and the cetacean community of northwest Scotland. *Biological Conservation* 124, 477–483.

MacNair, M.R. (1981) Tolerance of higher plants to toxic metals. In: Bishop, J.A. and Cook, L.M. (eds) *Genetic Consequences of Man-Made Change*. Academic Press, London, pp. 177–208.

Maestre, F.T., Bradford, M.A. and Reynolds, J.F. (2005) Soil nutrient heterogeneity interacts with elevated $CO_2$ and nutrient availability to determine species and assemblage responses in a model grassland community. *New Phytologist* 168, 637–649.

Magurran, A. (2004) *Measuring Biodiversity*. Blackwell Publishing, Malden, Massachusetts.

Malchair, S., De Boeck, H.J., Lemmens, C., Ceulemans, R., Merckx, R., Nijs, I. and Carnol, M. (2010) Diversity–function relationship of ammonia-oxidizing bacteria in soils among functional groups of grassland species under climate warming. *Applied Soil Ecology* 44, 15–23.

Malcolm, J., Markham, A., Neilson, R. and Garaci, M. (2002) Estimated migration rates under scenarios of global climate change. *Journal of Biogeography* 29, 835–849.

Malcolm, J., Liu, C., Neilson, R., Hansen, L. and Hannah, L. (2006) Global warming and extinctions of endemic species from biodiversity hotspots. *Conservation Biology* 20, 538–548.

Malhi, Y., Roberts, J., Betts, R., Killeen, T., Li, W. and Nobre, C. (2008) Climate change, deforestation, and the fate of the Amazon. *Science* 319, 169–172.

Malmstrom, C.M. and Field, C.B. (1997) Virus-induced differences in the response of oat plants to elevated carbon dioxide. *Plant, Cell & Environment* 20, 178–188.

Manderscheid, R. and Weigel, H.J. (2007) Drought stress effects on wheat are mitigated by atmospheric $CO_2$ enrichment. *Agronomy for Sustainable Development* 27, 79–87.

Mann, M.E., Bradley, R.S. and Hughes, M.K. (1998) Global-scale temperature patterns and climate forcing over the past six centuries. *Nature* 392, 779–787.

Mann, M.E., Bradley, R.S. and Hughes, M.K. (1999) Northern hemisphere temperatures during the past millennium: inferences, uncertainties, and limitations. *Geophysical Research Letters* 26, 759–762.

Mann, M.E., Bradley, R.S. and Hughes, M.K. (2003) Optimal surface temperature reconstructions using terrestrial borehole data. *Journal of Geophysical Research* 108, 4203; doi:10.1029/2002JD002532.

Manning, W. (2005) Establishing a cause and effect relationship for ambient ozone exposure and tree growth in the forest: progress and an experimental approach. *Environmental Pollution* 137, 443–454.

Manning, W. and von Tiedemann, A. (1995) Climate change: potential effects of increased atmospheric carbon dioxide ($CO_2$), ozone ($O_3$), and ultraviolet-B (UV-B) radiation on plant diseases. *Environmental Pollution* 88, 219–245.

Markkola, A.M., Ohtonen, A., Ahonenjonnarth, U. and Ohtonen, R. (1996) Scots pine responses to $CO_2$ enrichment. 1. Ectomycorrhizal fungi and soil fauna. *Environmental Pollution* 94, 309–316.

Martin, P. and Klein, R. (1984) *Quaternary Extinctions: A Prehistoric Revolution.* University of Arizona Press, Tucson, Arizona.

Martin, T.E. (2007) Climate correlates of 20 years of trophic changes in a high-elevation riparian system. *Ecology* 88, 367–380.

Marx, M.C., Wood, M. and Jarvis, S.C. (2001) A microplate fluorimetric assay for the study of enzyme diversity in soils. *Soil Biology & Biochemistry* 33, 1633–1640.

Masek, J.G. (2001) Stability of boreal forest stands during recent climate change: evidence from landsat satellite imagery. *Journal of Biogeography* 28, 967–976.

Matesanz, S., Brooker, R.W., Valladares, F. and Klotz, S. (2009) Temporal dynamics of marginal steppic vegetation over a 26-year period of substantial environmental change. *Journal of Vegetation Science* 20, 299–310.

Matyssek, R. and Sandermann, H. (2003) Impact of ozone on trees: an ecophysiological perspective. *Progress in Botany* 64, 349–404.

Mayhew, P., Jenkins, G. and Benton, T. (2008) A long-term association between global temperature and biodiversity, origination and extinction in the fossil record. *Proceedings. Biological Sciences/The Royal Society* 275, 47–53.

Maynard-Smith, J. (1989) *Evolutionary Genetics.* Oxford University Press, Oxford, UK.

Mayr, C., Miller, M. and Insam, H. (1999) Elevated $CO_2$ alters community-level physiological profiles and enzyme activities in alpine grassland. *Journal of Microbiological Methods* 36, 35–43.

McCarthy, H.R., Oren, R., Finzi, A.C. and Johnsen, K.H. (2006) Canopy leaf area constrains [$CO_2$]-induced enhancement of productivity and partitioning among aboveground carbon pools. *Proceedings of the National Academy of Sciences USA* 103, 19356–19361.

McCarty, J. (2001) Review: Ecological consequences of recent climate change. *Conservation Biology* 15, 320–331.

McComas, W.F. (1998) The principal elements of the nature of science: dispelling the myths. In: McComas, W.F.

(ed.) *The Nature of Science in Science Education: Rationales and Strategies.* Kluwer Academic Publishers, Dortecht, The Netherlands, pp. 53–70.

McDonald, A., Riha, S., Ditommaso, A. and Degaetano, A. (2009) Climate change and the geography of weed damage: analysis of US maize systems suggests the potential for significant range transformations. *Agriculture, Ecosystems & Environment* 130, 131–140.

McGeoch, M.A., Le Roux, P.C., Hugo, E.A. and Chown, S.L. (2006) Species and community responses to short-term climate manipulation: microarthropods in the sub-Antarctic. *Austral Ecology* 31, 719–731.

McIntyre, S. and McKitrick, R. (2005a) Hockey sticks, principal components, and spurious significance. *Geophysical Research Letters* 32, L03710; doi:10.1029/2004GL021750.

McIntyre, S. and McKitrick, R. (2005b) The M & M critique of the MBH98 northern hemisphere climate index: update and implications. *Energy and Environment* 16, 69–100.

McKee, D., Hatton, K., Eaton, J.W., Atkinson, D., Atherton, A., Harvey, I. and Moss, B. (2002) Effects of simulated climate warming on macrophytes in freshwater microcosm communities. *Aquatic Botany* 74, 71–83.

McKenney, D.W., Pedlar, J.H., Lawrence, K., Campbell, K. and Hutchinson, M.F. (2007) Potential impacts of climate change on the distribution of North American trees. *Bioscience* 57, 939–948.

McLachlan, J., Clark, J. and Manos, P. (2005) Molecular indicators of tree migration capacity under rapid climate change. *Ecology* 86, 2088–2098.

McLachlan, J., Hellmann, J. and Schwartz, M. (2007) A framework for debate of assisted migration in an era of climate change. *Conservation Biology* 21, 297–302.

McLaughlin, J., Hellmann, J., Boggs, C. and Ehrlich, P. (2002) Climate change hastens population extinctions. *Proceedings of the National Academy of Sciences USA* 99, 6070–6074.

Meehl, G.A., Stocker, T.F., Collins, W.D., Friedlingstein, P., Gaye, A.T., Gregory, J.M., Kitoh, A., Knutti, R., Murphy, J.M., Noda, A., Raper, S.C.B., Watterson, I.G., Weaver, A.J. and Zhao, Z.-C. (eds) (2007) *Global Climate Projections.* Cambridge University Press, Cambridge, UK.

Melotto, M., Underwood, W. and He, S. (2008) Role of stomata in plant innate immunity and foliar bacterial diseases. *Annual Review of Phytopathology* 46, 101–122.

Mendelsohn, R., Nordhaus, W. and Shaw, D. (1994) The impact of global warming on agriculture: a Ricardian analysis. *American Economic Review* 84, 753–771.

Menon, S., Soberon, J., Li, X. and Peterson, A. (2010) Preliminary global assessment of terrestrial biodiversity consequences of sea-level rise mediated by climate change. *Biodiversity and Conservation* 19, 1599–1609.

Merrill, R.T., McElhinny, M.W. and McFadden, P.L. (1990) *The Magnetic Field of the Earth: Paleomagnetism, the Core, and the Deep Mantle.* Academic Press, San Diego, California.

Mika, A.M., Weiss, R.M., Olfert, O., Hallett, R.H. and Newman, J.A. (2008) Will climate change be beneficial or detrimental to the invasive swede midge in North America? Contrasting predictions using climate projections from different general circulation models. *Global Change Biology* 14, 1721–1733.

Mikkelsen, T., Beier, C., Jonasson, S., Holmstrup, M., Schmidt, I., Ambus, P., Pilegaard, K., Michelsen, A., Albert, K. and Andresen, L. (2008) Experimental design of multifactor climate change experiments with elevated $CO_2$, warming and drought: the CLIMAITE project. *Functional Ecology* 22, 185–195.

Millar, C., Stephenson, N. and Stephens, S. (2007) Climate change and forests of the future: managing in the face of uncertainty. *Ecological Applications* 17, 2145–2151.

Millennium Ecosystem Assessment (2005) *Ecosystems and Human Well-Being: Biodiversity Synthesis.* World Resources Institute, Washington, DC.

Millien, V. (2006) Morphological evolution is accelerated among island mammals. *PLoS Biology* 4, 1863–1868.

Mitchell, E.A.D., Gilbert, D., Buttler, A., Amblard, C., Grosvernier, P. and Gobat, J.M. (2003) Structure of microbial communities in sphagnum peatlands and effect of atmospheric carbon dioxide enrichment. *Microbial Ecology* 46, 187–199.

Moise, E.R.D. and Henry, H.A.L. (2010) Like moths to a street lamp: exaggerated animal densities in plot-level global change field experiments. *Oikos* 119, 791–795.

Mokany, K., Raison, R.J. and Prokushkin, A.S. (2006) Critical analysis of root:shoot ratios in terrestrial biomes. *Global Change Biology* 12, 84–96.

Molau, U. (1996) Climatic impacts on flowering, growth, and vigour in an arctic-alpine cushion plant, *Diapensia lapponica*, under different snow cover regimes. *Ecological Bulletins* 45, 210–219.

Monnin, E., Indermühle, A., Dällenbach, A., Flückiger, J., Stauffer, B., Stocker, T.F., Raynaud, D. and Barnola, J.-M. (2001) Atmospheric $CO_2$ concentrations over the last glacial termination. *Science* 291, 112–114.

Montealegre, C.M., Van Kessel, C., Blumenthal, J.M., Hur, H.G., Hartwig, U.A. and Sadowsky, M.J. (2000) Elevated atmospheric $CO_2$ alters microbial population structure in a pasture ecosystem. *Global Change Biology* 6, 475–482.

Montealegre, C.M., Van Kessel, C., Russelle, M.P. and Sadowsky, M.J. (2002) Changes in microbial activity and composition in a pasture ecosystem exposed to elevated atmospheric carbon dioxide. *Plant and Soil* 243, 197–207.

Mooney, H., Larigauderie, A., Cesario, M., Elmquist, T., Hoegh-Guldberg, O., Lavorel, S., Mace, G., Palmer, M., Scholes, R. and Yahara, T. (2009) Biodiversity, climate change, and ecosystem services. *Current Opinion in Environmental Sustainability* 1, 46–54.

Moorcroft, P., Pacala, S. and Lewis, M. (2006) Potential role of natural enemies during tree range expansions following climate change. *Journal of Theoretical Biology* 241, 601–616.

Morecroft, M.D., Masters, G.J., Brown, V.K., Clarke, I.P., Taylor, M.E. and Whitehouse, A.T. (2004) Changing precipitation patterns alter plant community dynamics and succession in an ex-arable grassland. *Functional Ecology* 18, 648–655.

Morin, P.J. (1999) *Community Ecology.* Blackwell Science Inc., Oxford.

Morin, X. and Thuiller, W. (2009) Comparing niche-and process-based models to reduce prediction uncertainty in species range shifts under climate change. *Ecology* 90, 1301–1313.

Moritz, C., Patton, J.L., Conroy, C.J., Parra, J.L., White, G.C. and Beissinger, S.R. (2008) Impact of a century of climate change on small-mammal communities in Yosemite National Park, USA. *Science* 322, 261–264.

Morris, W., Pfister, C., Tuljapurkar, S., Haridas, C., Boggs, C., Boyce, M., Bruna, E., Church, D., Coulson, T. and Doak, D. (2008) Longevity can buffer plant and animal populations against changing climatic variability. *Ecology* 89, 19–25.

Morrison, L.W., Korzukhin, M.D. and Porter, S.D. (2005) Predicted range expansion of the invasive fire ant, *Solenopsis invicta*, in the eastern United States based on the VEMAP global warming scenario. *Diversity & Distributions* 11, 199–204.

Mouthon, J. and Daufresne, M. (2006) Effects of the 2003 heatwave and climatic warming on mollusc communities of the Saóne: a large lowland river and of its two main tributaries (France). *Global Change Biology* 12, 441–449.

Mrosovsky, N. (1994) Sex ratios of sea turtles. *Journal of Experimental Zoology* 270, 16–27.

Mueller, J. and Hellmann, J. (2008) An assessment of invasion risk from assisted migration. *Conservation Biology* 22, 562–567.

Mulder, C., Uliassi, D. and Doak, D. (2001) Physical stress and diversity–productivity relationships: the role of positive interactions. *Proceedings of the National Academy of Sciences USA* 98, 6704–6708.

Myers, N., Mittermeier, R., Mittermeier, C., Da Fonseca, G. and Kent, J. (2000) Biodiversity hotspots for conservation priorities. *Nature* 403, 853–858.

Myers, P., Lundrigan, B.L., Hoffman, S.M.G., Haraminac, A.P. and Seto, S.H. (2009) Climate-induced changes in the small mammal communities of the Northern Great Lakes Region. *Global Change Biology* 15, 1434–1454.

Nabuurs, G.-J., Pussinen, A., Karjalainen, T., Erhard, M. and Kramer, K. (2002) Stemwood volume increment changes in European forests due to climate change – a simulation study with the EFISCEN model. *Global Change Biology* 8, 304–316.

Nadelhoffer, K., Emmett, B., Gundersen, P., Kjonaas, O., Koopmans, C., Schleppi, P., Tietema, A. and Wright, R. (1999) Nitrogen deposition makes a minor contribution to carbon sequestration in temperate forests. *Nature* 398, 145–148.

Nadkarni, N.M. and Solano, R. (2002) Potential effects of climate change on canopy communities in a tropical cloud forest: an experimental approach. *Oecologia* 131, 580–586.

Naeem, S., Lawton, J., Crawley, M. and Thompson, L. (1996) Biodiversity and plant productivity in a model assemblage of plant species. *Oikos* 76, 259–264.

Nagy, P., Bakonyi, G., Peli, E., Sonnemann, I. and Tuba, Z. (2008) Long-term response of the nematode community to elevated atmospheric $CO_2$ in a temperate dry grassland soil. *Community Ecology* 9, 167–173.

Nakicenovic, N. and Swart, R. (eds) (2000) *Emissions Scenarios: A Special Report of Working Group III of the Intergovernmental Panel on Climate Change.* Cambridge University Press, Cambridge, UK.

NAS (2006) *Surface Temperature Reconstructions for the Last 2,000 Years.* Board on Atmospheric Sciences and Climate, Division on Earth and Life Studies, National Research Council of the National Academies of Science, Washington, DC.

Navas, M.L. (1998) Individual species performance and response of multispecific communities to elevated $CO_2$: a review. *Functional Ecology* 12, 721–727.

Neher, D.A., Weicht, T.R., Moorhead, D.L. and Sinsabaugh, R.L. (2004) Elevated $CO_2$ alters functional attributes of nematode communities in forest soils. *Functional Ecology* 18, 584–591.

Nemani, R., Keeling, C., Hashimoto, H., Jolly, W., Piper, S., Tucker, C., Myneni, R. and Running, S. (2003) Climate-driven increases in global terrestrial net primary production from 1982 to 1999. *Science* 300, 1560–1563.

Newman, J.A. (2004) Climate change and cereal aphids: the relative effects of increasing $CO_2$ and temperature on aphid population dynamics. *Global Change Biology* 10, 5–15.

Newman, J.A. (2005) Climate change and the fate of cereal aphids in Southern Britain. *Global Change Biology* 11, 940–944.

Newman, J.A. (2006) Using the output from global circulation models to predict changes in the distribution and abundance of cereal aphids in Canada: a mechanistic modeling approach. *Global Change Biology* 12, 1634–1642.

Newman, J.A., Bergelson, J. and Grafen, A. (1997) Blocking factors and hypothesis tests in ecology: is your statistics text wrong? *Ecology* 78, 1312–1320.

Newman, J.A., Gibson, D.J., Parsons, A.J. and Thornley, J.H.M. (2003) How predictable are aphid population responses to elevated $CO_2$? *Journal of Animal Ecology* 72, 556–566.

Newton, P.C.D., Carran, R.A., Edwards, G.R. and Niklaus, P.A. (eds) (2007) *Agroecosystems in a Changing Climate.* Taylor and Francis Group, Boca Raton, Florida.

Niinisto, S., Silvola, J. and Kellomaki, S. (2004) Soil $CO_2$ efflux in a boreal pine forest under atmospheric $CO_2$ enrichment and air warming. *Global Change Biology* 10, 1363–1376.

Niklaus, P.A. and Körner, C. (2004) Synthesis of a six-year study of calcareous grassland responses to *in situ* $CO_2$ enrichment. *Ecological Monographs* 74, 491–511.

Niklaus, P.A., Alphei, D., Ebersberger, D., Kampichler, C., Kandeler, E. and Tscherko, D. (2003) Six years of *in situ* $CO_2$ enrichment evoke changes in soil structure and soil biota of nutrient-poor grassland. *Global Change Biology* 9, 585–600.

Niklaus, P.A., Alphei, J., Kampichler, C., Kandeler, E., Körner, C., Tscherko, D. and Wohlfender, M. (2007) Interactive effects of plant species diversity and elevated $CO_2$ on soil biota and nutrient cycling. *Ecology* 88, 3153–3163.

Niu, S., Wu, M., Han, Y.I., Xia, J., Zhang, Z.H.E., Yang, H. and Wan, S. (2010) Nitrogen effects on net ecosystem carbon exchange in a temperate steppe. *Global Change Biology* 16, 144–155.

Niu, S.L. and Wan, S.Q. (2008) Warming changes plant competitive hierarchy in a temperate steppe in northern China. *Journal of Plant Ecology* 1, 103–110.

Nobel, P. (1991) Achievable productivities of certain CAM plants: basis for high values compared with $C_3$ and $C_4$ plants. *New Phytologist* 119, 183–205.

Nobel, P. (2000) Crop ecosystem responses to climatic change: crassulacean acid metabolism crops. In: Reddy, K.R. and Hodges, H.F. (eds) *Climate Change and Global Crop Productivity.* CAB International, Wallingford, UK, pp. 315–331.

Norby, R.J. and Luo, Y. (2004) Evaluating ecosystem responses to rising atmospheric $CO_2$ and global warming in a multi-factor world. *New Phytologist* 162, 281–293.

Norby, R.J., Cotrufo, M.F., Ineson, P., O'Neill, E.G. and Canadell, J.G. (2001) Elevated $CO_2$, litter chemistry, and decomposition: a synthesis. *Oecologia* 127, 153–165.

Norby, R.J., Ledford, J., Reilly, C.D., Miller, N.E. and O'Neill, E.G. (2004) Fine-root production dominates response of a deciduous forest to atmospheric $CO_2$ enrichment. *Proceedings of the National Academy of Sciences USA* 101, 9689–9693.

Norby, R.J., Delucia, E.H., Gielen, B., Calfapietra, C., Giardina, C.P., King, J.S., Ledford, J., McCarthy, H.R., Moore, D.J.P., Ceulemans, R., De Angelis, P.,

Finzi, A.C., Karnosky, D.F., Kubiske, M.E., Lukac, M., Pregitzer, K.S., Scarascia-Mugnozza, G.E., Schlesinger, W.H. and Oren, R. (2005) Forest response to elevated $CO_2$ is conserved across a broad range of productivity. *Proceedings of the National Academy of Sciences USA* 102, 18052–18056.

Norby, R.J., Rustad, L.E., Dukes, J.S., Ojima, D.S., Parton, W.J., Del Grosso, S.J., McMurtrie, R.E. and Pepper, D.A. (2007) *Ecosystem Responses to Warming and Interacting Global Change Factors*. Springer-Verlag, Berlin.

Novak, K., Cherubini, P., Saurer, M., Fuhrer, J., Skelly, J., Krauchi, N. and Schaub, M. (2007) Ozone air pollution effects on tree-ring growth, $\delta^{13}C$, visible foliar injury and leaf gas exchange in three ozone-sensitive woody plant species. *Tree Physiology* 27, 941–949.

Nowak, R., Ellsworth, D. and Smith, S. (2004) Functional responses of plants to elevated atmospheric $CO_2$: do photosynthetic and productivity data from face experiments support early predictions? *New Phytologist* 162, 253–280.

Oberbauer, S.F., Tweedie, C.E., Welker, J.M., Fahnestock, J.T., Henry, G.H.R., Webber, P.J., Hollister, R.D., Walker, M.D., Kuchy, A., Elmore, E. and Starr, G. (2007) Tundra $CO_2$ fluxes in response to experimental warming across latitudinal and moisture gradients. *Ecological Monographs* 77, 221–238.

O'Connor, D. (2008) Governing the global commons: linking carbon sequestration and biodiversity conservation in tropical forests. *Global Environmental Change* 18, 368–374.

Oechel, W. and Billings, W. (1992) Effects of global change on the carbon balance of arctic plants and ecosystems. In: *Arctic Ecosystems in a Changing Climate: An Ecophysiological Perspective*, pp. 139–168. Academic Press, San Francisco, California.

Oechel, W., Hastings, S., Vourlitis, G., Jenkins, M., Riechers, G. and Grulke, N. (1993) Recent change of Arctic tundra ecosystems from a net carbon dioxide sink to a source. *Nature* 361, 520–523.

Oksanen, L. (2001) Logic of experiments in ecology: is pseudoreplication a pseudoissue? *Oikos* 94, 27–38.

O'Lear, H.A. and Blair, J.M. (1999) Responses of soil microarthropods to changes in soil water availability in tallgrass prairie. *Biology and Fertility of Soils* 29, 207–217.

Olofsson, J., Oksanen, L., Callaghan, T., Hulme, P.E., Oksanen, T. and Suominen, O. (2009) Herbivores inhibit climate-driven shrub expansion on the tundra. *Global Change Biology* 15, 2681–2693.

O'Malley, M.A. (2007) The nineteenth century roots of 'everything is everywhere'. *Nature Reviews Microbiology* 5, 647–651.

Ooi, M.K.J., Auld, T.D. and Denham, A.J. (2009) Climate change and bet-hedging: interactions between increased soil temperatures and seed bank persistence. *Global Change Biology* 15, 2375–2386.

Oren, R., Ellsworth, D.S., Johnsen, K.H., Phillips, N., Ewers, B.E., Maier, C., Schafer, K.V.R., McCarthy, H., Hendrey, G., McNulty, S.G. and Katul, G.G. (2001) Soil fertility limits carbon sequestration by forest ecosystems in a $CO_2$-enriched atmosphere. *Nature* 411, 469–472.

Örstan, A. (2009) Will assisted colonization be a viable option to save terrestrial gastropods threatened by climate change? *Tentacle* 17, 14–16.

Owen-Smith, N. (1987) Pleistocene extinctions: the pivotal role of megaherbivores. *Paleobiology* 13, 351–362.

Paine, R., Tegner, M. and Johnson, E. (1998) Compounded perturbations yield ecological surprises. *Ecosystems* 1, 535–545.

Parker, B.L., Skinner, M., Gouli, S., Ashikaga, T. and Teillon, H.B. (1999) Low lethal temperature for hemlock woolly adelgid (Homoptera: Adelgidae). *Environmental Entomology* 28, 1085–1091.

Parmesan, C. (2006) Ecological and evolutionary responses to recent climate change. *Annual Review of Ecology, Evolution, and Systematics* 37, 637–669.

Parmesan, C. (2007) Influences of species, latitudes and methodologies on estimates of phenological response to global warming. *Global Change Biology* 13, 1860–1872.

Parmesan, C. and Yohe, G. (2003) A globally coherent fingerprint of climate change impacts across natural systems. *Nature* 421, 37–42.

Parmesan, C., Ryrholm, N., Stefanescu, C., Hill, J., Thomas, C., Descimon, H., Huntley, B., Kaila, L., Kullberg, J. and Tammaru, T. (1999) Poleward shifts in geographical ranges of butterfly species associated with regional warming. *Nature* 399, 579–583.

Parolo, G. and Rossi, G. (2008) Upward migration of vascular plants following a climate warming trend in the Alps. *Basic and Applied Ecology* 9, 100–107.

Parry, M. (2007) The implications of climate change for crop yields, global food supply and risk of hunger. *SAT eJournal | ejournal.icrisat.org* 4, issue 1; available at http://www.icrisat.org/journal/Special Project/sp14.pdf

Parton, W.J., Stewart, J.W.B. and Cole, C.V. (1988) Dynamics of C, N, P and S in grassland soils: a model. *Biogeochemistry* 5, 109–131.

Parton, W.J., Morgan, J.A., Wang, G.M. and Del Grosso, S. (2007) Projected ecosystem impact of the prairie heating and $CO_2$ enrichment experiment. *New Phytologist* 174, 823–834.

Pearce, F. (2006) Climate: the great hockey stick debate. *New Scientist* 18 March, issue 2543, 40.

Pearson, R.G. (2006) Climate change and the migration capacity of species. *Trends in Ecology & Evolution* 21, 111–113.

Pearson, R.G. and Dawson, T.P. (2003) Predicting the impacts of climate change on the distribution of species: are bioclimate envelope models useful? *Global Ecology and Biogeography* 12, 361–371.

Pearson, R.G. and Dawson, T.P. (2005) Long-distance plant dispersal and habitat fragmentation: identifying conservation targets for spatial landscape planning under climate change. *Biological Conservation* 123, 389–401.

Pearson, R.G., Thuiller, W., Araujo, M.B., Martinez-Meyer, E., Brotons, L., McClean, C., Miles, L., Segurado, P., Dawson, T.P. and Lees, D.C. (2006) Model-based uncertainty in species range prediction. *Journal of Biogeography* 33, 1704–1711.

Pelini, S.L., Dzurisin, J.D.K., Prior, K.M., Williams, C.M., Marsico, T.D., Sinclair, B.J. and Hellmann, J.J. (2009) Translocation experiments with butterflies reveal limits to enhancement of poleward populations under climate change. *Proceedings of the National Academy of Sciences USA* 106, 11160–11165.

Peñuelas, J., Prieto, P., Beier, C., Cesaraccio, C., De Angelis, P., De Dato, G., Emmett, B.A., Estiarte, M., Garadnai, J., Gorissen, A., Lang, E.K., Kroel-Dulay, G., Llorens, L., Pellizzaro, G., Riis-Nielsen, T., Schmidt, I.K., Sirca, C., Sowerby, A., Spano, D. and Tietema, A. (2007) Response of plant species richness and primary productivity in shrublands along a north–south gradient in Europe to seven years of experimental warming and drought: reductions in primary productivity in the heat and drought year of 2003. *Global Change Biology* 13, 2563–2581.

Peñuelas, J., Hunt, J.M., Ogaya, R.A. and Jump, A.S. (2008) Twentieth century changes of tree-ring $\delta^{13}C$ at the southern range-edge of *Fagus sylvatica*: increasing water-use efficiency does not avoid the growth decline induced by warming at low altitudes. *Global Change Biology* 14, 1076–1088.

Percy, K.E., Awmack, C.S., Lindroth, R.L., Kubiske, M.E., Kopper, B.J., Isebrands, J.G., Pregitzer, K.S., Hendrey, G.R., Dickson, R.E., Zak, D.R., Oksanen, E., Sober, J., Harrington, R. and Karnosky, D.F. (2002) Altered performance of forest pests under atmospheres enriched by $CO_2$ and $O_3$. *Nature* 420, 403–407.

Pergams, O. and Ashley, M. (2001) Microevolution in island rodents. *Genetica* 112, 245–256.

Perry, D. and Borchers, J. (1990) Climate change and ecosystem responses. *Northwest Environmental Journal* 6, 293–313.

Peters, R. and Darling, J. (1985) The greenhouse effect and nature reserves. *BioScience* 35, 707–717.

Peterson, A., Ball, J., Luo, Y., Field, C., Reich, P., Curtis, P., Griffin, K., Gunderson, C., Norby, R. and Tissue, D. (1999) The photosynthesis–leaf nitrogen relationship at ambient and elevated atmospheric carbon dioxide: a meta-analysis. *Global Change Biology* 5, 331–346.

Petit, J.R., Jouzel, J., Raynaud, D., Barkov, N.I., Barnola, J.-M., Basile, I., Bender, M., Chappellaz, J., Davisk, M., Delaygue, G., Delmotte, M., Kotlyakov, V.M.,

Legrand, M., Lipenkov, V.Y., Lorius, C., Pe'Pin, L., Ritz, C., Saltzmank, E. and Stievenard, M. (1999) Climate and atmospheric history of the past 420,000 years from the Vostok ice core, Antarctica. *Nature* 399, 429–436.

Phillips, O.L., Aragão, L.E., Lewis, S.L., Fisher, J.B., Lloyd, J., López-González, G., Malhi, Y., Monteagudo, A., Peacock, J., Quesada, C.A., van der Heijden, G., Almeida, S., Amaral, I., Arroyo, L., Aymard, G., Baker, T.R., Bánki, O., Blanc, L., Bonal, D., Brando, P., Chave, J., de Oliveira, A.C., Cardozo, N.D., Czimczik, C.I., Feldpausch, T.R., Freitas, M.A., Gloor, E., Higuchi, N., Jiménez, E., Lloyd, G., Meir, P., Mendoza, C., Morel, A., Neill, D.A., Nepstad, D., Patiño, S., Peñuela, M.C., Prieto, A., Ramírez, F., Schwarz, M., Silva, J., Silveira, M., Thomas, A.S., Steege, H.T., Stropp, J., Vásquez, R., Zelazowski, P., Alvarez Dávila, E., Andelman, S., Andrade, A., Chao, K.J., Erwin, T., Di Fiore, A., Honorio, C.E., Keeling, H., Killeen, T.J., Laurance, W.F., Peña Cruz, A., Pitman, N.C., Núñez Vargas, P., Ramírez-Angulo, H., Rudas, A., Salamão, R., Silva, N., Terborgh, J. and Torres-Lezama, A. (2009) Drought sensitivity of the Amazon rainforest. *Science* 323, 1344–1346.

Pickett, S.T.A. and White, P.S. (1985) Patch dynamics: a synthesis. In: Pickett, S.T.A. and Whitem, P.S. (eds) *The Ecology Of Natural Disturbance and Patch Dynamics*. Academic Press, New York, pp. 371–384

Pigliucci, M. (2002) *Denying Evolution: Creationism, Scientism and the Nature of Science*. Sinauer Associates, Sunderland, Massachusetts.

Pinto, D., Blande, J., Nykänen, R., Dong, W., Nerg, A. and Holopainen, J. (2007) Ozone degrades common herbivore-induced plant volatiles: does this affect herbivore prey location by predators and parasitoids? *Journal of Chemical Ecology* 33, 683–694.

Piou, C., Berger, U., Hildenbrandt, H. and Feller, I. (2008) Testing the intermediate disturbance hypothesis in species-poor systems: a simulation experiment for mangrove forests. *Journal of Vegetation Science* 19, 417–424.

Plum, N.M. and Filser, J. (2005) Floods and drought: response of earthworms and potworms (Oligochaeta: Lumbricidae, Enchytraeidae) to hydrological extremes in wet grassland. *Pedobiologia* 49, 443–453.

Pollack, H.N., Huang, S. and Shen, P.-Y. (1998) Climate change record in subsurface temperatures: a global perspective. *Science* 282, 279–281.

Poorter, H. and Navas, M. (2003) Plant growth and competition at elevated $CO_2$: on winners, losers and functional groups. *New Phytologist* 157, 175–198.

Poorter, H. and Pérez-Soba, M. (2001) The growth response of plants to elevated $CO_2$ under non-optimal environmental conditions. *Oecologia* 129, 1–20.

Poorter, H., Van Berkel, T., Baxter, R., Den Hertog, J., Dijkstra, P. and Gifford, R.M. (1997) The effect of elevated $CO_2$ on the chemical composition and construction costs of leaves of 27 C$_3$ plants. *Plant, Cell & Environment* 20, 472–482.

Pope, V. (2007) Models 'key to climate forecasts'. http://newsbbc.co.uk/2/hi/science/nature/6320515.stm (accessed 5 June 2010).

Post, E. and Forchhammer, M. (2008) Climate change reduces reproductive success of an arctic herbivore through trophic mismatch. *Philosophical Transactions of the Royal Society of London. Series B: Biological Sciences* 363, 2367–2375.

Post, E. and Pedersen, C. (2008) Opposing plant community responses to warming with and without herbivores. *Proceedings of the National Academy of Sciences USA* 105, 12353–12358.

Pounds, J.A. and Crump, M.L. (1994) Amphibian declines and climate disturbance: the case of the golden toad and the harlequin frog. *Conservation Biology* 8, 72–85.

Pounds, J.A., Fogden, M.P.L., Savage, J.M. and Gorman, G.C. (1997) Tests of null models for amphibian declines on a tropical mountain. *Conservation Biology* 11, 1307–1322.

Pounds, J.A., Fogden, M.P.L. and Campbell, J.H. (1999) Biological response to climate change on a tropical mountain. *Nature* 398, 611–615.

Pounds, J.A., Bustamante, M.R., Coloma, L., Consuegra, J.A., Fogden, M., Foster, P.N., La Marca, E., Masters, K.L., Merino-Viteri, A., Puschendorf, R., Sanchez-Azofeifa, G.A., Still, C.J., Ron, S.R. and Young, B. (2006) Widespread amphibian extinctions from epidemic disease driven by global warming. *Nature* 439, 161–167.

Prideaux, G., Roberts, R., Megirian, D., Westaway, K., Hellstrom, J. and Olley, J. (2007) Mammalian responses to pleistocene climate change in south-eastern Australia. *Geology* 35, 33–36

Prieto, P., Peñuelas, J., Lloret, F., Llorens, L. and Estiarte, M. (2009) Experimental drought and warming decrease diversity and slow down post-fire succession in a Mediterranean shrubland. *Ecography* 32, 623–636.

Primo, A.L., Azeiteiro, U.M., Marques, S.C., Martinho, F. and Pardal, M.A. (2009) Changes in zooplankton diversity and distribution pattern under varying precipitation regimes in a southern temperate estuary. *Estuarine Coastal and Shelf Science* 82, 341–347.

Prospero, J.M. (1979) Mineral and sea salt aerosol concentrations in various ocean regions. *Journal of Geophysical Research* 84, 725–731.

Pyke, C. (2004) Habitat loss confounds climate change impacts. *Frontiers in Ecology and the Environment* 2, 178–182.

Pyke, C. (2005) Interactions between habitat loss and climate change: implications for fairy shrimp in the central valley ecoregion of California, USA. *Climatic Change* 68, 199–218.

Pyke, C. and Marty, J. (2005) Cattle grazing mediates climate change impacts on ephemeral wetlands. *Conservation Biology* 19, 1619–1625.

Rahel, F. and Olden, J. (2008) Assessing the effects of climate change on aquatic invasive species. *Conservation Biology* 22, 521–533.

Ramankutty, N. and Foley, J.A. (1998) Characterizing patterns of global land use: an analysis of global croplands data. *Global Biogeochemical Cycles* 12, 667–685.

Randall, D.A., Wood, R.A., Bony, S., Colman, R., Fichefet, T., Fyfe, J., Kattsov, V., Pitman, A., Shukla, J., Srinivasan, J., Stouffer, R.J., Sumi, A. and Taylor, K.E. (2007) Cilmate models and their evaluation. In: Solomon, S., Qin, D., Manning, M., Chen, Z., Marquis, M., Averyt, K.B., Tignor, M. and Miller, H.L. (eds) *Climate Change 2007: The Physical Science Basis. Contribution of Working Group I to the Fourth Assessment Report of the Intergovernmental Panel on Climate Change*. Cambridge University Press, Cambridge, UK, pp. 589–663.

Randerson, J.T., Chapin, F.S., Harden, J.W., Neff, J.C. and Harmon, M.E. (2002) Net ecosystem production: a comprehensive measure of net carbon accumulation by ecosystems. *Ecological Applications* 12, 937–947.

Rasmussen, L., Beier, C. and Bergstedt, A. (2002) Experimental manipulations of old pine forest ecosystems to predict the potential tree growth effects of increased CO$_2$ and temperature in a future climate. *Forest Ecology and Management* 158, 179–188.

Raup, D. (1986) Biological extinction in earth history. *Science* 231, 1528–1533.

Raup, D. (1992) Large-body impact and extinction in the phanerozoic. *Paleobiology* 18, 80–88.

Raup, D. and Sepkoski, J. Jr (1982) Mass extinctions in the marine fossil record. *Science* 215, 1501–1503.

Rea, D.K., Leinen, M. and Janecek, T.R. (1985) Geologic approach to the long-term history of atmospheric circulation. *Science* 227, 721–725.

Réale, D., McAdam, A., Boutin, S. and Berteaux, D. (2003) Genetic and plastic responses of a northern mammal to climate change. *Proceedings. Biological Sciences/The Royal Society* 270, 591–596.

Regehr, E., Hunter, C., Caswell, H., Amstrup, S. and Stirling, I. (2009) Survival and breeding of polar bears in the southern Beaufort Sea in relation to sea ice. *Journal of Animal Ecology* 79, 117–127.

Reich, P.B., Hobbie, S.E., Lee, T., Ellsworth, D.S., West, J.B., Tilman, D., Knops, J.M.H., Naeem, S. and Trost, J. (2006) Nitrogen limitation constrains sustainability of ecosystem response to CO$_2$. *Nature* 440, 922–925.

Rejmanek, M. and Richardson, D. (1996) What attributes make some plant species more invasive? *Ecology* 77, 1655–1661.

Reusch, T. and Wood, T. (2007) Molecular ecology of global change. *Molecular Ecology* 16, 3973–3992.

Reusch, T., Ehlers, A., Hämmerli, A. and Worm, B. (2005) Ecosystem recovery after climatic extremes enhanced by genotypic diversity. *Proceedings of the National Academy of Sciences USA* 102, 2826–2831.

Reyes, A.V., Wiles, G.C., Smith, D.J., Barclay, D.J., Allen, S., Jackson, S., Larocque, S., Laxton, S., Lewis, D., Calkin, P.E. and Clague, J.J. (2006) Expansion of alpine glaciers in Pacific North America in the first millennium AD. *Geology* 34, 57–60.

Rhemtulla, J., Hall, R., Higgs, E. and MacDonald, S. (2002) Eighty years of change: vegetation in the montane ecoregion of Jasper National Park, Alberta, Canada. *Canadian Journal of Forest Research* 32, 2010–2021.

Ricciardi, A. and Simberloff, D. (2009a) Assisted colonization is not a viable conservation strategy. *Trends in Ecology & Evolution* 24, 248–253.

Ricciardi, A. and Simberloff, D. (2009b) Assisted colonization: good intentions and dubious risk assessment. *Trends in Ecology & Evolution* 24, 476–477.

Richardson, S.J., Press, M.C., Parsons, A.N. and Hartley, S.E. (2002) How do nutrients and warming impact on plant communities and their insect herbivores? A 9-year study from a sub-Arctic heath. *Journal of Ecology* 90, 544–556.

Rillig, M.C. and Allen, M.F. (1999) What is the role of arbuscular mycorrhizal fungi in plant-to-ecosystem responses to elevated atmospheric $CO_2$? *Mycorrhiza* 9, 1–8.

Rillig, M.C., Wright, S.F., Shaw, M.R. and Field, C.B. (2002) Artificial climate warming positively affects arbuscular mycorrhizae but decreases soil aggregate water stability in an annual grassland. *Oikos* 97, 52–58.

Rinnan, R., Michelsen, A., Baath, E. and Jonasson, S. (2007) Fifteen years of climate change manipulations alter soil microbial communities in a subarctic heath ecosystem. *Global Change Biology* 13, 28–39.

Rinnan, R., Michelsen, A. and Jonasson, S. (2008) Effects of litter addition and warming on soil carbon, nutrient pools and microbial communities in a subarctic heath ecosystem. *Applied Soil Ecology* 39, 271–281.

Rinnan, R., Rousk, J., Yergeau, E., Kowalchuk, G.A. and Baath, E. (2009a) Temperature adaptation of soil bacterial communities along an Antarctic climate gradient: predicting responses to climate warming. *Global Change Biology* 15, 2615–2625.

Rinnan, R., Stark, S. and Tolvanen, A. (2009b) Responses of vegetation and soil microbial communities to warming and simulated herbivory in a subarctic heath. *Journal of Ecology* 97, 788–800.

Ripley, B.S., Gilbert, M.E., Ibrahim, D.G. and Osborne, C.P. (2007) Drought constraints on $C_4$ photosynthesis: stomatal and metabolic limitations in $C_3$ and $C_4$ subspecies of *Alloteropsis semialata*. *Journal of Experimental Botany* 58, 1351–1363.

Roberts, R., Flannery, T., Ayliffe, L., Yoshida, H., Olley, J., Prideaux, G., Laslett, G., Baynes, A., Smith, M. and Jones, R. (2001) New ages for the last Australian megafauna: continent-wide extinction about 46,000 years ago. *Science* 292, 1888–1892.

Robertson, T.R., Zak, J.C. and Tissue, D.T. (2010) Precipitation magnitude and timing differentially affect species richness and plant density in the sotol grassland of the Chihuahuan Desert. *Oecologia* 162, 185–197.

Robinson, C.H., Wookey, P.A., Lee, J.A., Callaghan, T.V. and Press, M.C. (1998) Plant community responses to simulated environmental change at a high arctic polar semi-desert. *Ecology* 79, 856–866.

Robinson, R.A., Crick, H.Q.P., Learmonth, J.A., MacLean, I.M.D., Thomas, C.D., Bairlein, F., Forchhammer, M.C., Francis, C.M., Gill, J.A., Godley, B.J., Harwood, J., Hays, G.C., Huntley, B., Hutson, A.M., Pierce, G.J., Rehfisch, M.M., Sims, D.W., Begona Santos, M., Sparks, T.H., Stroud, D.A. and Visser, M.E. (2009) Travelling through a warming world: climate change and migratory species. *Endangered Species Research* 7, 87–99.

Rodenhouse, N., Christenson, L., Parry, D. and Green, L. (2009) Climate change effects on native fauna of northeastern forests. *Canadian Journal of Forest Research* 39, 249–263.

Rodriguez-Zaragoza, S. (1994) Ecology of free-living amebas. *Critical Reviews in Microbiology* 20, 225–241.

Rogers, A., Ainsworth, E.A. and Leakey, A.D.B. (2009) Will elevated carbon dioxide concentration amplify the benefits of nitrogen fixation in legumes? *Plant Physiology* 151, 1009–1016.

Ronn, R., Ekelund, F. and Christensen, S. (2003) Effects of elevated atmospheric $CO_2$ on protozoan abundance in soil planted with wheat and on decomposition of wheat roots. *Plant and Soil* 251, 13–21.

Root, T., Price, J., Hall, K., Schneider, S., Rosenzweig, C. and Pounds, J. (2003) Fingerprints of global warming on wild animals and plants. *Nature* 421, 57–60.

Rose, G. (2005) Capelin (*Mallotus villosus*) distribution and climate: a sea 'canary' for marine ecosystem change. *ICES Journal of Marine Science* 62, 1524–1530.

Rosenzweig, C. and Hillel, D. (1998) *Climate Change and the Global Harvest*. Oxford University Press, Oxford, UK.

Rosenzweig, C. and Iglesias, A. (2006) Potential impacts of climate change on world food supply: data sets from a major crop modeling study. http://sedac.ciesin.columbia.edu/giss_crop_study/ (accessed 30 October 2009).

Rosenzweig, C. and Tubiello, F. (2007) Adaptation and mitigation strategies in agriculture: an analysis of potential synergies. *Mitigation and Adaptation Strategies for Global Change* 12, 855–873.

Rosenzweig, C., Parry, M., Fischer, G. and Frohberg, K. (1993) *Climate Change and World Food Supply. Research Report No. 3*. University of Oxford, Environmental Change Unit, Oxford, UK.

Roth, V. (1992) Inferences from allometry and fossils: dwarfing of elephants on islands. *Oxford Surveys in Evolutionary Biology* 8, 259–259.

Roy, B.A., Gusewell, S. and Harte, J. (2004) Response of plant pathogens and herbivores to a warming experiment. *Ecology* 85, 2570–2581.

Ruddiman, W. (2001) *Earth's Climate: Past and Future.* W.H. Freeman and Co., New York.

Ruess, L., Michelsen, A. and Jonasson, S. (1999a) Simulated climate change in subarctic soils: responses in nematode species composition and dominance structure. *Nematology* 1, 513–526.

Ruess, L., Michelsen, A., Schmidt, I.K. and Jonasson, S. (1999b) Simulated climate change affecting microorganisms, nematode density and biodiversity in subarctic soils. *Plant and Soil* 212, 63–73.

Running, S. (2008) Ecosystem disturbance, carbon, and climate. *Science* 321, 652–653.

Rupp, T., Chapin, F.S. and Starfield, A. (2001) Modeling the influence of topographic barriers on treeline advance at the forest–tundra ecotone in northwestern Alaska. *Climatic Change* 48, 399–416.

Rupp, T., Starfield, A., Chapin, F.S. and Duffy, P. (2002) Modeling the impact of black spruce on the fire regime of Alaskan boreal forest. *Climatic Change* 55, 213–233.

Rusek, J. (1998) Biodiversity of collembola and their functional role in the ecosystem. *Biodiversity and Conservation* 7, 1207–1219.

Rustad, L. (2008) The response of terrestrial ecosystems to global climate change: towards an integrated approach. *Science of the Total Environment* 404, 222–235.

Rustad, L.E., Campbell, J.L., Marion, G.M., Norby, R.J., Mitchell, M.J., Hartley, A.E., Cornelissen, J.H.C. and Gurevitch, J. (2001) A meta-analysis of the response of soil respiration, net nitrogen mineralization, and aboveground plant growth to experimental ecosystem warming. *Oecologia* 126, 543–562.

Ryan, G., Rasmussen, S. and Newman, J.A. (2010) Global atmospheric change and trophic interactions: are there any general responses? In: Ninkovic, V. (ed.) *Plant Communication From an Ecological Perspective.* Springer-Verlag, Berlin, pp. 179–214.

Ryan, M. (1991) Effects of climate change on plant respiration. *Ecological Applications* 1, 157–167.

Sadowsky, M.J. and Schortemeyer, M. (1997) Soil microbial responses to increased concentrations of atmospheric $CO_2$. *Global Change Biology* 3, 217–224.

Saether, B., Tufto, J., Engen, S., Jerstad, K., Rostad, O. and Skatan, J. (2000) Population dynamical consequences of climate change for a small temperate songbird. *Science* 287, 854–856.

Safranyik, L. and Linton, D. (1991) Unseasonably low fall and winter temperatures affecting mountain pine beetle and pine engraver beetle populations and damage in the British Columbia Chilcotin region. *Journal of the Entomological Society of British Columbia* 88, 17–21.

Safranyik, L. and Linton, D. (1998) Mortality of mountain pine beetle larvae, *Dendroctonus ponderosae* (Coleoptera: Scolytidae) in logs of lodgepole pine (*Pinus contorta* var. *latifolia*) at constant low temperatures. *Journal of the Entomological Society of British Columbia* 95, 81–87.

Sagan, C. and Mullen, G. (1972) Earth and Mars: evolution of atmospheres and surface temperatures. *Science* 177, 52–56.

Sage, R.F. (1996) Atmospheric modification and vegetation responses to environmental stress. *Global Change Biology* 2, 79–83.

Sage, R.F. (2004) The evolution of $C_4$ photosynthesis. *New Phytologist* 161, 341–370.

Saitoh, T., Cazelles, B., Vik, J., Viljugrein, H. and Stenseth, N. (2006) Effects of regime shifts on the population dynamics of the grey-sided vole in Hokkaido, Japan. *Climate Research* 32, 109–118.

Sala, O.E., Chapin, F.S., Armesto, J.J., Berlow, E., Bloomfield, J., Dirzo, R., Huber-Sanwald, E., Huenneke, L.F., Jackson, R.B., Kinzig, A., Leemans, R., Lodge, D.M., Mooney, H.A., Oesterheld, M., Poff, N.L., Sykes, M.T., Walker, B.H., Walker, M. and Wall, D.H. (2000) Global biodiversity scenarios for the year 2100. *Science* 287, 1770–1774.

Saleska, S.R., Harte, J. and Torn, M.S. (1999) The effect of experimental ecosystem warming on $CO_2$ fluxes in a montane meadow. *Global Change Biology* 5, 125–141.

Salinari, F., Giosu, S., Tubiello, F., Rettori, A., Rossi, V., Spanna, F., Rosenzweig, C. and Gullino, M.L. (2006) Downy mildew (*Plasmopara viticola*) epidemics on grapevine under climate change. *Global Change Biology* 12, 1299–1307.

Salt, D.T., Fenwick, P. and Whittaker, J.B. (1996) Interspecific herbivore interactions in a high $CO_2$ environment: root and shoot aphids feeding on cardamine. *Oikos* 77, 326–330.

Salzer, M.W., Hughes, M.K., Bunn, A.G. and Kipfmueller, K.F. (2009) Recent unprecedented tree-ring growth in bristlecone pine at the highest elevations and possible causes. *Proceedings of the National Academy of Sciences USA* 106, 20348–20353.

Sarmiento, J. and Toggweiler, J. (1984) A new model for the role of the oceans in determining atmospheric $pCO_2$. *Nature* 308, 621–624.

Sasaki, T., Okubo, S., Okayasu, T., Jamsran, U., Ohkuro, T. and Takeuchi, K. (2009) Management applicability of the intermediate disturbance hypothesis across Mongolian rangeland ecosystems. *Ecological Applications* 19, 423–432.

Sax, D. and Gaines, S. (2003) Species diversity: from global decreases to local increases. *Trends in Ecology & Evolution* 18, 561–566.

Sax, D. and Gaines, S. (2008) Species invasions and extinction: the future of native biodiversity on islands. *Proceedings of the National Academy of Sciences USA* 105, 11490–11497.

Sax, D., Smith, K. and Thompson, A. (2009) Managed relocation: a nuanced evaluation is needed. *Trends in Ecology & Evolution* 24, 472–473.

Schaphoff, S., Lucht, W., Gerten, D., Sitch, S., Cramer, W. and Prentice, I.C. (2006) Terrestrial biosphere carbon storage under alternative climate projections. *Climatic Change* 74, 97–122.

Scheller, R.M. and Mladenoff, D.J. (2005) A spatially interactive simulation of climate change, harvesting, wind, and tree species migration and projected changes to forest composition and biomass in northern Wisconsin, USA. *Global Change Biology* 11, 307–321.

Schiel, D.R., Steinbeck, J.R. and Foster, M.S. (2004) Ten years of induced ocean warming causes comprehensive changes in marine benthic communities. *Ecology* 85, 1833–1839.

Schiermeier, Q. (2009) Ocean fertilization: dead in the water? Study casts doubt on iron-induced carbon sequestration. *Nature* 457, 520–521.

Schmidhuber, J. and Tubiello, F. (2007) Global food security under climate change. *Proceedings of the National Academy of Sciences USA* 104, 19703–19708.

Schmidt, N. and Jensen, P. (2003) Changes in mammalian body length over 175 years – adaptations to a fragmented landscape? *Conservation Ecology* 7, 6, http://www.consecol.org/vol7/iss2/art6/.

Schnyder, H., Mächler, F. and Nösberger, J. (1984) Influence of temperature and $O_2$ concentration on photosynthesis and light activation of ribulosebisphosphate carboxylase oxygenase in intact leaves of white clover (*Trifolium repens* L.). *Journal of Experimental Botany* 35, 147–156.

Schortemeyer, M., Hartwig, U.A., Hendrey, G.R. and Sadowsky, M.J. (1996) Microbial community changes in the rhizospheres of white clover and perennial ryegrass exposed to free air carbon dioxide enrichment (FACE). *Soil Biology & Biochemistry* 28, 1717–1724.

Schuur, E.A.G., Bockheim, J., Canadell, J.G., Euskirchen, E., Field, C.B., Goryachkin, S.V., Hagemann, S., Kuhry, P., Lafleur, P.M., Lee, H., Mazhitova, G., Nelson, F.E., Rinke, A., Romanovsky, V.E., Shiklomanov, N., Tarnocai, C., Venevsky, S., Vogel, J.G. and Zimov, S.A. (2008) Vulnerability of permafrost carbon to climate change: implications for the global carbon cycle. *BioScience* 58, 701–714.

Schwartz, M., Ahas, R. and Aasa, A. (2006) Onset of spring starting earlier across the northern hemisphere. *Global Change Biology* 12, 343–351.

Schweiger, O., Biesmeijer, J.C., Bommarco, R., Hickler, T., Hulme, P.E., Klotz, S., Kühn, I., Moora, M., Nielsen, A., Ohlemüller, R., Petanidou, T., Potts, S.G., Pyšek, P., Stout, J.C., Sykes, M.T., Tscheulin, T., Vilà, M., Walther, G.-R., Westphal, C., Winter, M., Zobel, M. and Settele, J. (2010) Multiple stressors on biotic interactions: how climate change and alien species interact to affect pollination. *Biological Reviews* 85, 777–795.

Segurado, P. and Araujo, M.B. (2004) An evaluation of methods for modelling species distributions. *Journal of Biogeography* 31, 1555–1568.

Sekercioglu, C., Schneider, S., Fay, J. and Loarie, S. (2008) Climate change, elevational range shifts, and bird extinctions. *Conservation Biology* 22, 140–150.

Shafer, S., Bartlein, P. and Thompson, R. (2001) Potential changes in the distributions of western North America tree and shrub taxa under future climate scenarios. *Ecosystems* 4, 200–215.

Shaver, G.R., Canadell, J., Chapin, F.S., Gurevitch, J., Harte, J., Henry, G., Ineson, P., Jonasson, S., Melillo, J., Pitelka, L. and Rustad, L. (2000) Global warming and terrestrial ecosystems: a conceptual framework for analysis. *BioScience* 50, 871–882.

Shaw, J. (1989) *Heavy Metal Tolerance in Plants: Evolutionary Aspects*. CRC Press, Boca Raton, Florida.

Shaw, M.R., Zavaleta, E.S., Chiariello, N.R., Cleland, E.E., Mooney, H.A. and Field, C.B. (2002) Grassland responses to global environmental changes suppressed by elevated $CO_2$. *Science* 298, 1987–1990.

Sheldon, N.D. (2006) Precambrian paleosols and atmospheric $CO_2$ levels. *Precambrian Research* 147, 148–155.

Shen, K. and Harte, J. (eds) (2000) *Ecosystem Climate Manipulations*. Springer-Verlag, Berlin.

Shi, L.B., Guttenberger, M., Kottke, I. and Hampp, R. (2002) The effect of drought on mycorrhizas of beech (*Fagus sylvatica* L.): changes in community structure, and the content of carbohydrates and nitrogen storage bodies of the fungi. *Mycorrhiza* 12, 303–311.

Shirey, P. and Lamberti, G. (2009) Assisted colonization under the US Endangered Species Act. *Conservation Letters* 3, 45–52.

Siegenthaler, U., Stocker, T.F., Monnin, E., Lüthi, D., Schwander, J., Stauffer, B., Raynaud, D., Barnola, J.-M., Fischer, H., Masson-Delmotte, V. and Jouzel, J. (2005) Stable carbon cycle–climate relationship during the Late Pleistocene. *Science* 310, 1313–1317.

Silliman, B.R., Van De Koppel, J., Bertness, M.D., Stanton, L.E. and Mendelssohn, I.A. (2005) Drought, snails, and large-scale die-off of southern US salt marshes. *Science* 310, 1803–1806.

Silva, L.C.R., Anand, M., Oliveria, J.M. and Pillar, V.D. (2009) Past century changes in *Araucaria angustifolia* (Bertol.) Kuntze water use efficiency and growth in forest and grassland ecosystems of southern Brazil: implications for forest expansion. *Global Change Biology* 15, 2387–2396.

Silva, L.C.R., Anand, M. and Leithead, M.D. (2010) Recent widespread tree growth decline despite increasing atmospheric $CO_2$. *PLoS One* 5, e11543.

Silvertown, J., Poulton, P., Johnston, E., Edwards, G., Heard, M. and Biss, P. (2006) The Park Grass Experiment 1856–2006: its contribution to ecology. *Journal of Ecology* 94, 801–814.

Simmons, A. and Thomas, C. (2004) Changes in dispersal during species' range expansions. *The American Naturalist* 164, 378–395.

Simmons, B.L., Wall, D.H., Adams, B.J., Ayres, E., Barrett, J.E. and Virginia, R.A. (2009) Long-term experimental warming reduces soil nematode populations in the McMurdo Dry Valleys, Antarctica. *Soil Biology & Biochemistry* 41, 2052–2060.

Sinclair, B.J. (2002) Effects of increased temperatures simulating climate change on terrestrial invertebrates on Ross Island, Antarctica. *Pedobiologia* 46, 150–160.

Singer, M. and Thomas, C. (1996) Evolutionary responses of a butterfly metapopulation to human- and climate-caused environmental variation. *The American Naturalist* 148, 9–39.

Sjursen, H., Michelsen, A. and Jonasson, S. (2005a) Effects of long-term soil warming and fertilisation on microarthropod abundances in three sub-Arctic ecosystems. *Applied Soil Ecology* 30, 148–161.

Sjursen, H.S., Michelsen, A. and Holmstrup, M. (2005b) Effects of freeze–thaw cycles on microarthropods and nutrient availability in a sub-Arctic soil. *Applied Soil Ecology* 28, 79–93.

Skelly, D., Joseph, L., Possingham, H., Freidenburg, L., Farrugia, T., Kinnison, M. and Hendry, A. (2007) Evolutionary responses to climate change. *Conservation Biology* 21, 1353–1355.

Skerratt, L., Berger, L., Speare, R., Cashins, S., McDonald, K., Phillott, A., Hines, H. and Kenyon, N. (2007) Spread of chytridiomycosis has caused the rapid global decline and extinction of frogs. *EcoHealth* 4, 125–134.

Smith, J.R., Fong, P. and Ambrose, R.F. (2006) Dramatic declines in mussel bed community diversity: response to climate change? *Ecology* 87, 1153–1161.

Smith, M., Seabloom, E., Corvallis, O., Meche, G., Lafayette, L., Knops, J., Ritchie, L. and Syracuse, N. (2007) Does species diversity limit productivity in natural grassland communities? *Ecology Letters* 10, 680–689.

Smith, M., Knapp, A. and Collins, S. (2009) A framework for assessing ecosystem dynamics in response to chronic resource alterations induced by global change. *Ecology* 90, 3279–3289.

Smol, J.P., Last, W.M. and Birks, H.B. (2002a) *Tracking Environmental Change Using Lake Sediments: Terrestrial, Algal, and Siliceous Indicators.* Kluwer Academic Publishers, Dordrecht, The Netherlands.

Smol, J.P., Last, W.M. and Birks, H.B. (2002b) *Tracking Environmental Change Using Lake Sediments: Zoological Indicators.* Kluwer Academic Publishers, Dordrecht, The Netherlands.

Smol, J.P., Wolfe, A.P., Birks, H.J.B., Douglas, M.S.V., Jones, V.J., Korhola, A., Pienitz, R., Rühland, K., Sorvari, S., Antoniades, D., Brooks, S.J., Fallu, M.-A., Hughes, M., Keatley, B.E., Laing, T.E., Michelutti, N.,

Nazarova, L., Nyman, M., Paterson, A.M., Perren, B., Quinlan, R., Rautio, M., Saulnier-Talbot, É., Siitonen, S., Solovieva, N. and Weckström, J. (2005) Climate-driven regime shifts in the biological communities of arctic lakes. *Proceedings of the National Academy of Sciences USA* 102, 4397–4402.

Smulders, M., Cobben, M., Arens, P. and Verboom, J. (2009) Landscape genetics of fragmented forests: anticipating climate change by facilitating migration. *iForest: Biogeosciences and Forestry* 2, 128–132.

Snaydon, R. (1970) Rapid population differentiation in a mosaic environment. I. The response of *Anthoxanthum odoratum* populations to soils. *Evolution* 24, 257–269.

Sohlenius, B. and Bostrom, S. (1999) Effects of global warming on nematode diversity in a Swedish tundra soil – a soil transplantation experiment. *Nematology* 1, 695–709.

Sohngen, B., Mendelsohn, R. and Sedjo, R.A. (2001) A global model of climate change impacts on timber markets. *Journal of Agricultural and Resource Economics* 26, 326–343.

Sonnemann, I. and Wolters, V. (2005) The microfood web of grassland soils responds to a moderate increase in atmospheric $CO_2$. *Global Change Biology* 11, 1148–1155.

Sowerby, A., Emmett, B., Beier, C., Tietema, A., Peñuelas, J., Estiarte, M., Van Meeteren, M.J.M., Hughes, S. and Freeman, C. (2005) Microbial community changes in heathland soil communities along a geographical gradient: interaction with climate change manipulations. *Soil Biology & Biochemistry* 37, 1805–1813.

Speer, J.H. (2010) *Fundamentals of Tree Ring Research.* University of Arizona Press, Tucson, Arizona.

Springer, C. and Ward, J. (2007) Flowering time and elevated atmospheric $CO_2$. *New Phytologist* 176, 243–255.

Srivastava, D.S. and Vellend, M. (2005) Biodiversity–ecosystem function research: is it relevant to conservation? *Annual Review of Ecology, Evolution, and Systematics* 36, 267–294.

St Clair, S., Sharpe, W. and Lynch, J. (2008) Key interactions between nutrient limitation and climatic factors in temperate forests: a synthesis of the sugar maple literature. *Canadian Journal of Forest Research* 38, 401–414.

St Clair, S., Sudderth, E., Castanha, C., Torn, M. and Ackerly, D. (2009) Plant responsiveness to variation in precipitation and nitrogen is consistent across the compositional diversity of a California annual grassland. *Journal of Vegetation Science* 20, 860–870.

Staddon, P.L. and Fitter, A.H. (1998) Does elevated atmospheric carbon dioxide affect arbuscular mycorrhizas? *Trends in Ecology & Evolution* 13, 455–458.

Staddon, P.L., Gregersen, R. and Jakobsen, I. (2004a) The response of two *Glomus* mycorrhizal fungi and a fine endophyte to elevated atmospheric $CO_2$, soil warming and drought. *Global Change Biology* 10, 1909–1921.

Staddon, P.L., Jakobsen, I. and Blum, H. (2004b) Nitrogen input mediates the effect of free-air $CO_2$ enrichment on mycorrhizal fungal abundance. *Global Change Biology* 10, 1678–1688.

Staley, J.T., Mortimer, S.R., Morecroft, M.D., Brown, V.K. and Masters, G.J. (2007) Summer drought alters plant-mediated competition between foliar- and root-feeding insects. *Global Change Biology* 13, 866–877.

Staley, J.T., Mortimer, S.R. and Morecroft, M.D. (2008) Drought impacts on above–belowground interactions: do effects differ between annual and perennial host species? *Basic and Applied Ecology* 9, 673–681.

Stamp, N. (2003) Out of the quagmire of plant defense hypotheses. *The Quarterly Review of Biology* 78, 23–55.

Stearns, S. (1989) The evolutionary significance of phenotypic plasticity. *BioScience* 39, 436–445.

Stefansdottir, L., Solmundsson, J., Marteinsdottir, G., Kristinsson, K. and Jonasson, J.P. (2010) Groundfish species diversity and assemblage structure in Icelandic waters during recent years of warming. *Fisheries Oceanography* 19, 42–62.

Stevens, C., Thompson, K., Grime, J., Long, C. and Gowing, D. (2010) Contribution of acidification and eutrophication to declines in species richness of calcifuge grasslands along a gradient of atmospheric nitrogen deposition. *Functional Ecology* 24, 478–484.

Stewart, C. (2004) *Genetically Modified Planet: Environmental Impacts of Genetically Engineered Plants.* Oxford University Press, New York.

Stewart, M. (1995) Climate driven population fluctuations in rain forest frogs. *Journal of Herpetology* 29, 437–446.

Sticht, C., Schrader, S., Giesemann, A. and Weigel, H.J. (2006) Effects of elevated atmospheric $CO_2$ and N fertilization on abundance, diversity and C-isotopic signature of collembolan communities in arable soil. *Applied Soil Ecology* 34, 219–229.

Sticht, C., Schrader, S., Giesemann, A. and Weigel, H.J. (2009) Sensitivity of nematode feeding types in arable soil to free air $CO_2$ enrichment (FACE) is crop specific. *Pedobiologia* 52, 337–349.

Stiling, P. and Cornelissen, T. (2007) How does elevated carbon dioxide ($CO_2$) affect plant–herbivore interactions? A field experiment and meta-analysis of $CO_2$-mediated changes on plant chemistry and herbivore performance. *Global Change Biology* 13, 1823–1842.

Stiling, P., Rossi, A.M., Hungate, B., Dijkstra, P. and Hinkle, C.R. (1999) Decreased leaf-miner abundance in elevated $CO_2$: reduced leaf quality and increased parasitoid attack. *Ecological Applications* 9, 240–244.

Still, C., Foster, P. and Schneider, S. (1999) Simulating the effects of climate change on tropical montane cloud forests. *Nature* 398, 608–610.

Stocklin, J., Schweizer, K. and Korner, C. (1998) Effects of elevated $CO_2$ and phosphorus addition on productivity and community composition of intact monoliths from calcareous grassland. *Oecologia* 116, 50–56.

Stohlgren, T., Owen, A. and Lee, M. (2000) Monitoring shifts in plant diversity in response to climate change: a method for landscapes. *Biodiversity and Conservation* 9, 65–86.

Stokes, M.A. and Smiley, T.L. (1968) *An Introduction to Tree-Ring Dating.* University of Chicago Press, Chicago. Illinois.

Strecker, A.L., Cobb, T.P. and Vinebrooke, R.D. (2004) Effects of experimental greenhouse warming on phytoplankton and zooplankton communities in fishless alpine ponds. *Limnology and Oceanography* 49, 1182–1190.

Strom, A., Francis, R.C., Mantua, N.J., Miles, E.L. and Peterson, D.L. (2004) North Pacific climate recorded in growth rings of geoduck clams: a new tool for paleoenvironmental reconstruction. *Geophysical Research Letters* 31, L06206; doi:10.1029/2004GL019440.

Su, H.X., Sang, W.G., Wang, Y.X. and Ma, K.P. (2007) Simulating *Picea schrenkiana* forest productivity under climatic changes and atmospheric $CO_2$ increase in Tianshan Mountains, Xinjiang Autonomous Region, China. *Forest Ecology and Management* 246, 273–284.

Sulkava, P. and Huhta, V. (2003) Effects of hard frost and freeze–thaw cycles on decomposer communities and N mineralisation in boreal forest soil. *Applied Soil Ecology* 22, 225–239.

Sutherst, R.W. (2000) Climate change and invasive species: a conceptual framework. In: Mooney, H.A. and Hobbs, R.J. (eds) *Invasive Species in a Changing World.* Island Press, Washington, DC, pp. 211–240.

Sutherst, R.W., Floyd, R. and Maywald, G.F. (1996) The potential geographical distribution of the cane toad, *Bufo marinus* L. in Australia. *Conservation Biology* 10, 294–299.

Svensson, J., Lindegarth, M. and Pavia, H. (2009) Equal rates of disturbance cause different patterns of diversity. *Ecology* 90, 496–505.

Syvertsen, J.P. and Graham, J.H. (1999) Phosphorus supply and arbuscular mycorrhizas increase growth and net gas exchange responses of two *Citrus* spp. grown at elevated [$CO_2$]. *Plant and Soil* 208, 209–219.

Talhelm, A.F., Pregitzer, K.S. and Zak, D.R. (2009) Species-specific responses to atmospheric carbon dioxide and tropospheric ozone mediate changes in soil carbon. *Ecology Letters* 12, 1219–1228.

Taub, D., Miller, B. and Allen, H. (2008) Effects of elevated $CO_2$ on the protein concentration of food crops: a meta-analysis. *Global Change Biology* 14, 565–575.

Taylor, A.R., Schroter, D., Pflug, A. and Wolters, V. (2004) Response of different decomposer communities to the manipulation of moisture availability: potential effects of changing precipitation patterns. *Global Change Biology* 10, 1313–1324.

Temple, S.A. (1977) Plant–animal mutualism: coevolution with dodo leads to near extinction of plant. *Science* 197, 885–886.

Teyssonneyre, F., Picon-Cochard, C., Falcimagne, R. and Soussana, J.F. (2002) Effects of elevated $CO_2$ and cutting frequency on plant community structure in a temperate grassland. *Global Change Biology* 8, 1034–1046.

Thomas, C.D., Singer, M.C. and Boughton, D. (1996) Catastrophic extinction of population sources in a butterfly metapopulation. *The American Naturalist* 148, 957–975.

Thomas, C.D., Cameron, A., Green, R.E., Bakkenes, M., Beaumont, L.J., Collingham, Y.C., Erasmus, B.F.N., De Sigueira, M.F., Grainger, A., Hannah, L., Hughes, L., Huntley, B., Van Jaarsveld, A.S., Midgley, G.F., Miles, L., Ortega-Huerta, M.A., Peterson, A.T., Phillips, O.L. and Williams, S.E. (2004) Extinction risk from climate change. *Nature* 427, 145–148.

Thomas, C.D., Franco, A.M.A. and Hill, J.K. (2006) Range retractions and extinction in the face of climate warming. *Trends in Ecology & Evolution* 21, 415–416.

Thomas, R.B., Richter, D.D., Ye, H., Heine, P.R. and Strain, B.R. (1991) Nitrogen dynamics and growth of seedlings of an N-fixing tree (*Gliricidia sepium* (Jacq.) Walp.) exposed to elevated atmospheric carbon-dioxide. *Oecologia* 88, 415–421.

Thompson, J. (1991) Phenotypic plasticity as a component of evolutionary change. *Trends in Ecology & Evolution* 6, 246–249.

Thompson, P. and Ollason, J. (2001) Lagged effects of ocean climate change on fulmar population dynamics. *Nature* 413, 417–420.

Thonon, I. and Klok, C. (2007) Impact of a changed inundation regime caused by climate change and floodplain rehabilitation on population viability of earthworms in a lower River Rhine floodplain. *Science of the Total Environment* 372, 585–594.

Thornley, J.H.M. (1998) *Grassland Dynamics: An Ecosystem Simulation Model.* CAB International, Wallingford, UK.

Thornley, J.H.M. and Cannell, M.G.R. (1996) Temperate forest responses to carbon dioxide, temperature and nitrogen: a model analysis. *Plant, Cell & Environment* 19, 1331–1348.

Thornley, J.H.M. and Johnson, I.R. (1990) *Plant and Crop Modelling: A Mathematical Approach to Plant and Crop Physiology.* Clarendon Press, Oxford, UK.

Thuiller, W., Lavorel, S., Araújo, M.B., Sykes, M.T. and Prentice, I.C. (2005) Climate change threats to plant diversity in Europe. *Proceedings of the National Academy of Sciences USA* 102, 8245–8250.

Thuiller, W., Richardson, D.M. and Midgley, G.F. (2007) Will climate change promote alien plant invasions? In: Nentwig, W. (ed.) *Biological Invasions.* Springer-Verlag, Berlion, pp. 197–211.

Tilman, D. and Downing, J.A. (1994) Biodiversity and stability in grasslands. *Nature* 367, 363–365.

Tilman, D., Wedin, D. and Knops, J. (1996) Productivity and sustainability influenced by biodiversity in grassland ecosystems. *Nature* 379, 718–720.

Toberman, H., Freeman, C., Evans, C., Fenner, N. and Artz, R.R.E. (2008) Summer drought decreases soil fungal diversity and associated phenol oxidase activity in upland calluna heathland soil. *FEMS Microbiology Ecology* 66, 426–436.

Tobin, P.C., Nagarkatti, S., Loeb, G. and Saunders, M.C. (2008) Historical and projected interactions between climate change and insect voltinism in a multivoltine species. *Global Change Biology* 14, 951–957.

Trappe, J.M. (1987) Phylogenetic and ecologic aspects of mycotrophy in the angiosperms from an evolutionary standpoint. In: Safir, G.R. (ed.) *Ecophysiology of VA Mycorrhizal Plants.* CRC Press, Boca Raton, Florida, pp. 2–25.

Travis, J. (2003) Climate change and habitat destruction: a deadly anthropogenic cocktail. *Proceedings. Biological Sciences/The Royal Society* 270, 467–473.

Treonis, A.M. and Lussenhop, J.F. (1997) Rapid response of soil protozoa to elevated $CO_2$. *Biology and Fertility of Soils* 25, 60–62.

Treseder, K.K. (2004) A meta-analysis of mycorrhizal responses to nitrogen, phosphorus, and atmospheric $CO_2$ in field studies. *New Phytologist* 164, 347–355.

Treseder, K.K., Egerton-Warburton, L.M., Allen, M.F., Cheng, Y.F. and Oechel, W.C. (2003) Alteration of soil carbon pools and communities of mycorrhizal fungi in chaparral exposed to elevated carbon dioxide. *Ecosystems* 6, 786–796.

Trivers, R. and Willard, D. (1973) Natural selection of parental ability to vary the sex ratio of offspring. *Science* 179, 90–92.

Trubina, M. (2009) Species richness and resilience of forest communities: combined effects of short-term disturbance and long-term pollution. *Plant Ecology* 201, 339–350.

Trumble, J. and Butler, C. (2009) Climate change will exacerbate California's insect pest problems. *California Agriculture* 63, 73–78.

Tsiafouli, M.A., Kallimanis, A.S., Katana, E., Stamou, G.P. and Sgardelis, S.P. (2005) Responses of soil microarthropods to experimental short-term manipulations of soil moisture. *Applied Soil Ecology* 29, 17–26.

Tubiello, F., Rosenzweig, C., Goldberg, R., Jagtap, S. and Jones, J. (2002) Effects of climate change on US crop production: simulation results using two different GCM scenarios. Part I: Wheat, potato, maize, and citrus. *Climate Research* 20, 259–270.

Tungate, K.D., Israel, D.W., Watson, D.M. and Rufty, T.W. (2007) Potential changes in weed competitiveness in an agroecological system with elevated temperatures. *Environmental and Experimental Botany* 60, 42–49.

Turner, N. (1999) Time to burn: traditional use of fire to enhance resource production by aboriginal peoples in British Columbia. In: Boyd, R. (ed.) *Indians, Fire and the Land in the Pacific Northwest*. Oregon State University Press, Corvallis, Oregon, pp. 185–218.

Ungerer, M., Ayres, M. and Lombardero, M. (1999) Climate and the northern distribution limits of *Dendroctonus frontalis* Zimmermann (Coleoptera: Scolytidae). *Journal of Biogeography* 26, 1133–1145.

Van der Veken, S., Hermy, M., Vellend, M., Knapen, A. and Verheyen, K. (2008) Garden plants get a head start on climate change. *Frontiers in Ecology and the Environment* 6, 212–216.

Van Mantgem, P., Stephenson, N., Byrne, J., Daniels, L., Franklin, J., Fule, P., Harmon, M., Larson, A., Smith, J., Taylor, A. and Veblen, T. (2009) Widespread increase of tree mortality rates in the Western United States. *Science* 323, 521–524.

Van Turnhout, C., Foppen, R., Leuven, R., Van Strien, A. and Siepel, H. (2010) Life-history and ecological correlates of population change in Dutch breeding birds. *Biological Conservation* 143, 173–181.

Vanriper, C., Vanriper, S.G., Goff, M.L. and Laird, M. (1986) The epizootiology and ecological significance of malaria in Hawaiian land birds. *Ecological Monographs* 56, 327–344.

Vegvari, Z., Bokony, V., Barta, Z. and Kovacs, G. (2010) Life history predicts advancement of avian spring migration in response to climate change. *Global Change Biology* 16, 1–11.

Veron, J. (2008) *A Reef in Time: The Great Barrier Reef From Beginning to End*. Belknap Press, Cambridge, Massachusetts.

Vestgarden, L.S. and Austnes, K. (2009) Effects of freeze–thaw on C and N release from soils below different vegetation in a montane system: a laboratory experiment. *Global Change Biology* 15, 876–887.

Villalpando, S.N., Williams, R.S. and Norby, R.J. (2009) Elevated air temperature alters an old-field insect community in a multifactor climate change experiment. *Global Change Biology* 15, 930–942.

Vincent, G., De Foresta, H. and Mulia, R. (2009) Co-occurring tree species show contrasting sensitivity to ENSO-related droughts in planted dipterocarp forests. *Forest Ecology and Management* 258, 1316–1322.

Virkkala, R., Heikkinen, R., Leikola, N. and Luoto, M. (2008) Projected large-scale range reductions of northern-boreal land bird species due to climate change. *Biological Conservation* 141, 1343–1353.

Virtanen, T., Neuvonen, S., Nikula, A., Varama, M. and Niemelä, P. (1996) Climate change and the risks of *Neodiprion sertifer* outbreaks on scots pine. *Silva Fennica* 30, 169–177.

Visser, M.E. and Both, C. (2005) Shifts in phenology due to global climate change: the need for a yardstick. *Proceedings. Biological Sciences/The Royal Society* 272, 2561–2569.

Vitousek, P., Mooney, H., Lubchenco, J. and Melillo, J. (1997) Human domination of Earth's ecosystems. *Science* 277, 494–499.

Vors, L. and Boyce, M. (2009) Global declines of caribou and reindeer. *Global Change Biology* 15, 2626–2633.

Vu, J.C.V. (2005) Rising atmospheric $CO_2$ and $C_4$ photosynthesis. In: Pessarakli, M. (ed.) *Handbook of Photosynthesis*, 2nd edn. Taylor and Francis, Boca Raton, Florida, pp. 315–326.

Wahl, E.R. and Ammann, C.M. (2007) Robustness of the Mann, Bradley, Hughes reconstruction of northern hemisphere surface temperatures: examination of criticisms based on the nature and processing of proxy climate evidence. *Climatic Change* 85, 33–69.

Waldrop, M.P. and Firestone, M.K. (2006) Response of microbial community composition and function to soil climate change. *Microbial Ecology* 52, 716–724.

Walker, I.R. (2001) Midges: *Chironomidae* and related diptera. In: Smol, J.P., Birks, H.J.B. and Last, W.M. (eds) *Tracking Environmental Change using Lake Sediments*. Vol. 4. *Zoological Indicaters*. Kluwer Academic Publishers, Dordrecht, The Netherlands, pp. 43–66.

Walker, M.D., Wahren, C.H., Hollister, R.D., Henry, G.H.R., Ahlquist, L.E., Alatalo, J.M., Bret-Harte, M.S., Calef, M.P., Callaghan, T.V., Carroll, A.B., Epstein, H.E., Jonsdottir, I.S., Klein, J.A., Magnusson, B., Molau, U., Oberbauer, S.F., Rewa, S.P., Robinson, C.H., Shaver, G.R., Suding, K.N., Thompson, C.C., Tolvanen, A., Totland, O., Turner, P.L., Tweedie, C.E., Webber, P.J. and Wookey, P.A. (2006) Plant community responses to experimental warming across the tundra biome. *Proceedings of the National Academy of Sciences USA* 103, 1342–1346.

Wallace, A.R. (1876) *The Geographical Distribution of Animals*. Harper and Brothers, New York.

Walther, G., Post, E., Convey, P., Menzel, A., Parmesan, C., Beebee, T., Fromentin, J., Hoegh-Guldberg, O. and Bairlein, F. (2002) Ecological responses to recent climate change. *Nature* 416, 389–395.

Wand, S., Midgley, G., Jones, M. and Curtis, P. (1999) Responses of wild $C_4$ and $C_3$ grass (Poaceae) species to elevated atmospheric $CO_2$ concentration: a meta-analytic test of current theories and perceptions. *Global Change Biology* 5, 723–741.

Wang, B. and Qiu, Y.L. (2006) Phylogenetic distribution and evolution of mycorrhizas in land plants. *Mycorrhiza* 16, 299–363.

Warren, P.H., Law, R. and Weatherby, A.J. (2003) Mapping the assembly of protist communities in microcosms. *Ecology* 84, 1001–1011.

Watson, R. and Team, C. (2001) *Climate Change 2001: Synthesis Report. Contribution of Working Groups I, II, and III to the Third Assessment Report of the Intergovernmental Panel on Climate Change.* Cambridge University Press, Cambridge, UK.

Webb, T. III and Bartlein, P.J. (1992) Global changes during the last 3 million years: climatic controls and biotic responses. *Annual Reviews of Ecology and Systematics* 23, 141–173.

Weintraub, M.N. and Schimel, J.P. (2005) Nitrogen cycling and the spread of shrubs control changes in the carbon balance of Arctic tundra ecosystems. *BioScience* 55, 408–415.

West, J.B., Hillerislambers, J., Lee, T.D., Hobbie, S.E. and Reich, P.B. (2005) Legume species identity and soil nitrogen supply determine symbiotic nitrogen-fixation responses to elevated atmospheric [$CO_2$]. *New Phytologist* 167, 523–530.

West-Eberhard, M. (1989) Phenotypic plasticity and the origins of diversity. *Annual Review of Ecology and Systematics* 20, 249–278.

Westerling, A.L., Hidalgo, H.G., Cayan, D.R. and Swetnam, T.W. (2006) Warming and earlier spring increase western US forest wildfire activity. *Science* 313, 940–943.

Whitford, W.G. and Sobhy, H.M. (1999) Effects of repeated drought on soil microarthropod communities in the northern Chihuahuan Desert. *Biology and Fertility of Soils* 28, 117–120.

Whitney, K.D. and Gabler, C.A. (2008) Rapid evolution in introduced species, 'invasive traits' and recipient communities: challenges for predicting invasive potential. *Diversity & Distributions* 14, 569–580.

Wiens, J.A., Rotenberry, J.T. and Vanhorne, B. (1986) A lesson in the limitations of field experiments – shrubsteppe birds and habitat alteration. *Ecology* 67, 365–376.

Wiles, G.C., McAllister, R.P., Davi, N.K. and Jacoby, G.C. (2003) Eolian response to Little Ice Age climate change, tana dunes, Chugach Mountains, Alaska, USA. *Arctic, Antarctic, and Alpine Research* 35, 67–73.

Wilf, P. and Labandeira, C.C. (1999) Response of plant–insect associations to Paleocene–Eocene warming. *Science* 284, 2153–2156.

Williams, D. and Liebhold, A. (1995) Forest defoliators and climatic change: potential changes in spatial distribution of outbreaks of western spruce budworm (Lepidoptera: Tortricidae) and gypsy moth (Lepidoptera: Lymantriidae). *Environmental Entomology* 24, 1–9.

Williams, D. and Liebhold, A. (2002) Climate change and the outbreak ranges of two North American bark beetles. *Agricultural and Forest Entomology* 4, 87–99.

Williams, J., Shuman, B., Webb, T. III, Bartlein, P. and Leduc, P. (2004) Late-Quaternary vegetation dynamics in North America: scaling from taxa to biomes. *Ecological Monographs* 74, 309–334.

Williams, M.A. (2007) Response of microbial communities to water stress in irrigated and drought-prone tallgrass prairie soils. *Soil Biology & Biochemistry* 39, 2750–2757.

Williams, S.E., Bolitho, E.E. and Fox, S. (2003) Climate change in Australian tropical rainforests: an impending environmental catastrophe. *Proceedings. Biological Sciences/The Royal Society* 270, 1887–1892.

Williams, W.J., Eldridge, D.J. and Alchin, B.M. (2008) Grazing and drought reduce cyanobacterial soil crusts in an Australian acacia woodland. *Journal of Arid Environments* 72, 1064–1075.

Willig, M., Kaufman, D. and Stevens, R. (2003) Latitudinal gradients of biodiversity: pattern, process, scale, and synthesis. *Annual Review of Ecology, Evolution, and Systematics* 34, 273–309.

Willig, M., Bloch, C., Brokaw, N., Higgins, C., Thompson, J. and Zimmermann, C. (2007) Cross-scale responses of biodiversity to hurricane and anthropogenic disturbance in a tropical forest. *Ecosystems* 10, 824–838.

Willis, C.G., Ruhfel, B., Primack, R.B., Miller-Rushing, A.J. and Davis, C.C. (2008) Phylogenetic patterns of species loss in Thoreau's Woods are driven by climate change. *Proceedings of the National Academy of Sciences USA* 105, 17029–17033.

Willis, S., Hill, J., Thomas, C., Roy, D., Fox, R., Blakeley, D. and Huntley, B. (2009) Assisted colonization in a changing climate: a test-study using two UK butterflies. *Conservation Letters* 2, 46–52.

Wilsey, B.J. (2001) Effects of elevated $CO_2$ on the response of *Phleum pratense* and *Poa pratensis* to aboveground defoliation and root-feeding nematodes. *International Journal of Plant Sciences* 162, 1275–1282.

Wilson, R., Gutierez, D., Gutierez, J. and Monserrat, V. (2007) An elevational shift in butterfly species richness and composition accompanying recent climate change. *Global Change Biology* 13, 1873–1887.

Wilson, S. and Nilsson, C. (2009) Arctic alpine vegetation change over 20 years. *Global Change Biology* 15, 1676–1684.

Winkel, W. and Hudde, H. (1997) Long-term trends in reproductive traits of tits (*Parus major, P. caeruleus*) and pied flycatchers (*Ficedula hypoleuca*). *Journal of Avian Biology* 28, 187–190.

Winkler, J.B. and Herbst, M. (2004) Do plants of a semi-natural grassland community benefit from long-term

$CO_2$ enrichment? *Basic and Applied Ecology* 5, 131–143.

Wipf, S., Stoeckli, V. and Bebi, P. (2009) Winter climate change in alpine tundra: plant responses to changes in snow depth and snowmelt timing. *Climatic Change* 94, 105–121.

Wood, R. and Bishop, J. (1981) Insecticide resistance: populations and evolution. In: Wood, R., Bishop, J. and Cook, L. (eds) *Genetic Consequences of Man-Made Change*. Academic Press, London, pp. 97–127.

Woodall, C., Oswalt, C., Westfall, J., Perry, C., Nelson, M. and Finley, A. (2009) An indicator of tree migration in forests of the eastern United States. *Forest Ecology and Management* 257, 1434–1444.

Woodhouse, C.A. and Overpeck, J.T. (1998) 2000 years of drought variability in the central United States. *Bulletin of the American Meteorological Society* 79, 2693–2714.

Woodward, F. (1987) Stomatal numbers are sensitive to increases in $CO_2$ from pre-industrial levels. *Nature* 327, 617–618.

Wookey, P.A., Aerts, R., Bardgett, R.D., Baptist, F., Brathen, K.A., Cornelissen, J.H.C., Gough, L., Hartley, I.P., Hopkins, D.W., Lavorel, S. and Shaver, G.R. (2009) Ecosystem feedbacks and cascade processes: understanding their role in the responses of arctic and alpine ecosystems to environmental change. *Global Change Biology* 15, 1153–1172.

Wray, S. and Strain, B. (1986) Response of two old field perennials to interactions of $CO_2$ enrichment and drought stress. [*Aster pilosus* and *Andropogon virginicus* L]. *American Journal of Botany* 73, 1486–1491.

Wright, S. (1931) Evolution in Mendelian populations. *Genetics* 16, 97–159.

Wright, S.J. (2005) Tropical forests in a changing environment. *Trends in Ecology & Evolution* 20, 553–560.

Wu, W.B., Shibasaki, R., Yang, P., Tan, G.X., Matsumura, K.I. and Sugimoto, K. (2007) Global-scale modelling of future changes in sown areas of major crops. *Ecological Modelling* 208, 378–390.

Xiong, Z., Pan, Y. and Nakagoshi, N. (2007) Forest NPP change under the climate change in northeastern China. *Hikobia* 15, 11–21.

Yarie, J. and Billings, S. (2002) Carbon balance of the taiga forest within Alaska: present and future. *Canadian Journal of Forest Research* 32, 757–767.

Yeates, G.W. and Newton, P.C.D. (2009) Long-term changes in topsoil nematode populations in grazed pasture under elevated atmospheric carbon dioxide. *Biology and Fertility of Soils* 45, 799–808.

Yeates, G.W., Tate, K.R. and Newton, P.C.D. (1997) Response of the fauna of a grassland soil to doubling of atmospheric carbon dioxide concentration. *Biology and Fertility of Soils* 25, 307–315.

Yeates, G.W., Newton, P.C.D. and Ross, D.J. (1999) Response of soil nematode fauna to naturally elevated $CO_2$ levels influenced by soil pattern. *Nematology* 1, 285–293.

Yeates, G.W., Newton, P.C.D. and Ross, D.J. (2003) Significant changes in soil microfauna in grazed pasture under elevated carbon dioxide. *Biology and Fertility of Soils* 38, 319–326.

Yuan, J., Himanen, S., Holopainen, J., Chen, F. and Stewart, C. (2009) Smelling global climate change: mitigation of function for plant volatile organic compounds. *Trends in Ecology & Evolution* 24, 323–331.

Zak, D.R., Pregitzer, K.S., Curtis, P.S. and Holmes, W.E. (2000a) Atmospheric $CO_2$ and the composition and function of soil microbial communities. *Ecological Applications* 10, 47–59.

Zak, D.R., Pregitzer, K.S., King, J.S. and Holmes, W.E. (2000b) Research review. Elevated atmospheric $CO_2$, fine roots and the response of soil microorganisms: a review and hypothesis. *New Phytologist* 147, 201–222.

Zaller, J.G. and Arnone, J.A. (1997) Activity of surface-casting earthworms in a calcareous grassland under elevated atmospheric $CO_2$. *Oecologia* 111, 249–254.

Zaller, J.G. and Arnone, J.A. (1999) Interactions between plant species and earthworm casts in a calcareous grassland under elevated $CO_2$. *Ecology* 80, 873–881.

Zaller, J.G., Caldwell, M.M., Flint, S.D., Ballare, C.L., Scopel, A.L. and Sala, O.E. (2009) Solar UVB and warming affect decomposition and earthworms in a fen ecosystem in Tierra del Fuego, Argentina. *Global Change Biology* 15, 2493–2502.

Zavaleta, E.S., Shaw, M.R., Chiariello, N.R., Mooney, H.A. and Field, C.B. (2003) Additive effects of simulated climate changes, elevated $CO_2$, and nitrogen deposition on grassland diversity. *Proceedings of the National Academy of Sciences USA* 100, 7650–7654.

Zhang, W., Parker, K.M., Luo, Y., Wan, S., Wallace, L.L. and Hu, S. (2005) Soil microbial responses to experimental warming and clipping in a tallgrass prairie. *Global Change Biology* 11, 266–277.

Zheng, J.Q., Han, S.J., Ren, F.R., Zhou, Y.M., Zheng, X.B. and Wang, Y. (2009) Effects of long-term $CO_2$ fumigation on fungal communities in a temperate forest soil. *Soil Biology & Biochemistry* 41, 2244–2247.

Zhou, X., Perry, J., Woiwod, I., Harrington, R., Bale, J. and Clark, S. (1997) Temperature change and complex dynamics. *Oecologia* 112, 543–550.

Zhou, X., Sherry, R., An, Y., Wallace, L. and Luo, Y. (2006) Main and interactive effects of warming, clipping, and doubled precipitation on soil $CO_2$ efflux in a grassland ecosystem. *Global Biogeochemical Cycles* 20, GB1003; doi:10.1029/2005GB002526.

Zhu, X., Portis, A. and Long, S. (2004) Would transformation of $C_3$ crop plants with foreign Rubisco increase productivity? A computational analysis extrapolating from kinetic properties to canopy photosynthesis. *Plant, Cell & Environment* 27, 155–165.

Ziska, L.H. (2003a) Evaluation of the growth response of six invasive species to past, present and future atmospheric carbon dioxide. *Journal of Experimental Botany* 54, 395–404.

Ziska, L.H. (2003b) Evaluation of yield loss in field sorghum from a $C_3$ and $C_4$ weed with increasing $CO_2$. *Weed Science* 51, 914–918.

Ziska, L.H. and Bunce, J.A. (2007) Predicting the impact of changing $CO_2$ on crop yields: some thoughts on food. *New Phytologist* 175, 607–617.

Ziska, L.H., Morris, C.F. and Goins, E.W. (2004) Quantitative and qualitative evaluation of selected wheat varieties released since 1903 to increasing atmospheric carbon dioxide: can yield sensitivity to carbon dioxide be a factor in wheat performance? *Global Change Biology* 10, 1810–1819.

Ziska, L.H., Faulkner, S. and Lydon, J. (2009) Changes in biomass and root:shoot ratio of field-grown Canada thistle (*Cirsium arvense*), a noxious, invasive weed, with elevated $CO_2$: implications for control with glyphosate. *Weed Science* 52, 584–588.

Zogg, G.P., Zak, D.R., Ringelberg, D.B., MacDdonald, N.W., Pregitzer, K.S. and White, D.C. (1997) Compositional and functional shifts in microbial communities due to soil warming. *Soil Science Society of America Journal* 61, 475–481.

Zuckerberg, B., Woods, A. and Porter, W. (2009) Poleward shifts in breeding bird distributions in New York State. *Global Change Biology* 15, 1866–1883.

Zvereva, E. and Kozlov, M. (2006) Consequences of simultaneous elevation of carbon dioxide and temperature for plant–herbivore interactions: a meta-analysis. *Global Change Biology* 12, 27–41.

# Index

Page numbers in **bold** refer to figures, tables and boxes.

experimental methods
    design and interpretation 53–56, **58–59**,
        150–152
    manipulation of variables
        carbon dioxide **1**, 56–59, **60**, **61**, **62**
        precipitation 59, 61, **64**
        temperature 59, **63**, **64**
external validity 54, 57
extinction
    causes 203, 205
    interaction of multiple drivers 115
    mass events, past and present 203, **205**
    predictions 206, **207**
    species vulnerable to climate change
        endemics 203, 211
        sea turtles **154–155**
extirpations
    definition 203
    impact on population range 95
    risk factors 203, 205–206
extreme events
    agricultural impacts 187, **187**
    altering net carbon sinks to sources 221
    impact on community succession 115, **115**


FACE (Free Air Carbon dioxide Enrichment)
    crop phenology experiments 83, 85
    design and operation 57, **61**
    replication and representativeness 57, **62**,
        67–68, 127
fecundity (birth rate) 89–90
feedback systems
    climate change and forest fires 178–179, 225
    photosynthetic rate and $CO_2$ concentration 78, **79**
    plant–soil 123–124, 130, 168–171, **169**
    positive and negative 220–221
    between rock and hydrological cycles 22
fertility 89
fertilization effect (of $CO_2$) 76
    experimental evidence 126–128, **128**, 184, **184**,
        225–226
    interaction with water use efficiency 82–83, 223
field experiments
    agricultural crops **157**, 184, **185**
    effects of herbivores 116–117, **117**, **118**
    time scale 108–109, 168–169
    validity and result significance 54, 57
    *see also* FACE
fire
    dependence, for regeneration 114, 176–177, **177**
    effects on vegetation, short- and long-term 225
    frequency, effect of climate change 114, **178**,
        178–179, 224
    human management regimes 51, **232**
food security, global **192**
food webs, soil **167**, 167–168, 171

forests
    climate-induced changes 82–83
        carbon allocation in tree growth 175–176,
            **176**
        disturbance regimes 176–182, **178**
        interactions with human management 223,
            **223**, **232–233**
        ozone damage 229
    economic and environmental value 172
    modelling 67, 172–173, **173**, **174**, 174–175
    productivity, interacting control factors 127–129,
        **134–135**, 135, 172–175, 182
fossil fuels 119, **192**, 227
fractionation, isotopic **14**, 14–15
functional traits 91, **94**, 238


Garry oak (*Quercus garryana*) ecosystems **232**
General Circulation Models (GCMs)
    centres and model names 33–34, **34**
    range and robustness of predictions 35–36, **36**
    spatial and temporal resolution 35, **35**
    structure 32–33
    validation 38–41, **39**, **40**
genetics
    climate-related traits, correlation 152–153, **153**
    genetic variation sources 143–144
    genetically modified organisms 189, **190**
    of populations, at range edges 99–100, **100**
glaciation
    consequences and causes 9, **16–17**, 23
    continental ice shelf and polar bear decline 90
    dating by dendrochronology 8
    impacts shown in pollen record 53, **53**
    Quaternary Ice Ages and interglacials 24–27, **26**
    *see also* ice cores
Global Circulation (Climate) Models *see* General
        Circulation Models
global warming
    contribution of varying solar activity 24
    correlated with poleward range shifts 97, 97–98, **98**
    effects on permafrost and decomposition
        130–131, **131**
    effects on spring phenological events 83
    effects on vertebrate populations 89–90
    prehistoric climate, causes **16–17**, 25–26
    scientific and popular status 242
    surface temperature predictions (from models)
        **36**, 45
    and upward elevation (altitude) shifts 98–99, **99**
    *see also* temperature
grassland communities
    earthworm activity 171, **171**
    fire impacts 114–115, 225
    multiple stress interactions 113, 114, 229
    responses to water availability 110, **111**
    *see also* OCCAM experiment

recruitment 89, 112, 181–182
refugia 99–100, 209–210
Regional Climate Models (RCMs) 36, 38
Reid's paradox **102**
research *see* study methods
reserves, design and placement **211**, 211–212, **212**
resilience to disturbance
    forest fire 181–182
    grassland drought 110, **111**
resistance, rapid evolution of 147, 149
respiration *see* heterotrophic respiration;
    photorespiration
restoration ecology 210
rhizobia (root nodules) 165–166, **166**
ribulose-1,5-bisphosphate (RuBP) 74, **74**, 76, 77
roots, plant
    fine root turnover, trees 175, **176**
    mutualism with mycorrhizae and rhizobia
        164–166, **165**, **166**
    root:shoot ratios 80, **81**, 82, **82**
    root herbivores, impact of climate 166–167
Rubisco (ribulose-1,5-bisphosphate carboxylase oxygenase)
    nitrogen investment in, regulation 78, 80
    role in carbon fixation 73, **74**, 75
    sensitivity to $CO_2$:$O_2$ ratio 74, 77, **77**, **190**

salinity, community effects 113–114
sample size (experimental replicates) 55, **55**
scenarios
    interpretation by modelling frameworks **31**, 32
    purpose and construction 29–30, **30**
    qualitative storylines 30–31, **32**
sclerochronology 20–21, **21**
sea levels
    coastal habitat loss and extinctions 206
    palaeological rises and falls 24
sediments
    freshwater 9–13, 52–53
    ocean 9
selection experiments (evolution) 150, **151–152**, 152
sequestration *see* carbon sequestration
sex ratios (sea turtles), environmental control **154–155**
Shannon diversity (H) 201
shells, growth increment records 4, 20–21, **21**
Simpson's diversity index 202
simulations *see* models
soils
    carbon sequestration
        effect of soil community changes 161, 168
        global reservoir 129, 130
        organic matter, Century Model **48–49**, 67
    food webs 159, **167**, 167–168, 171
    freeze–thaw cycle testing
        effects of snow cover 163, **164**
        pseudoreplication 56
    organisms **160**
        interactions with plant roots 164–167
        mesofauna and macrofauna 163–164
        microbes 160–162, **161**
        protozoa 162, **162**
    plant response feedbacks 168–171, **169**
    respiration 121–122, 130, 226
    temperature control, experimental 59
solar brightness 23–24
spatial scale
    biodiversity patterns 202–203, **204**
    ease of study access 108
    resolution of models 35, **35**, 36, 38
    study methods 95
    *see also* distribution
species richness, definition 201
species-specific responses
    climate impact modelling 61, 63–67, **65**, **66**
    individual poleward migration rates 104, **105**
    as limitation to experimental conclusions
        238–239, **239**
    past, reconstructed 53, **102**
    related to functional traits 91
species–area relationship (SAR) 202–203, 206
SRES (Special Report on Emissions Scenarios,
    IPCC) 29, **36**
stability
    erratic fluctuations, population effects 95
    related to diversity 200–201, 206, 208
statistical power **54**, 54–55, **55**, 68–69
    community experiments 109–110
stomata
    conductance at elevated $CO_2$ 78, 80, 122
    density and pathogens 195
    regulation, to control water loss 75, 78, 79
stress
    global ecological change components 219, **220**
    habitat loss and human land use 229–231, **232–233**
    impacts of episodic disturbance 221–225
    interaction of multiple environmental stressors **217**,
        225–229
    synergy and feedbacks 219–221
study methods
    bias towards important/easily studied taxa 89,
        91, 108
    choice of technique 50, 69–70, 88–89
    experimental 53–61, **62**, 239–240
    geographical imbalance of research 57, 98–99,
        126, 182
    negative results, reporting of 68–69, 117
    observational 50–53
    'natural experiments' 50, 59, 89
    statistical meta-analysis 67–69, **70**
    theoretical models 61, 63–67, **68–69**
succession (communities) 114–115, **115**, 182
Sun
    distance from Earth, orbital eccentricity 17, 22–23
    solar brightness variation 23–24